LIBRARY
Gordon College
Wenham, Mass. 01984
W9-AGV-094
574.524
As 63
DISCARDED
JENKS LRC
GORDON COLLEGE

Aspects of

The Biology of Symbiosis

Proceedings of a symposium entitled THE BIOLOGY OF SYMBIOSIS *held in Boston, Massachusetts, under the joint auspices of*

The American Society of Zoologists
The American Association for the Advancement of Science
The Society for Invertebrate Pathology

organized by Thomas C. Cheng, Professor of Biology, Lehigh University, and Sidney J. Townsley, Professor of Zoology, University of Hawaii

Aspects of

The Biology of Symbiosis

EDITED BY

THOMAS C. CHENG

Professor of Biology
and
Director, Institute for Pathobiology
Lehigh University
Bethlehem, Pennsylvania

UNIVERSITY PARK PRESS

BALTIMORE

BUTTERWORTHS

LONDON

Library of Congress Catalog Card Number 72-148457

International Standard Book Number 0-8391-0039-6

Printed in Japan
by General Printing Co., Ltd., Yokohama

Contents

Contributors

GEORGE T. BARTHALMUS. *Department of Biology, The Pennsylvania State University, University Park, Pennsylvania.*

PETER CASTRO. *Department of Biological Sciences, University of Puerto Rico, Rio Piedras, Puerto Rico.*

THOMAS C. CHENG. *Department of Biology and Institute for Pathobiology, Lehigh University, Bethlehem, Pennsylvania.*

LEONARD G. EPP. *Department of Biology, The Pennsylvania State University, University Park, Pennsylvania.*

LARRY G. HARRIS. *Department of Zoology, University of New Hampshire, Durham, New Hampshire.*

SIRO KAWAGUTI. *Department of Biology, Faculty of Science, Okayama University, Okayama, Japan.*

DAVID R. LINCICOME. *Department of Zoology, Howard University, Washington, D.C.*

GEORGE S. LOSEY, Jr. *Hawaii Institute of Marine Biology, University of Hawaii, Kaneohe, Hawaii.*

CHARLES F. LYTLE. *Department of Zoology, North Carolina State University, Raleigh, North Carolina.*

RICHARD N. MARISCAL. *Department of Biological Science, Florida State University, Tallahassee, Florida.*

ERIK RIFKIN. *Department of Parasitology, Naval Medical Research Institute, Naval Medical Center, Bethesda, Maryland.*

M. A. STIREWALT. *Department of Parasitology, Naval Medical Research Institute, Naval Medical Center, Bethesda, Maryland.*

M. R. TRIPP. *Department of Biological Sciences, University of Delaware, Newark, Delaware.*

F. JOHN VERNBERG. *Belle W. Baruch Coast Research Institute, University of South Carolina, Columbia, South Carolina.*

WINONA B. VERNBERG. *Department of Biology, University of South Carolina, Columbia, South Carolina.*

Introduction

In recent years considerable interest has been placed on the "model" approach to the understanding of a number of biological phenomena. The study of heterospecific relationships is no exception. As a consequence, those individuals interested in parasitism, commensalism, and mutualism now recognize that these subcategories of symbiosis share several common denominators. In retrospect, this realization is understandable, for during the course of evolution these common phenomena have become established either as the result of parallel evolution or sequentially, depending upon the specific partnership. Regardless of the evolutionary mechanism, these common denominators must be interpreted as being of survival value not only for the organisms involved but also for the symbioses. In view of this concept, it appeared appropriate to bring together a small group of investigators with interests in a number of aspects of the dynamic equilibrium between host and symbiont so that we could share and profit from their experiences. These proceedings are the results of that symposium.

It is now generally recognized that a successful symbiotic relationship must include several phases: (1) contact between the two heterospecific organisms, including actual penetration in some instances, (2) establishment of the symbiont on or in the host, and (3) successful escape of the symbiont or its germ-cell-bearing progeny so that similar associations can be perpetuated.

Several aspects of the communication between host and symbiont are considered in the first group of contributions in this volume.

Within the second group of articles, consideration is given to a number of aspects of physiological interactions between the host and the symbiont after the latter has successfully entered the host. Furthermore, as Dr. C. F. Lytle and his associates have pointed out, such interactions may be still influenced by a number of ambient factors.

Within the third and final selection of articles, attention is directed toward an extremely important aspect of symbiosis; that is, how certain hosts prevent the establishment of symbionts by employing defense mechanisms and how certain successful symbionts are capable of protecting themselves from their hosts' defensive barriers.

As far as I have been able to determine, no one is currently investigating escape mechanisms among symbionts. Thus this interesting aspect of symbiosis must regretfully remain unrepresented in this volume. Perhaps at some later gathering of a similar nature, circumstances will be such that this information will form a part of the program.

It was not the intention of the organizers of this symposium to include authors dealing with all the known factors contributing to symbiosis and it is obvious that many other important and interesting factors have been omitted. It was also not our intention to imply that the authors are the only or the most prominent investigators in this area of biology. Many widely recognized authorities were not with us.

Finally, I wish to thank Dr. Sidney J. Townsley of the University of Hawaii and Dr. Erik Rifkin of the Naval Medical Research Institute for consenting to serve as session chairmen and for helping to organize this symposium. Without their assistance these proceedings would not have been possible. I also wish to thank the Invertebrate Zoology Division of the American Society of Zoologists, the American Association for the Advancement of Science, and the Society for Invertebrate Pathology under whose joint auspices this symposium was held on December 28 and 29, 1969. With all its faults, I sincerely hope that this symposium and the resulting proceedings will serve a useful function.

THOMAS C. CHENG, Editor
Bethlehem, Pennsylvania

Penetration Stimuli for Schistosome Cercariae

M. A. STIREWALT

Naval Medical Research Institute
Bethesda, Maryland

INTRODUCTION

The host-endoparasite association begins with the parasite's entry into the host's body. For many parasites, this happens passively, when some stage is ingested or inhaled by the host or injected into its tissues by an infected vector. For many others, the process is an active one, as the parasite burrows into host tissue, either from inside the body, or from outside, often through intact skin. This article is concerned with a parasite which penetrates the intact skin of its hosts.

Little is known about the infective behavior patterns of these parasites in general, their routes of entry into skin, or their mechanisms of penetration and migration through the skin of living organisms. This subject has been reviewed by Stirewalt (1966). It is strange that so few investigators have attempted to relate the behavior, morphology, and physiology of the parasites to the structure of the skin which they invade. The interactions are, it appears, highly complex and widely different in the various skin-parasite models.

A variety of experimental models is available. Skin, which is almost never exactly the same among different species of vertebrates, may be that of rodents, carnivores, birds, domestic animals, or primates. The parasite models one may choose from include the filariform larvae of many species of hookworms and related genera such as *Nippostrongylus* or *Strongyloides*; infective larvae of some of the filarial worms; some insect larvae; and schistosome cercariae, to cite a few.

From Bureau of Medicine and Surgery, Navy Department, Research Task MF12 524 009 1004.

The experiments reported herein were conducted according to the principles enunciated in "Guide for Laboratory Animal Facilities and Care" prepared by the Committee on the Guide for Laboratory Animal Resources, NAS-NRC.

The opinions and assertions contained herein are those of the author and are not to be construed as official or as reflecting the views of the Navy Department or the Naval service at large.

In view of our ignorance about the skin-parasite interactions in these models, it is difficult to decide where to direct the first experimental probes. Because of the circumstances of our position and competence, we have chosen the interaction between rodent skin and *Schistosoma mansoni* cercaria as our first model. It is our hope to compare later the skin of other hosts and other cercariae and nematode larvae.

A brief review of some of the salient features of the complex life cycle of *S. mansoni* may be helpful. Adult male and female worms live in certain mesenteric venules of man and other definitive hosts. Many of the eggs laid by the female worms reach fresh water, hatch, and free the contained miracidia. If they find and infect suitable snails, miracidia develop to produce large numbers of cercariae which are infective to man.

We will focus here on the cercarial invasion of skin of definitive hosts. In an artist's conception of a cercaria creeping on skin (Fig. 1), the cercaria lies along a wrinkle near the base of several fine hairs. Cercariae explore skin for varying lengths of time, select penetration sites at surface irregularities (Stirewalt and Hackey, 1956), and enter by a combination of muscular activity and secretion from the acetabular glands.

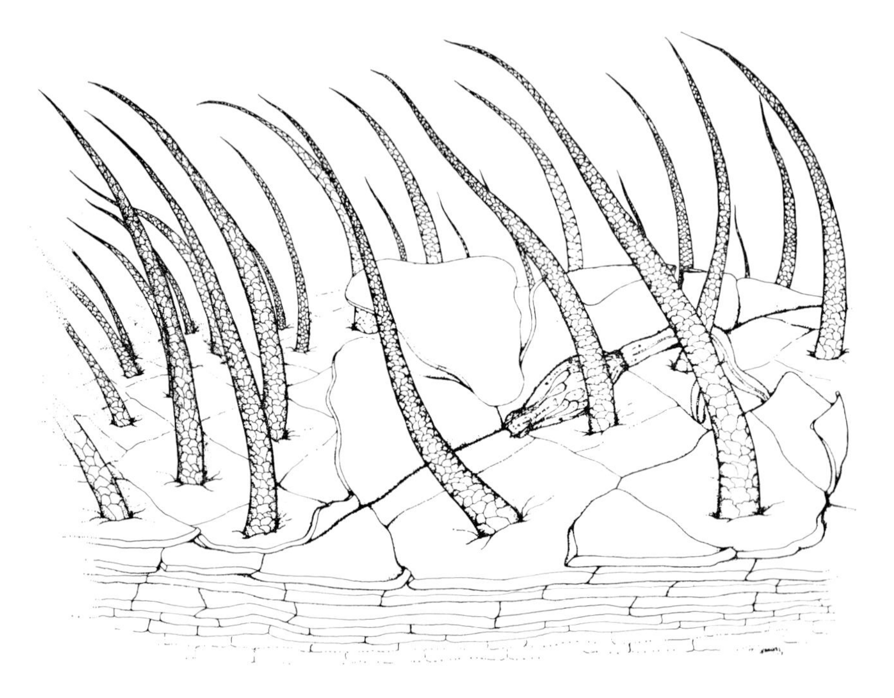

Fig. 1. Diagram of a cercaria in exploring position on skin.

For the past several years, we have been studying penetration mechanisms with the rodent skin – *S. mansoni* cercaria model. While this approach has not been exhaustively explored, we turn our attention here to an analysis of some of the results in another context, namely, identification of factors which serve as stimuli to cercariae to penetrate skin.

REVIEW OF THE LITERATURE

Almost as soon as it was proved that penetration of intact host skin by helminths was a route of infection of definitive hosts, speculation about penetration pathways and penetration mechanisms began (Fülleborn, 1914). Conditions which might facilitate the parasites' host-finding and skin penetration, however, were not considered as a separate issue until later (Fülleborn, 1924; Goodey, 1925).

These investigators were concerned with various hookworm and *Strongyloides* larvae. The conditions of interest were depth of the water medium, exsheathment, and temperature. Water depth was not interpreted as a penetration stimulus, of course, but a shallow water drop did provide for close contact of parasites and skin and perhaps for leverage for easier entry. It was reported that penetration by *Strongyloides fülleborni* larvae from the baboon was not under geotropic influence, but was rather under thermotropic influence. On the contrary, temperature did not appear to be an important factor in the penetration of other *Strongyloides* larvae into the skin of frogs, mice or guinea pigs (Kosugi, 1924).

Another factor, the attractiveness of skin, was mentioned by Abadie (1963), who said simply that the infective larvae of *Strongyloides ratti* were attracted to the rat's tail.

Penetration by schistosome cercariae began to receive especial attention in this regard in the 1950's. Neuhaus (1952) observed that the cercariae of *Trichobilharzia szidati* demonstrated phototaxis and geotaxis and, in addition, perceived and responded to thermal, mechanical, and chemical changes in their immediate environments. He suggested that these taxes served these cercariae as host-finding aids. Under the title of host recognition, Bolwig (1955) listed rat toe, scraped skin particles, sebaceous gland secretion, and a warm temperature point in cooler water as stimuli which caused cercariae of *Schistosoma haematobium* and *Schistosoma bovis* to attach to objects and cast their tails. He found no stimulation by blood, keratin, sweat, urea, lactic acid, cholesterol crystals, butter, sodium chloride, or carbon dioxide.

Otherwise, infectivity has been measured in more general terms. It has been expressed primarily as the ratio of number of mature worms recovered at autopsy of experimentally infected hosts to number of available infective larvae at their exposure (DeWitt, 1965; Purnell, 1966a,b; Olivier, 1966; among the most recent).

Obviously, the number of mature worms harbored by a host has been influenced by a multitude of factors. Developmental hazards of all kinds have taken their toll of invading cercariae and migrating schistosomules and worms. The host's natural resistance and acquired immune reactions, the parasite's long wandering through host tissues, and the hazards of its penetration into skin, all doubtless reduced in varying degrees the number of mature worms as compared with the number of cercariae at exposure.

It is true that several investigators have eliminated some of these factors by reporting results of their infectivity studies in terms of the ratio of the number of penetrating cercariae to the number of available cercariae (Stirewalt, 1953; Stirewalt and Fregeau, 1965, 1968; Stirewalt, *et al.*, 1965; Olivier, 1966; Warren and Peters, 1967a,b, 1968). The pertinent conditions examined as influential on penetration included the type of penetration membrane, ambient temperature, type of cercarial suspending medium, exposure time, number and dispersion of cercariae, developmental history and postemergence age of cercariae. All of these factors, however, were in the broad context of infection. For most, the effect was a general one on the cercaria as a total organism.

Actually, the ideas as well as the terms used (for example, infectivity and worm burden), were distorted by the fact that all aspects of the invasive process were considered in one final static picture. Examination of the dynamics involved requires attention to each specific phase of the invasive process alone.

Unfortunately, investigators have been slow to sort out specific aspects of the invasive process, such as penetration mechanisms, penetration stimuli, and cercarial responses, from the complex of influences which affect it. To achieve this sorting out it was necessary to study penetration alone *in vitro*, so one experimental condition at a time could be manipulated against a "steady state" background of otherwise optimal penetration conditions. Development of this capability began with the concept and practice of quantitating penetration alone *in vivo* (Stirewalt, 1953). This was accomplished by the simple maneuver of counting exactly the number of cercariae available to penetrate the tail skin of a mouse and then, at the end of the exposure period, recovering carefully and counting those not penetrating. Results with this procedure quickly made it evident that the environment played an important role in cercarial penetration of skin. Such factors as type and temperature of the water used for the cercarial suspension, length of the exposure period, and dispersion of the cercariae influenced the numbers of cercariae which penetrated (Stirewalt and Fregeau, 1965; Warren and Peters, 1967a). Likewise, many parasite-related conditions were influential: cercarial postemergence age (Oliver, 1966; Stirewalt and Fregeau, 1968), and developmental history in snail hosts, especially the maintenance temperature of infected snails and the point in the patency of snail infections at which the cercariae emerged (Stirewalt and Fregeau, 1968). In addition, a few host-related conditions also affected cercarial penetration; recent experience with penetration by homologous cercariae (Stirewalt, 1953), host species (Warren and Peters, 1967b), and mouse strain and age (Stirewalt *et al.*, 1965).

Review of these conditions, however, indicated that they probably influenced either the total penetration potential of cercariae or the barrier quality of skin. At any rate, they were not shown to serve specifically as stimuli to cercariae to penetrate. All in all, the cercaria–skin model with the skin *in situ* on a living host was not adaptable to the recognition of cercarial penetration stimuli.

The next step, examination of penetration of cercariae into skin of lipid-extracted and unextracted excised mouse ears (Wagner, 1959), was devised to demonstrate reduction of penetration by *Schistosomatium douthitti* cercariae after extraction of the skin with ether, and increased penetration again after residue of the skin extract or some free fatty acids were applied to the extracted skin. It did not lend itself to exact control of experimental conditions for additional investigation.

Development of a dried skin penetration membrane and a vessel for accommodating it followed (Fig. 2) (Stirewalt *et al.*, 1966). The vessel consisted

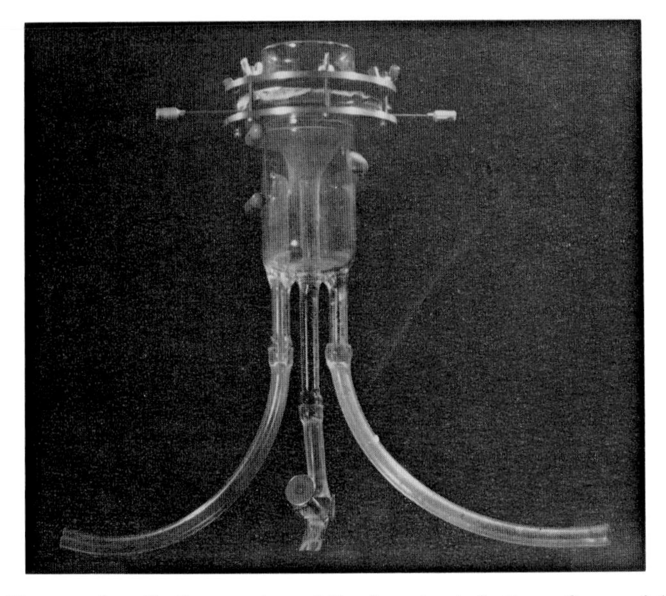

Fig. 2. Schistosomule collecting vessel consisting from top to bottom of: cercarial chamber, clamps enclosing skin penetration membrane, collecting funnel, water jacket for temperature control. (Reprinted with permission of *Transactions of the Royal Society of Tropical Medicine and Hygiene*, London).

of four essential parts: a top cercarial chamber whose floor was the skin membrane; the dried skin membrane; a collecting funnel immediately below, containing Hanks' balanced salt solution (HBSS); and a water jacket surrounding the collecting funnel for temperature control. With this equipment, membranes and conditions were tested for their comparative value for collecting schistosome larvae after they penetrated skin—we call them schistosomules. When this system was refined by the establishment of optimal conditions for collecting schistosomules (Stirewalt and Uy, 1969), most of the required conditions for testing penetration stimuli were satisfied. These investigators reported that, on the basis of the percentage of schistosomules collected after penetration of a thin, dried, rat skin membrane, several factors appeared to stimulate cercariae of *S. mansoni* to penetrate: a temperature differential providing a skin membrane warmer than the cercarial suspension; a light differential with the target skin bright and the cercarial suspension dark; and the presence of the skin membrane.

Substitution of the skin membrane with a Baudruche membrane from bovine intestine (Stirewalt and Uy, 1969) or of a hardened gelatin penetration membrane (Clegg, 1969) offered further information. Clegg (1969), with a gelatin penetration membrane hardened by chrome tanning and *Austrobilharzia terrigalensis* cercariae, showed that the surface lipid of chicken skin, the free sterol fraction of the extracted lipid, and a pure sample of cholesterol, the major sterol of bird skin, all stimulated these cercariae to penetrate. Other fractions had no stimulatory effect or negligible one. A 40°C temperature of the collecting medium in the penetration bottles also increased the percentage of schistosomules collected.

The final developmental step consists of the design of a system for quantitating cercarial penetration responses by observing cercarial behavior above penetration membranes, so that: (1) data on schistosomule collections can be verified, and (2) the assumption that the percentage of schistosomules collected equalled the percentage of cercariae attempting to penetrate can be avoided. It is described herein.

MATERIALS AND METHODS

Parasite

Both the schistosome, *Schistosoma mansoni*, and its snail host, *Biomphalaria glabrata*, were of Puerto Rican origin. Both have been reared in the laboratory for more than 25 years. Snails providing the cercariae had been individually exposed to 8–10 miracidia obtained from the macerated liver tissue of mice about two months after their individual exposure to approximately 100 cercariae. Exposed snails were kept at 26–28°C.

Cercariae were collected twice weekly in large numbers (500,000–1,000,000) from 300 to 500 snails. They were pooled by pouring them off the detritus and snail feces. The remaining contaminant material was picked from the cercarial suspension, and the cercariae concentrated in a suction Millepore filter (8μ pore size), to provide 1,000 cercariae per ml. Five ml were pipetted into each cercarial chamber.

Skin Extraction

Since the lipid film covering skin is the presenting surface to a cercaria, the obvious approach to testing the skin membrane for penetration stimuli was to remove it, and then collect schistosomules and observe cercarial behavior. This approach seemed especially promising since we knew that cercariae would swim to a glass surface and attempt to penetrate it only after skin lipids had been rubbed on it.

Many methods of removing the skin surface lipid were tried (Table 1). The most satisfactory, listed as "pentane and stripped," was as follows: the rat

TABLE 1. *Skin lipid removal techniques and their effect on* Schistosoma mansoni *schistosomule collections (three replications)*

Technique	*Schistosomule collection*	
	Av. %	*Range %*
Physohex washed	5	0.5–9
Ether extracted	9	6–11
Pentane extracted	23	13–39
Scotch tape stripped	28	14–48
Physohex and stripped	14	6–23
Pentane and stripped	4	1–9
Untreated control skin	51	32–68

abdomen was washed well with water at 35°C, air-dried, and immersed in pentane for 30 minutes. After air-drying again, the skin was excised and processed as usual (Stirewalt and Uy, 1969). It was then extracted with pentane again for 30 minutes, air-dried, and stripped 5 times with adhesive tape.

Penetration Membranes

Membranes which have been tested experimentally are listed in Table 2. The membrane of choice was the rat abdominal skin membrane from 50–60-day-old female Sprague-Dawley rats (NMRI:O(SD)). Rat skin membranes were prepared by drying, sanding, and plucking, as described by Stirewalt and

Uy (1969) except that drying took place *in vacuo* for 18 hours at room temperature of 23–24°C. Baudruche membranes (bovine intestinal membranes) were obtained from the Long and Long Co., Bellville, New Jersey.

TABLE 2. *Reaction of schistosome cercariae to membranes or surfaces other than normal skin* in situ

Membrane	*Cercarial penetration response*	*Penetration*
Glass	No	No
Glass+skin lipid	Yes	No
Baudruche membrane	No	No
Baudruche membrane+skin lipid	Yes	No
Chick eggshell membrane (Stirewalt *et al.*, 1966)	Yes	No
Chick chorioallantoic membrane 〃	Yes	Yes
Dried rat, mouse, hamster skin 〃	Yes	Yes
Mesentery (Stirewalt *et al.*, 1966)	—	Yes
Blood vessel wall 〃	—	No
Cat gall bladder 〃	—	No
Fish swim bladder 〃	—	No
Human horny layer sheets 〃	Yes	Yes
Green bean epidermis (Warren & Peters, 1968)	Yes	Yes
Gelatin membrane (Clegg, 1969)	Poor	Poor
Gelatin membrane+skin lipid (Clegg, 1969)	Yes	Yes

Schistosomule Collection

We believed, correctly as it turned out, that a skin membrane could be produced which would be thin enough to drop all the schistosomules into a collecting medium. Thus the number of schistosomules was equated with the number of cercariae penetrating, since no schistosomules were found to remain in the membrane. Here was a system for concentrating on penetration alone—actual entry into skin—uncomplicated by the effects of migration and schistosomule postpenetration development. Furthermore, this system made it possible to test the role of each potential penetration stimulus singly with the rest of the penetration environment optimal (Table 3) in a steady state.

The best penetration membrane was dried rat abdominal skin from female rats about 60 days old. HBSS was the collecting medium of choice. The highest percentage of schistosomules was recovered from 5,000 cercariae in a cercarial suspension 5 mm deep above 1,400 mm² of skin membrane. A light differential was optimal, with the cercarial chamber dark and a strong light below directed onto the under-surface of the skin through the collecting funnel. A temperature differential also was important with the temperature around the cercarial chambers maintained at about 28°C, that of the cercarial suspension at 34–35°C and that of the collecting medium at about 38°C (Stirewalt and Uy, 1969). Collecting vessels (Fig. 2) were described by Stirewalt and Uy (1969). Standard collecting conditions were established empirically (Table 3).

TABLE 3. *Optimal conditions for collecting schistosomules*

Condition	*Optimum*
Penetration membrane	Dried rat abdominal skin from about 60-day-old rats
Collecting medium	Hank's basic salt solution (HBSS)
Cercarial density	5000
Depth of cercarial suspension	5 mm
Light	Dark cercarial chamber and a light collecting funnel
Temperature	28° C room temperature 35° C cercarial chamber 38° C collecting medium

Cercarial Penetration Response

The results of this kind of testing, however, still did not give us the specific details we needed. They told us only that a certain percentage of the cercariae succeeded in penetrating. They did not distinguish between cercariae which unsuccessfully attempted to penetrate, and cercariae which made no attempt to penetrate. It is certainly possible that stimuli to penetrate might be active but some physiological condition of the cercariae or some barrier quality of the penetration membrane prevented passage. In other words, we were still mixing several aspects of the penetration process: the just-mentioned penetration capabilities of cercariae; host-finding; and penetration stimuli. So the next step was to add to our testing equipment a capability for observing cercarial behavior under the experimental conditions.

The vessels used for this purpose were essentially small schistosomule collection vessels, miniaturized to provide a diameter of 8 mm (Fig. 3). This reduced diameter was adopted to make it possible to observe cercarial behavior in the cercarial chamber through a stereomicroscope mounted laterally. The microscopic field covered the entire cercarial chamber so all cercariae could be observed. Except for size and the recording of results in terms of cercarial behavior instead of schistosomules collected, all aspects of the experiments were the same as with the schistosomule collecting apparatus.

One of two response patterns developed. Undisturbed, above an inert substrate such as a glass plate or a Baudruche membrane (treated bovine intes-

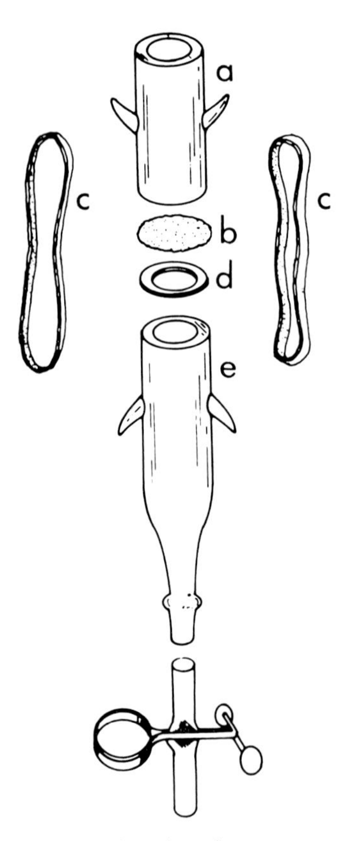

Fig. 3. Diagram of exploded vessel for observing cercarial penetration responses. *a*=cercarial observation chamber; *b*=penetration membrane; *c*=rubber bands which loop over hooks for holding vessel together; *d*=rubber gasket; *e*=schistosomule collecting chamber.

tinal membrane used as an insect feeding membrane), the distribution of cercariae in the water for several hours was essentially a water surface distribution (Fig. 4), though from time to time a few cercariae drifted slowly downward only to swim back up to the water surface before or immediately after contact with the substrate. Most cercariae never made contact with the substrate. This behavior constitutes absence of a penetration response.

By contrast, above a skin surface, within several minutes the behavior pattern of cercariae was as follows: most had drifted or swum to the skin and were creeping over the surface depositing mucus at each attachment of the oral sucker, or probing crevices, or they had assumed the penetration posture—attached orally, perpendicular or nearly so to the skin, and exhibiting the

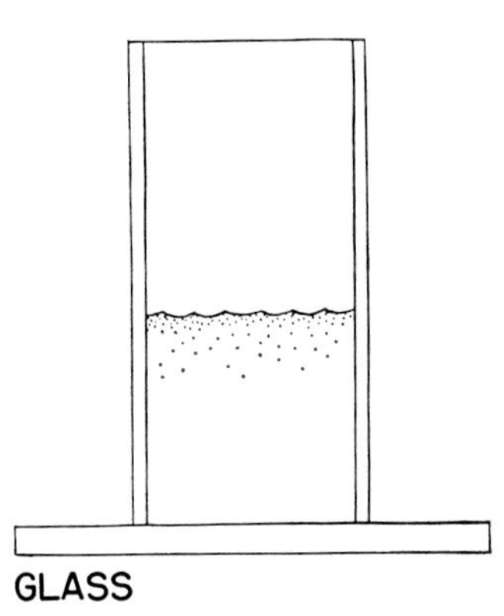

Fig. 4. Diagram representing absence of a penetration response. All cercariae are at or near the water surface.

intense general muscular activity associated with penetration (Fig. 5). This was a positive penetration response.

Responses were not always all-or-none. Then the number of cercariae making a positive response was recorded against time, and the results were treated quantitatively. An 8-minute reading was chosen as the experimental time.

Quantitative data of two kinds were thus used in the evaluation of penetration stimuli: (a) the percentage of schistosomules collected after penetration of the skin membrane, and (b) the percentage of cercariae observed to

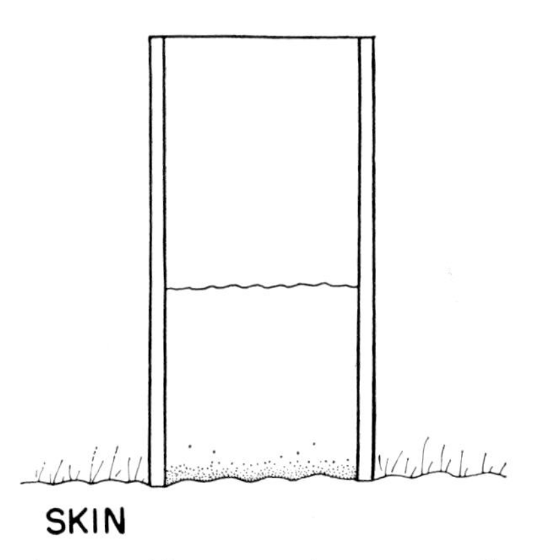

Fig. 5. Diagram representing a positive penetration response. Cercariae are at or near the skin surface.

make a positive penetration response within 8 minutes. In each case, the percentages from experimental vessels were compared with those from standard control vessels.

It was assumed that stimuli were perceived by cercariae and, if perceived, were expressed as penetration responses. Below is the sequence of steps describing the experimental protocol.

PENETRATION STIMULUS ⟶ CERCARIAL SENSORY RECEPTION ⟶ PENETRATION RESPONSE in terms of PERCENTAGE SCHISTOSOMULES COLLECTED / PERCENTAGE PENETRATION RESPONSES

Fig. 6. Diagram of sequence of steps describing the experimental protocol.

RESULTS

The results of testing factors for their capacity tc stimulate cercariae to approach skin, and to attach to it, explore it, and attempt to penetrate it, are recorded in Tables 4–8. These factors were tested concurrently by two parameters: schistosomule harvests through dried rat skin membranes; and cercarial penetration responses above penetration membranes.

During our search for optimal conditions for collecting schistosomules (Stirewalt and Uy, 1969) it was noted that water, saline solution, and HBSS to which serum had been added were not good schistosomule collecting media (Table 4). It did not seem reasonable, however, to interpret a collecting medium below the penetration membrane as a penetration stimulus.

When tested in terms of observed cercarial behavior above the skin membrane, the cercarial penetration response—100% positive within 8 minutes—denied the differential influence of any of the collecting media as a penetration stimulus. All of the cercariae were stimulated to penetrate, regardless of the collecting medium. Some other condition in our system interfered with their collection as schistosomules. This condition has not been identified.

TABLE 4. *Effect of the collecting medium on schistosomule collections and cercarial behavior (five replications)*

Collecting medium	*Schistosomule collection*		*Cercarial penetration response %*
	Av. %	*Range %*	
Water	<1	0–0.3	100
HBSS	52	50–56	100
Saline solution	0	0	100

Our next step was based on the fact that cercariae are positively phototactic. Light was tested as a penetration stimulus, first in terms of schistosomule harvests (Table 5). With otherwise optimal collecting conditions, uniform light, uniform darkness, and a light cercarial chamber over a dark collecting funnel gave essentially the same schistosomule collections. However, dark cercarial chamber over a strongly lighted collecting funnel—that is, a brightly lighted skin—greatly increased the percentage of schistosomules collected (Table 5).

TABLE 5. *Evaluation of light as a cercarial penetration stimulus (four replications)*

Lighting condition	*Schistosomule collection*		*Cercarial penetration response* %*
	Av. %	*Range* %	
Uniform room light	43	37–49	
Uniform darkness	44	36–56	—
Light/dark	37	27–50	80
Dark/light	74	69–80	100

* With uniform room temperature of 24°C, no cercarial penetration response was observed under any lighting conditions.

Cercarial behavior above a penetration membrane under different lighting conditions has not been sufficiently studied. As far as this behavior has been tested, it has not been clearly shown to be greatly influenced by light when other conditions were optimal (Table 5). It has not been assayed in the absence of a temperature stimulus.

Temperature, on the other hand, has been shown to play an important role in stimulating penetration. As tested by schistosomule collections, note that the lowest percentage of schistosomules was collected with a uniform temperature of 24°C (Table 6). Unfortunately, other uniform temperatures have not been evaluated. We also do not know what the cercarial sensitivity is to small temperature differences.

TABLE 6. *Evaluation of temperature as a cercarial penetration stimulus (seven replications)*

Temperature		*Schistosomule collection*		*Cercarial penetration response* %
Cercarial Chamber	*Collecting Medium*	*Av.* %	*Range* %	
24°C	24°C	5	3–7	0
	37°C	46	29–58	100
	42°C	21	11–29	—
35°C	38°C	60	56–64	100
24°C	18°C	—	—	0
	21°C	—	—	0

The optimal temperature of the collecting medium studied thus far was 38–39°C and that around the cercarial chambers was 28°C. Under these conditions the temperature of the cercarial suspension, i.e., in the darkened cercarial chamber, was 34–35°C. Changing either of these temperatures to the next level tested, reduced the numbers of schistosomules collected.

Table 6 shows that temperature was actually greatly influencing the cercarial penetration response. When a uniform room temperature was maintained throughout the system, no cercariae responded within 8 minutes, although about 50% had responded by penetrating at a 45-minute observation. With a target skin warmer than the cercarial suspension, all gave a penetration response. It is not clear from these data whether it is only the temperature differential which is necessary for the penetration response or whether some specific temperature is critical, but more recent data help to clarify this problem. When the temperature differential was reversed, and the cercarial chamber made warmer than the collecting medium, there was no penetration response at all. Obviously, we do not yet know all the answers, but a warm collecting medium, i.e., a warm skin target, was a strong stimulus to penetrate under the conditions used thus far.

It was apparent that even when light and temperature conditions were not optimal, some cercariae were still stimulated to penetrate. To check for other stimulating conditions, both a bright target skin and a warm target skin were eliminated from the system as penetration stimuli by using uniform light and temperature. Even so, in the presence of the skin membrane, from 4 to 13% of the cercariae penetrated and were collected as schistosomules (Table 7).

TABLE 7. *Effect of uniform room light and temperature (24°C) on the percentage of schistosomules collected (three replications)*

	Optimal conditions	*Uniform room light and temperature*
Average %	59	9
Range %	55–63	4–13

This fact was considered in the light of our knowledge that above a glass plate or a Baudruche membrane, even with conditions of light and temperature at the optimal stimulus levels, no cercariae exhibited a penetration response. It was clearly indicated that the skin membrane itself provided some penetration stimulus.

It is necessary now to try to present the skin surface as it must be sensed by host-finding cercariae. We will be concerned only with those skin structures which seem at present likely to be pertinent to cercarial entry into the host: the lipids covering the skin and hairs, the stratum corneum or horny layer immediately beneath it; and the projecting hairs.

A cercaria stimulated to penetrate skin or a skin membrane, comes in contact first with the lipid surface film. This is a fine homogeneous emulsion, half-aqueous and half-greasy (Rothman, 1954). It protects the skin against over-wetting, over-drying, abrupt temperature changes, and many infections. It comes from three sources: the sebaceous glands in the dermis around the hairs, sweat, and the cells of the keratinizing epidermis. It diffuses as a liquid over the skin surface, collecting especially along the skin sulci or wrinkles between the follicular orifices, and between the lamellae of the horny layer. At the surface temperature of skin, normally about 30°C, it solidifies into a thin layer similar to a hydrophylic ointment, following the contours and irregularities of the epidermis below it.

The environment thus provided must seem to a cercaria—slightly less than 1 mm long—to be a very rough one indeed. The topography of the surface consists of major plateaus separated by deep wrinkles, often deeper than the diameter of the cercaria (Kligman, 1964; Bernstein, 1969; Zelickson, 1967). Each plateau is secondarily crisscrossed with lesser wrinkles, forming smaller prominences and troughs. Further irregularities occur wherever underlying corneal scales are scuffed up (Fig. 1). Finally, hair shafts, several times the diameter of a cercaria, may project from mounds of tissue around the follicular orifices.

A few facts about the stratum corneum below the lipid film may be pertinent. This layer is comprised of thin flattened stratified dead cells. They are constantly pushed out and shed as flakes by the viable epidermal layers beneath, whose function it is to keep a supply of keratinized dead cells coming (Kligman, 1964).

The dead status of the horny layer is perhaps the prime factor in our development of an artificial penetration membrane of dried rat skin. The skin environment presented to a cercaria either *in situ* on a living host or *in vitro* after its excision and drying is probably essentially the same: a lipid surface film covering a laminated layer of dead cells composed of tough cell envelopes and closely packed fibrous keratin, all strongly bound together by a cementing substance which is considered to be mucopolysaccharide in composition.

Since the lipid film is the presenting surface to a cercaria, the obvious approach to testing the skin membrane for penetration stimuli was to remove it, and then to collect schistosomules and observe cercarial behavior. We considered this approach to be promising, since we knew that cercariae would give a positive penetration response to a glass surface only after skin lipids had been rubbed on to it.

The most satisfactory method of removing the surface lipid film, established empirically, was a combination of washing with warm water, pentane-extracting, and stripping with adhesive or scotch tape (Table 1). When the skin membrane had been so treated, only 4% of the cercariae were collected as schistosomules. Removal of the skin lipid by this method was also effective in eliminating from the skin membrane its capability of stimulating cercariae to penetrate. Only 0–2% of the cercariae made a penetration response (Table 8).

The critical test was to restore to the membrane the capacity for stimulating cercariae to penetrate. This was done with the pentane-treated skin membrane rubbed with human skin and hair lipids from the forehead and scalp. Under the conditions presented here, the penetration stimulating capacity was partially restored as recorded by the percentage of schistosomules collected. It was completely restored in terms of the cercarial penetration response (Table 8).

TABLE 8. *Evaluation of skin surface lipid as a cercarial penetration stimulus (four replications)*

Stimulus	*Schistosomule collection*		*Cercarial penetration*
	Av. %	*Range %*	*response %*
Untreated skin membrane	59	54–68	100
Membrane with skin lipid removed[a]	4	1–9	2
Treated membrane+skin lipid	18	10–25	100
Baudruche membrane	0	—	0
Baudruche+skin lipid	0	—	100

[a] Pentane and stripped (Table 1)

One of several explanations for the fact that restoration of the stimulating capacity of the skin membrane was only partial may be mentioned: (1) the difficulty of completely restoring the lipid film to the rough, irregular surface of the horny layer; (2) the possibility that the extraction treatment makes the skin membrane tougher to penetrate.

To test the cercarial response to skin lipids further, an inert membrane—the Baudruche (bovine intestinal) membrane mentioned earlier—was substituted for the skin membrane. To this substance cercariae made no penetration response at all. They hovered at the water surface, occasionally drifting downward in the water and then swimming back to the surface. Very few ever drifted far enough to touch the membrane, and these at once swam back up again. None was observed to remain on the membrane or explore the surface or attempt to penetrate. With skin lipid added, however, the penetration response was complete and unequivocal (Table 8), just as it is to a skin lipid film rubbed on a glass slide. Exploration was followed by attempts to penetrate which continued until the cercariae were exhausted, for they are unable to penetrate this membrane. After an hour, all were exhausted, lying on the membrane, unable to swim. The only motion discernible was spasmodic twitching.

The skin lipid film thus appears to provide an important stimulus. Furthermore, it or some other stimulus is effective not only upon contact but also remotely. The behavior of cercariae under optimal conditions indicated that some kind of penetration stimulus was perceived quickly by them even when they were at the surface of a cercarial suspension 5mm deep. Most of the cercariae under these conditions sank or swam directly to the skin membrane, attaching to the hair or skin within 8 minutes, and attempting to penetrate. When the depth of the cercarial suspension was increased to 15mm, however, they were farther from the skin membrane since cercariae tend to hang at the water surface. Whereas at the 8-minute observation all the cercariae in the suspension which was 5mm deep, had made the penetration response, only 76% of those in the 15mm deep suspension had done so. Even after 30 minutes only 93% had been stimulated to penetrate.

One other penetration stimulus has been recognized. We have always noted that when cercariae were penetrating a restricted area of skin, many entered at the same point—in a group. What happened was that a penetrating cercaria was often joined by others swimming or creeping in the vicinity, as the latter changed direction and moved to join the penetrant. As many as 15 have been seen collaborating to penetrate (Stirewalt and Hackey, 1956; Stirewalt *et al.*, 1966). The frequency and regularity with which this multiple entry has been observed suggests very strongly that some diffusible substance related to the penetration process is a penetration stimulus to nearby cercariae.

This phenomenon of group penetration has been observed regularly not only into skin (Stirewalt, 1959) and skin membranes, but also on other membranes studied, especially on Baudruche membranes and glass slides rubbed with skin lipids, and even during cercarial penetration of vaseline (Stirewalt and Strome, 1965). In consideration of the last statement, it could not be a response to some degradation product of the destroyed host tissues.

The penetration response is not made in the vicinity of swimming or creeping cercariae and so is not related to the mucus deposited by creeping cercariae from their postacetabular glands. It seems most likely that it is a response to preacetabular gland secretion.

DISCUSSION

It should be understood that even the refinements listed in Review of the Literature leave us with a picture which, albeit relatively simplified, may still be somewhat complex. The stimuli which bring cercariae to the skin may or may not be the same ones which cause them to attach, creep and explore, expend their post-acetabular gland secretion, begin entry, and secrete the contents of their preacetabular glands. For the moment it is our working assumption that, in the main, the stimuli *are* the same. In line with this assumption, we have included, for the present all of these responses in the single term, penetration response. This term was used no matter whether the response resulted in successful penetration or in an unsuccessful attempt, as happens when cercariae are stimulated to attack a Baudruche membrane or an eggshell membrane.

The need for distinguishing among these responses, as done by Bolwig (1955), as well as among the stimuli which elicit them, has already been demonstrated. Clegg (1969) found with his model that cercariae were not attracted to the cholesterol-treated gelatin film, but reacted by attaching and penetrating only after they had made contact by chance with the surface. To stimulate penetration, he found it necessary, therefore, to apply *Austrobilharzia terrigalensis* cercariae closely to the gelatin membrane surface. In contrast, observation of cercarial behavior with our model left no doubt that *Schistosoma mansoni* cercariae were indeed attracted to the rat skin membrane. The difference in their behavior above skin on the one hand, and above an inert Baudruche membrane or glass surface on the other, was striking. Under optimal conditions over a skin membrane, instead of remaining at or near the water surface as they did over glass, they dropped quickly to the skin or hair, to either of which they attached upon contact. Those which made contact first with a hair crept along it directly and unhesitatingly toward the skin, never away from it. These indications that some stimulus is perceived at a distance were strengthened by the finding that penetration responses were made most quickly and by the greatest percentage of cercariae when the depth of the suspension was least. The suggestion which immediately comes to mind, namely that the explanation is to be found in an increased number of random contacts of cercariae with the skin membrane when the suspension depth is less, is improbable for the following reason. The suspension depths used were such that cercariae above a nonstimulating glass or Baudruche membrane almost never dropped close enough to make contact with the surface.

The most likely stimuli perceivable by cercariae without contact, that is, remotely, were temperature and light: a warm target membrane and a bright light. The influence of these in terms of schistosomule harvests was dramatic. The effect of light alone on the observed cercarial penetration response, however, has not been demonstrated in the absence of a temperature stimulus. It is not probable that skin lipid is a remotely-sensed stimulus since there is no evidence of a water-soluble component in it which could diffuse through the cercarial suspension. Thus temperature seems to be the stimulus most likely to be remotely perceived. It may be mentioned in this regard, however, that once a few penetrants are in action, there is good evidence that a diffusible substance is present in the cercarial suspension. This substance appears to be a secretion from the preacetabular glands of the penetrants.

The possibility that a bright light on the membrane played a role in eliciting the penetration response is in line with the firmly-established positively phototactic character of cercariae. No experimental data are at hand on the comparative effect of reflected and transmitted light or of different types of light source. Only transmitted light was used in these experiments. The physics of light transmission and reflectance under water has not been related to this problem.

That a warm target was influential has been clearly demonstrated here for *S. mansoni* cercariae. *A. terrigalensis* cercariae reacted similarly under comparable conditions (Clegg, 1969). Temperature has also been shown to affect worm burdens of *S. mansoni* by DeWitt (1965) and of *Schistosoma haematobium* and *Schistosoma bovis* by Bolwig (1955). In addition, penetration of skin by *S. mansoni* was influenced in the studies of Stirewalt and Fregeau (1965). No temperature differentials were involved in either of these experiments. The differences in stimulating temperatures, 38–39° C for *S. mansoni* and 40° C for *A. terrigalensis,* are probably insignificant. If, however, the sensitivity to temperature of the collecting medium found under our experimental conditions (at 37° C, schistosomule collections averaging 46% compared with 60% at 38–39° C) can be confirmed, temperature differentials at this level do appear to be critical. In line with this, it should be noted that a collecting temperature of 42° C reduced the stimulating effect on *S. mansoni*, at least in terms of schistosomules collected.

It may be relevant in this context to cite the statement of Rothman (1954) that the skin temperature is measurably increased over large subcutaneous arteries. Cercariae observed over the chick chorioallantoic membrane usually collected to penetrate above the large underlying embryonic blood vessels. On the other hand, neither a warm temperature nor a bright light alone or in combination, stimulated cercariae to approach, attach, and creep over a Baudruche membrane. These conditions appear to act, then, as important but auxiliary stimuli to the stimulus from skin, and must be assessed in relationship to the penetration membrane used. Further study of the relative stimulating effect of the skin, light, and temperature is in progress.

It is necessary to consider here the penetration membranes or surfaces other than normal skin *in situ*, which have been tested with schistosome cercariae (Table 2). Most observers of cercariae in action have believed that they would attempt to penetrate any nonrepellant substratum presented to them. This is not true. If glass or Baudruche membranes are handled with gloves to prevent deposit on them of skin lipids, *S. mansoni* cercariae ignore their presence and make no attempt to penetrate them. The finding that some substances were attacked and others ignored must be interpreted as showing that the cercarial reaction to substratums is discriminating.

To identify with assurance, at this moment, a common characteristic of the stimulating substances is not possible. Skin lipids were present either naturally or experimentally on many membranes, but not on mesentery, eggshell or chorioallantoic membranes, green bean epidermis, or gelatin membranes. Eggshell membranes as well as the horny layers of skin, however, do contain keratin (Rothman, 1954), but it seems unlikely that this substance should be the penetration stimulus since it is covered with the surface lipid film. One important possibility must be considered: that is, unless the membranes just mentioned were handled specifically to prevent it, skin lipid may well have inadvertently been rubbed on them. Further study taking this into account must solve this problem.

Several unresolved problems should be brought up here and suggested for further study. The first is concerned with the methods for removing the skin surface lipid completely without leaving repellant residues or changing the susceptibility of the membrane to successful penetration. Of all the methods tested (Table 1), the one listed as "pentane and stripped" was considered to be the least likely to leave such residues and so was adopted as standard. Nothing is known about changes in penetrability of the extracted membrane,

The second problem is that of reapplication of skin lipids to the extracted rat skin membrane. We were never able to restore full capacity to the membrane for schistosomule collecting. It might be that it is impossible to replace this lipid in the deep wrinkles and around the irregularities where most *S. mansoni* cercariae find penetration sites. Or it might be that the lipid film applied was so thick that it provided an added barrier to penetration, but this hypothesis needs proof. It can only be said here that in terms of schistosomule harvests the restoration was only partial; in terms of cercarial penetration responses, the stimulating effect was fully restored (Table 8), which points to the actual process of penetration itself as being the phase of reaction affected.

The third problem is the comparative stimulating effect of sebum covering the hairs as opposed to the skin surface lipid film composed of sebum, lipid from the keratinizing epidermis and, in some animals, sweat. No information on this is available. The observation mentioned above that some cercariae approaching skin touched hairs first and crept along them to the skin surface has always indicated that contact with hair was random and that the hair shaft served only as a means of approach to the skin.

When all the data are considered it must be concluded, at this time, that skin lipid film appears to be a primary penetration stimulus. Its removal from mouse ear skin (Wagner, 1959) and from dried rat skin membranes as used in this study resulted in greatly decreased schistosomule harvests as well as reduced cercarial penetration responses (Table 8). Its restoration to extracted skin membranes restored the cercarial penetration responses completely and the schistosomule harvests partially. Similar results were reported with other cercariae and skin *in situ* (Bolwig, 1955; Wagner, 1959). Its addition to glass and Baudruche membranes incited cercariae to attack them, although they could not penetrate them. Clegg (1969) found the same with gelatin films. In addition, he showed that the active fraction of chicken skin lipid for *A. terrigalensis* cercariae was in the free sterol fraction, more specifically, cholesterol. With his model, fatty acids could not be proved to be stimulatory, although Wagner found them to be for *Schistosomatium douthitti* cercariae.

In closing, the suggestion seems appropriate that the times are propitious for elucidation of all phases of skin penetration mechanisms by parasites. Much is known about the composition of skin layers and many models of skin and parasite are available. In future, emphasis should be placed on sorting out and studying separately the various phases of what has been called here the penetration response: finding the host, selecting penetration sites, initiating entry and, for schistosome cercariae, secreting the contents of the postacetabular and preacetabular glands. Distinguishing between secretion from the two types of acetabular glands is of especial importance since these secretions have been shown to be different histochemically, to be secreted during different stages of the process of penetration, and to be under different muscular control (Stirewalt and Kruidenier, 1961).

SUMMARY

Several promising conditions were tested as penetration stimuli for cercariae of *Schistosoma mansoni.* Testing was performed *in vitor* using schistosomule collecting vessels and a system for observing cercarial behavior above a penetration membrane. Optimal conditions for collecting schistosomules through a dried rat skin membrane were maintained except that one condition at a time was changed to permit its evaluation as a penetration stimulus. Two kinds of data were recorded: the percentage of cercariae collected as schistosomules after their penetration of the skin membrane; and the percentage of cercariae observed to make a penetration response, that is, swim to the membrane, attach, creep and explore, and begin entry. The following conditions served as penetration stimuli: the skin surface lipid; a warm target membrane in a temperature-differentiated environment; possibly a brightly-lighted target membrane in a light-differentiated system; and diffusing secretion from the preacetabular glands of penetrating cercariae.

ACKNOWLEDGMENTS

The cooperation of other staff members of the Naval Medical Research Institute is gratefully acknowledged. Mildred Walters, Augusto Uy, and R. D. Arrieta of the Parasitology Division supplied materials and technical assistance. M. L. Yeager and N. J. Laxey, Graphic Arts, provided the illustrations. The members of the Photography Laboratory prepared the photographs.

REFERENCES

Abadie, S. H. 1963. The life cycle of *Strongyloides ratti*. J. Parasitol. 49:241–248.

Bernstein, E. O. 1969. Skin replication procedure for the scanning electron microscope. Science 166:252–253.

Bolwig, N. 1955. An experimental study of the behavior and host recognition in *Schistosoma* cercariae. S. Afr. J. Sci. 51:338–344.

Clegg, J. A. 1969. Skin penetration by cercariae of the bird schistosome *Austrobilharzia terrigalensis*: the stimulatory effect of cholesterol. Parasitology 59:973–989.

DeWitt, W. B. 1965. Effects of temperature on penetration of mice by cercariae of *Schistosoma mansoni*. Amer. J. Trop. Med. Hyg. 14:579–580.

Fülleborn, F. 1914. Untersuchungen über den Infektionsweg bei *Strongyloides* und *Ankylostomum* und die Biologie dieser Parasiten. Arch. f. Schiffs. Tropenhyg. 18:182–236.

Fülleborn, F. 1924. Über "Taxis" (Tropismus) bei *Strongyloides* und *Ankylostomen* larven. Arch. f. Schiffs. Tropenhyg. 28:144–165.

Goodey, T. 1925. Observations on certain conditions requisite for skin penetration by the infective larvae of *Strongyloides* and Ankylostomes. J. Helminthol. 3:51–62.

Kligman, A. M. 1964. The biology of the stratum corneum. In W. Montagna and W. C. Lobitz (eds.) The Epidermis. Academic Press, New York. 649 p.

Kosugi, I. 1924. Wie weit wirken bei dem Eindringen von Strongyloideslarven und anderen parasitischen Nematoden in das Gewebe spezifische Reize. Arch. f. Schiffs. Tropenhyg. 28:179–187.

Neuhaus, W. 1952. Biologie und Entwicklung von *Trichobilharzia szidati* n.sp. (Trematoda. Schistosomatidae), einem Erreger von Dermatitis beim Menschen. Zeit. Parasit. 15:203–266.

Olivier, L. J. 1966. Infectivity of *Schistosoma mansoni* cercariae. Amer. J. Trop. Med. Hyg. 15:882–885.

Purnell, R. E. 1966a. Host-parasite relationships in schistosomiasis. I. The effect of temperature on the infection of *Biomphalaria sudanica tanganyicensis* with *Schistosoma mansoni* miracidia and of laboratory mice with *Schistosoma mansoni* miracidia and of laboratory mice with *Schistosoma mansoni* cercariae. Ann. Trop. Med. Parasitol. 60:90–93.

Purnell, R. E. 1966b. Host-parasite relationships in schistosomiasis. II. The effects of age and sex on the infection of mice and hamsters with cercariae of *Schistosoma mansoni* and of hamsters with cercariae of *Schistosoma haematobium*. Ann. Trop. Med. Parasitol. 60:94–99.

Rothman, S. 1954. Physiology and Biochemistry of the Skin. Univ. Chicago Press, Chicago. 741 p.

Stirewalt, M. A. 1953. The influence of previous infection of mice with *Schistosoma mansoni* on a challenging infection with the homologous parasite. Amer. J. Trop. Med. Hyg. 2:867–882.

Stirewalt, M. A. 1959. Chronological analysis, pattern and rate of migration of cercariae of *Schistosoma mansoni* in body, ear and tail skin of mice. Ann. Trop. Med. Parasitol. 53:400–413.

Stirewalt, M. A. 1966. Skin penetration mechanisms of helminths. In E. J. L. Soulsby (ed.) The Biology of Parasites. Academic Press, New York. 354 p.

Stirewalt, M. A. and Fregeau, W. A. 1965. Effect of selected experimental conditions on penetration and maturation of *Schistosoma mansoni* in mice. I. Environmental. Exp. Parasitol. 17:168–179.

Stirewalt, M. A. and Fregeau, W. A. 1968. Effect of selected experimental conditions on penetration and maturation of *Schistosoma mansoni* in mice. II. Parasite-related conditions. Exp. Parasitol 22:73–95

Stirewalt, M. A. and Hackey, J. R. 1956. Penetration of host skin by cercariae of *Schistosoma mansoni*. I. Observed entry into skin of mouse, hamster, rat, monkey and man. J. Parasitol. 42:565–580.

Stirewalt, M. A. and Kruidenier, F. J. 1961. Activity of the acetabular secretory apparatus of cercariae of *Schistosoma mansoni* under experimental conditions. Exp. Parasitol. 11: 191–211.

Stirewalt, M. A., Minnick, D. R. and Fregeau, W. A. 1966. Definition and collection in quantity of schistosomules of *Schistosoma mansoni*. Exp. Parasitol. 60:352–360.

Stirewalt, M. A., Shepperson, J. R., and Lincicome, D. R. 1965. Comparison of penetration and maturation of *Schistosoma mansoni* in four strains of mice. Parasitology 55: 227–235.

Stirewalt, M. A. and Strome, C. P. 1965. Cercarial penetration of a skin-substitute medium in phase contrast cinemicrography. J. Parasitol. 51 (Sec. 2): 59.

Stirewalt, M. A. and Uy, A. 1969. *Schistosoma mansoni*: cercarial penetration and schistosomule collection in an *in vitro* system. Exp. Parasitol. 26:17–28.

Wagner, A. 1959. The initiation of penetration of *Schistosomatium douthitti* cercariae. J. Parasitol. 45 (Sec. 2): 16.

Warren, K. S. and Peters, P. A. 1967a. Quantitative aspects of exposure time and cercaria. dispersion on penetration and maturation of *Schistosoma mansoni* in mice. Ann. Tropl Med. Parasitol. 61:294–301.

Warren, K. S. and Peters, P. A. 1967b. Comparison of penetration and maturation of *Schistosoma mansoni* in the hamster, mouse, guinea pig, rabbit, and rat. Amer. J. Trop. Med. Hyg. 16:718–722.

Warren, K. S. and Peters, P. A. 1968. Cercariae of *Schistosoma mansoni* and plants: attempt to penetrate *Phaseolus vulgaris* and *Hedychium coronarium* produces a cercaricide. Nature 217:647–648.

Zelickson, A. S. 1967. Ultrastructure of Normal and Abnormal Skin. Lea and Febiger, Philadelphia. 431 p.

Interaction Between *Schistosoma mansoni* Schistosomules and Penetrated Mouse Skin at the Ultrastructural Level

ERIK RIFKIN

Naval Medical Research Institute
Bethesda, Maryland

INTRODUCTION AND REVIEW OF THE LITERATURE

At the host-parasite interface structural and physiological interactions between heterospecific organisms take place. In the case of the helminth parasite, the tegument (surface) of the worm is in intimate contact with the fluids and tissues of the host. Likewise, the host cells are exposed to modified tegumental structures in addition to potentially harmful elements elaborated by the parasite.

This region of host-parasite interaction can effectively be interpreted at the ultrastructural level. Although the electron microscope is becoming a standard tool in the study of biological materials, a review of the literature will show that it has been infrequently used in studying helminth-host interactions. Specifically in regard to the Trematoda, although the literature is replete with ultrastructural studies of the tegument, the worms had been removed from their host prior to being observed. In spite of this, these studies have shed considerable light on the structural nature of the trematode surface, and will be reviewed here.

Hyman (1951) summarized the early work on the structure of the trematode "cuticle" in the following manner.

> The trematodes lack an epidermis and are clothed instead with a resistant cuticle. The homology and origin of this cuticle have long been

From Bureau of Medicine and Surgery, Navy Department, Research Task MF12 524 009 1004.

The experiments reported herein were conducted according to the principles enunciated in "Guide for Laboratory Animal Facilities and Care" prepared by the Committee on the Guide for Laboratory Animal Resources, NAS-NRC.

The opinions and assertions contained herein are those of the author and are not to be construed as official or as reflecting the views of the Navy Department or the Naval service at large.

disputed, and several theories have been advanced: (1) that the cuticle is an altered and degenerated epidermis, (2) that it is the basement membrane of the former epidermis, (3) that it is the outer layer of an insunk epidermis, the cells and nuclei of which have sunk beneath the subcuticular musculature, (4) that the cells in question are not epidermal but are mesenchymal (parenchymal) cells that secrete the cuticle, and (5) that the cuticle is secreted by the ordinary mesenchyme, not by special cells.

With the advent of the electron microscope it was found that the "cuticle" did not consist of an inert, acellular covering but of a cytoplasmic syncytium (external tegument) bordered by a unit membrane. Threadgold (1963) and Björkman and Thorsell (1964) demonstrated that this cytoplasmic layer was 15–21 μ thick on the adult *Fasciola hepatica*. Within this syncytium were found mitochondria, rough and smooth endoplasmic reticulum, and numerous dense rod-shaped bodies 20–30 mμ in greatest diameter. Small vesicles and membrane-bound vacuoles were also present. This external syncytium was connected by cytoplasmic bridges to cytons sunken among the parenchyma. Unlike the syncytial layer, these cytons were nucleated and contained dictyosomes. These cytons represent a portion of the epidermis and are collectively designated as the internal tegument. Projecting above the surface of the external tegument, yet covered by the limiting unit membrane, were electron-dense spines.

Burton (1964) studied the ultrastructure of the adult tegument of *Haematoloechus medioplexus* (a lung fluke of amphibians) and found it to be similar to that of *F. hepatica*. The syncytial layer contained vesicles, fibrous elements, and dense, membrane-bound, oval bodies. These dense bodies were also located in the sunken epidermal cells. Burton postulated that these bodies, formed in the inner tegument, migrated to the syncytial surface or external tegument via the cytoplasmic bridges and added to the formation of the fibrous elements and/or the spines.

Burton (1966) also studied the tegument of the frog bladder fluke, *Gorgoderina* sp., and observed a cytoplasmic syncytium with perinuclear cytoplasmic units lying in the parenchyma. He concluded that two types of secretion bodies were formed in the perinuclear cytoplasm (internal tegument) and were transported through cytoplasmic bridges to the cytoplasmic syncytium. Just beneath the limiting membrane of the syncytium was a dense zone of particulate material. No spines were present. Although he found no evidence of exchange between the surface and environment in *Gorgoderina* sp., he concluded that structural evidence did indicate such exchange in *H. medioplexus*. In comparing the tegument of the lung and bladder flukes, Burton (1966) stated "that the integument of the lung fluke is in greater communication with its surroundings than that of the bladder fluke. This is not surprising, considering the dissimilar environments of the worms, and also considering that the bladder probably has less to offer, from a metabolic standpoint, than the lung."

The tegument of adult *Schistosoma mansoni* was first studied at the ultrastructural level by Senft *et al.* (1961). These investigators concluded that the worms were covered by an acellular amorphous cuticle. Supposedly this layer was secreted by large cells embedded in the parenchyma. Cytoplasmic connections between these cells and the "cuticle" were not seen.

Morris and Threadgold (1968), in their study of adult *S. mansoni*, described the "cuticle" as an outer syncytial portion of a tegumental unit. They observed a folded syncytial surface which was connected to nucleated tegumental cells embedded in the parenchyma. A few mitochondria, electron-dense rod-like inclusions, and spherical electron-lucid inclusions were described in both the inner and outer teguments. Smith *et al.* (1969), as part of a larger study, reported structural variations between cercariae, schistosomules, and male and female adult schistosomes. The basic design of the body surfaces of all three stages once again consisted of an outer syncytial layer connected to sunken cytons by cytoplasmic processes. Tubular infoldings of the outer plasma membrane became more complex with age in all three stages. The outer tegument of the cercaria was covered by a fibrillar envelope. In the adult, ribosomes and dictyosomes were found in the cytons, and the external tegument included mitochondria, multilaminate vesicles, discoid granules, spines, and dense bodies. It was concluded that the cell organelles associated with protein synthesis were in the cytons, while those dealing with absorption and protection were in the syncytial external tegument.

The present study is concerned with the interaction of *S. mansoni* schistosomules and penetrated mouse epidermal cells. Few examples of ultrastructural studies concerned with the host-parasite interface can be found in the literature. Rifkin *et al.* (1969) studied the constituents of encapsulating cysts in the American oyster, *Crassostrea virginica*, formed in response to the larva of the cestode *Tylocephalum* sp. The capasule elaborated by the host in response to the parasite consisted of three cell types and two types of extracellular fibers. The major type of extracellular fiber in the vicinity of the host-parasite interface was found attached to the outer surface of the unit membrane covering the parasite's microvilli. Spines protruding from the worm's tegument were commonly embedded in the fibroblasts of host origin which made up the inner wall of the capsule. Morseth and Soulsby (1969b) demonstrated in vitro that rabbit or guinea pig polymorphonuclear leukocytes showed firm Ab-mediated adherence to the cuticle of *Ascaris suum* larvae. In the areas of leukocyte-parasite contact there was wrinkling and dissolution of the cuticle. In an earlier study (Morseth and Soulsby, 1969a), these workers showed that adhering pyrinophils were associated with little or no evidence of alteration of the cuticular integrity.

MATERIALS AND METHODS

The *Schistosoma mansoni* cercariae used in this study were from laboratory-maintained *Biomphalaria glabrata* colonies originally from Puerto Rico. The cercariae were collected from shedding snails and concentrated in small vessels prior to use. The skin to which the cercariae would eventually be exposed was excised from the belly of five-day-old Swiss albino mice (NIH/P1 strain) weighing an average of 2–3 grams each. The skin was cut into pieces $9 cm^2$, and securely placed over a cork, the center of which (7 cm) had been removed.

One hundred and fifty cercariae were placed on the skin and the penetration process was observed under a dissection microscope. After 20 minutes, the skin was minced into blocks approximately $1 mm^3$ and subsequently fixed.

The fixative employed was cold (4° C) glutaraldehyde in 0.1 M phosphate buffer maintained at pH 7.4. After fixation for 2 hours, the tissues were washed for 90 minutes in phosphate buffer and post-fixed in a phosphate-buffered 1 % osmium tetroxide solution. This was followed by washing in 0.1 M sodium acetate solution for 90 minutes, stained with uranyl acetate (2 %) for 2 hours, and finally washed again in sodium acetate solution prior to dehydration in an ethanol series. A low viscosity Epon (Bailey, personal communication) was used as the embedding medium.

Sections, 3μ thick, were cut on an AO microtome, heated, and stained with basic toluidine blue, in order to locate the parasites. The blocks were then placed on a Porter Blum MT-2 ultramicrotome and ultrathin sections approximately 700–1000 Å (gold and silver interference colors) were obtained and stained with lead citrate (Reynolds, 1963). The sections were placed on non-coated copper grids and examined on a Siemens 1A electron microscope operated at 80 Kv.

RESULTS

Two types of tissue are in close contact as schistosomules move into and through host skin: (1) tegument of the parasite, and (2) host epidermal cells surrounding it. The tegument of the schistosomule will be considered first.

Parasite Tegument

The tegument of *Schistosoma mansoni* schistosomules has not been described in detail. Like that of the cercaria and adult, it consists of what may arbitrarily be defined as external and internal tegumentary levels (Fig. 1). The external level consists of an anucleate, cytoplasmic syncytium delimited on its outer surface by a unit membrane. A fibrous basal lamina situated mediad to the external tegument separates this layer from the underlying subsurfacial musculature. Cytoplasmic bridges extend from the external tegument, through the musculature, and connect the syncytium with its cytons. These nucleated cell bodies lie below the musculature and are juxtaposed with the cytons of the circular and longitudinal muscle complex.

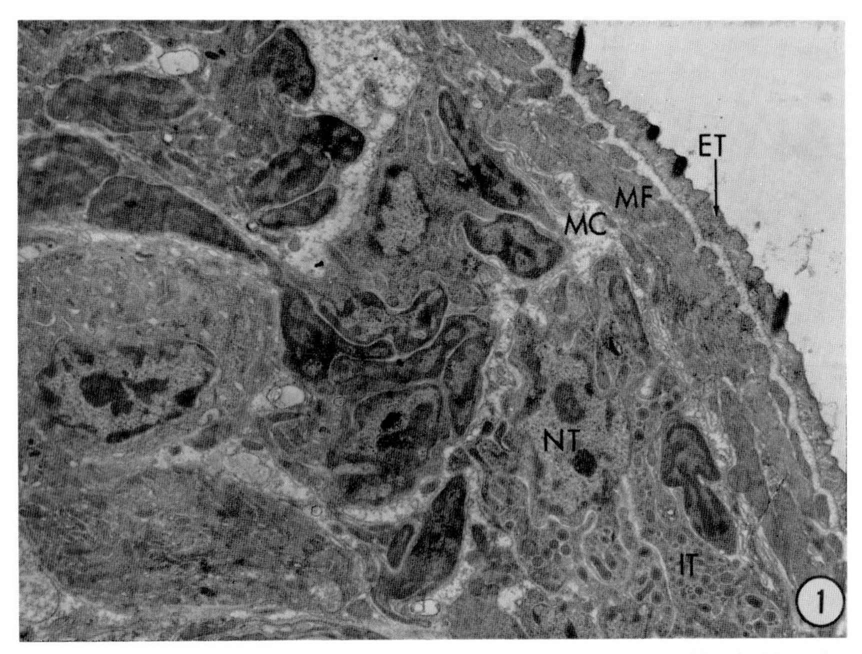

Fig. 1. Electron micrograph showing the external and internal tegumental levels. Note the underlying subsurfacial muscle fibers and cytons of the internal tegument. (×7,100.)

External Tegument.—The outer limiting unit membrane that surrounds the schistosomule is approximately 100 Å in diameter (Figs. 3,4) and follows the contour of the external tegument. The outer surface of the syncytium is irregular and includes infolded areas which produce deep pockets (Figs. 3,6).

Lamellated, electron-dense, tegumentry spines are embedded within the syncytium with their apices protruding from its surface (Figs. 3, 6, 8). These spines, which vary considerably in size (average $1.1 \times 0.2\mu$), are covered on their outer surfaces by the surfacial unit membrane. The proximal surfaces of the spines are contained by the plasma membrane delimiting the inner surface of the syncytium. Electron-dense regions also appear just beneath the outer limiting unit membrane and above the inner plasma membrane (Figs. 2, 6).

The most prominent organelle in the external tegument as well as in the cytoplasmic bridges and the cytons, is the lamellated body. When observed in cytons, these bodies are round or dumbbell-shaped, with an electron-dense core from which emanates a myelin-like substructure. When observed in the external tegument, the electron-dense core is diffuse, and in many instances disappears entirely (Figs. 3, 4). Furthermore, the myelin-like substructure expands, giving the entire body the appearance of an amorphous group of membranes. One of the prominent features of the external tegument is its extensive membranous network (Fig. 7). Mitochondria are present, although rarely, and their profiles average 0.25μ in greatest diameter (Fig. 8). Endoplasmic reticulum and dictyosomes are conspicuously absent.

The basal lamina, averaging 0.5μ in width, is composed of a fibrous reticular network embedded in a homogenous matrix (Figs. 3, 6). Fine filamentous projections from the basal lamina extend into the syncytial external tegument thereby increasing the surface area at this junction (Fig. 3).

Cytoplasmic Bridges.—The cytoplasmic bridges, averaging 100 mμ in width, connect the cytons comprising the internal tegument with the external syncytial tegument (Figs. 2, 3, 4, 6). Lamellated bodies are quite numerous in these processes and their appearance is intermediate between similar organelles found in the external and internal levels of the tegument (Figs. 2, 3, 4, 6). Infrequently, mitochondria with tubular cristae are found within these cytoplasmic bridges (Fig. 6). Running parallel to the long axis of these extensive processes (the bridges) are microtubules (Figs. 2, 3, 4, 6). These tubules, measuring 25 mμ in diameter, traverse the entire length of the process and are contained toward its periphery (Fig. 3). The processes are bordered externally by an outer limiting membrane.

Cytons.—The nucleated cell bodies lie mediad to the longitudinal subsurfacial muscles (Fig. 5). Anastomosing cytoplasmic projections extend from these cytons into the adjacent extracellular spaces. The cytoplasm of each cyton is replate with the dumbbell-shaped lamellated bodies described earlier. These bodies are membrane bound, 0.6μ in diameter at their ends, and include lamellae which are spaced 150 Å apart. The electron-dense core is 0.3μ in cross-section. The cytons contain mitochondria, the profiles of which measure 0.7μ in greatest diameter. Glycogen granules are scattered throughout the cytoplasm.

Host Epidermis

From 15 to 20 minutes after application of cercariae to the skin, the strata spinosum and granulosum of the host's epidermis were observed to be in con-

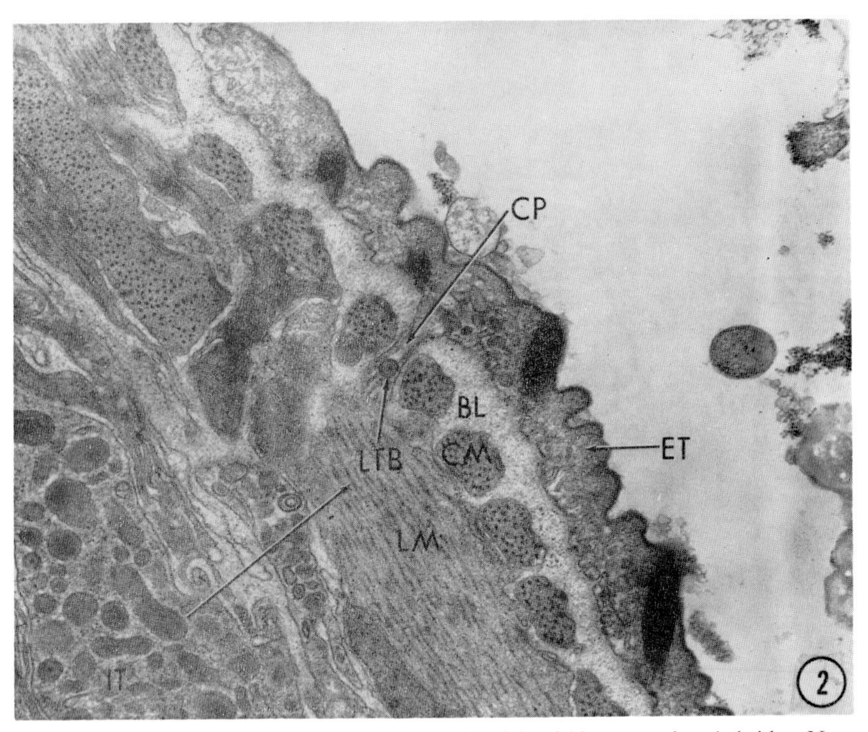

Fig. 2. Electron micrograph showing a lamellated vesicle within a cytoplasmic bridge. Note the lamellated vesicles in the cyton of the internal tegument. (×25,000.) *BL*=basal lamina, *CM*=circular muscle, *CP*=connective process (cytoplasmic bridge), *ET*= external tegument, *IT*=internal tegument, *LM*=longitudinal muscle, *LTB*=lamellated tegumental body, *MC*=muscle cell, *MF*=muscle fibers, *NT*=tegument nucleus.

tact with the tegument of the schistosomules (Figs. 8, 9, 10). These host cells each contain a nucleus varying in shape from ovoid to elliptical which is surrounded by a folded nuclear membrane (Fig. 11). The chromatin is most densely concentrated toward the periphery of the nucleus bordering the inner leaf of the nuclear membrane. A prominent nucleolus is readily apparent and measures approximately 0.5μ in diameter.

Desmosomes frequently occur at irregular intervals along the limiting unit membrane (Figs. 10, 11). Three filament-like elements run within and are parallel to the long axis of the attachment plaque. The cell membranes in the vicinity of the desmosomes are highly convoluted.

Tonofibrils are scattered throughout the cytoplasm but it is difficult to distinguish their tonofilament substructure (Figs. 10, 11). These tonofibrils in many instances are in juxtaposition with the desmosomes or are attached directly to them.

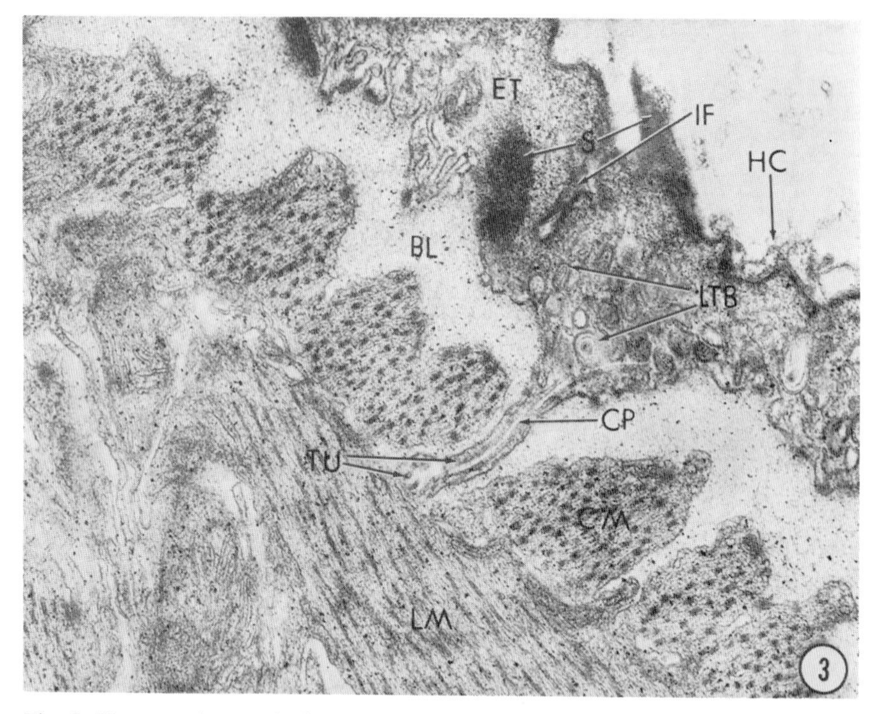

Fig. 3. Electron micrograph showing microtubules peripherally located in a cytoplasmic bridge. Note the membranous lamellated vesicles in the syncytium. Segments of a hirsute coat can be seen adhering to the membrane of the external tegument. (×40,500.)

The most distinguishing feature of the stratum granulosum is the keratohyalin granules (Figs. 9, 10, 11). Although they characteristically display a dense granular appearance, their number, size (average 1μ) and distribution vary from cell to cell. The granules coalesce with the tonofibrils, and careful examination shows them to be bordered by a trilamellar unit membrane.

Endoplasmic reticula are randomly dispersed throughout the cytoplasm but distyosomes are not clearly defined. The cytoplasm contains many ribosomes as well as lamellated bodies measuring 0.2μ in diameter (Figs. 9, 10, 11). Mitochondria (0.5μ profile) with elongate tubular cristae are dispersed throughout the cytoplasm but are more concentrated in the perinuclear region.

Host-Parasite Interaction

Any description of the schistosomule-host epidermal interaction should be preceded by a brief account of schistosomular migration through the horny layers of the host's skin into the strata granulosum and spinosum. The horny layer, or stratum corneum, consists of layered, dead, epidermal cells that provide a protective covering for the animal. These flattened cell layers are held together by a cohesive intercellular cement (Kligman, 1964).

The cercaria orients itself perpendicular to the skin surface and enters the stratum corneum by muscular activity of the tail and elongation and contraction of the body. The latter produces an enlarging of the site of entry (Gordon and Griffiths, 1951). The parasite turns and follows the stratum lucidum or keratogenous zone just beneath the stratum corneum (Stirewalt, 1959). Secretions from the preacetabular glands are now released into the skin (Stirewalt and Kruidenier, 1961), and after a testing period (Gordon and Griffiths, 1951), the schistosomule, as the parasite is now called, enters the living epidermal layers.

The apices of the spines come into direct contact with the host's epithelium as does the unit membrane surrounding the syncytial external tegument (Figs. 8, 9). The abrasive action of the spines against the host cells causes extensive

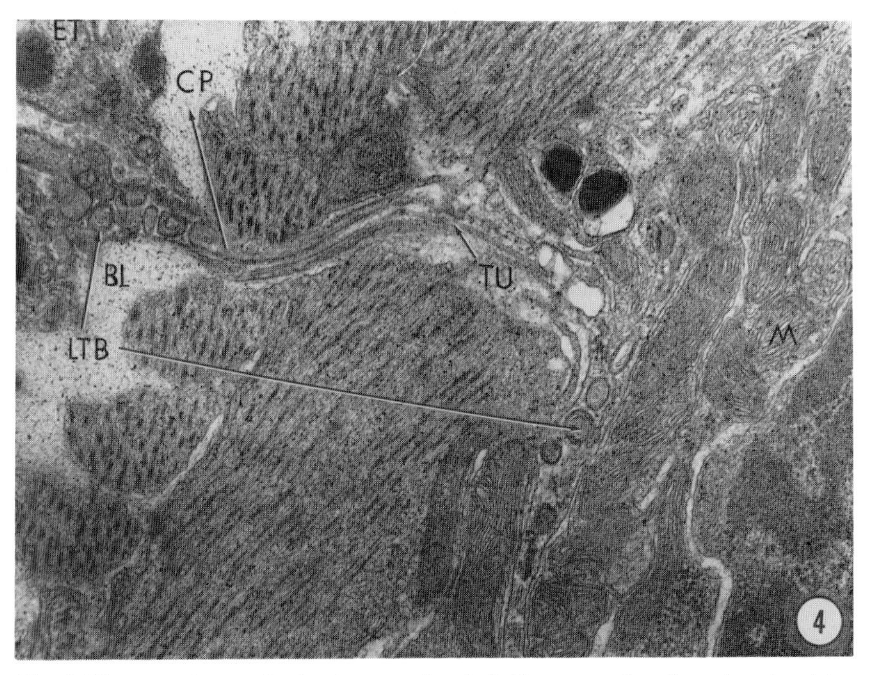

Fig. 4. Electron micrograph showing cytoplasmic bridge connecting the external and internal levels of the tegument. Note the lamellated tegumental bodies in both levels of the tegument. (×39,000.). *BL*=basal lamina, *CM*=circular muscle, *CP*=connective process (cytoplasmic bridge), *ET*=external tegument, *HC*=hirsute coat, *IF*=infolded tegument, *LM*=longitudinal muscle, *LTB*=lamellated tegumental body, *M*=mitochondria, *S*=spines, *TU*=tubules.

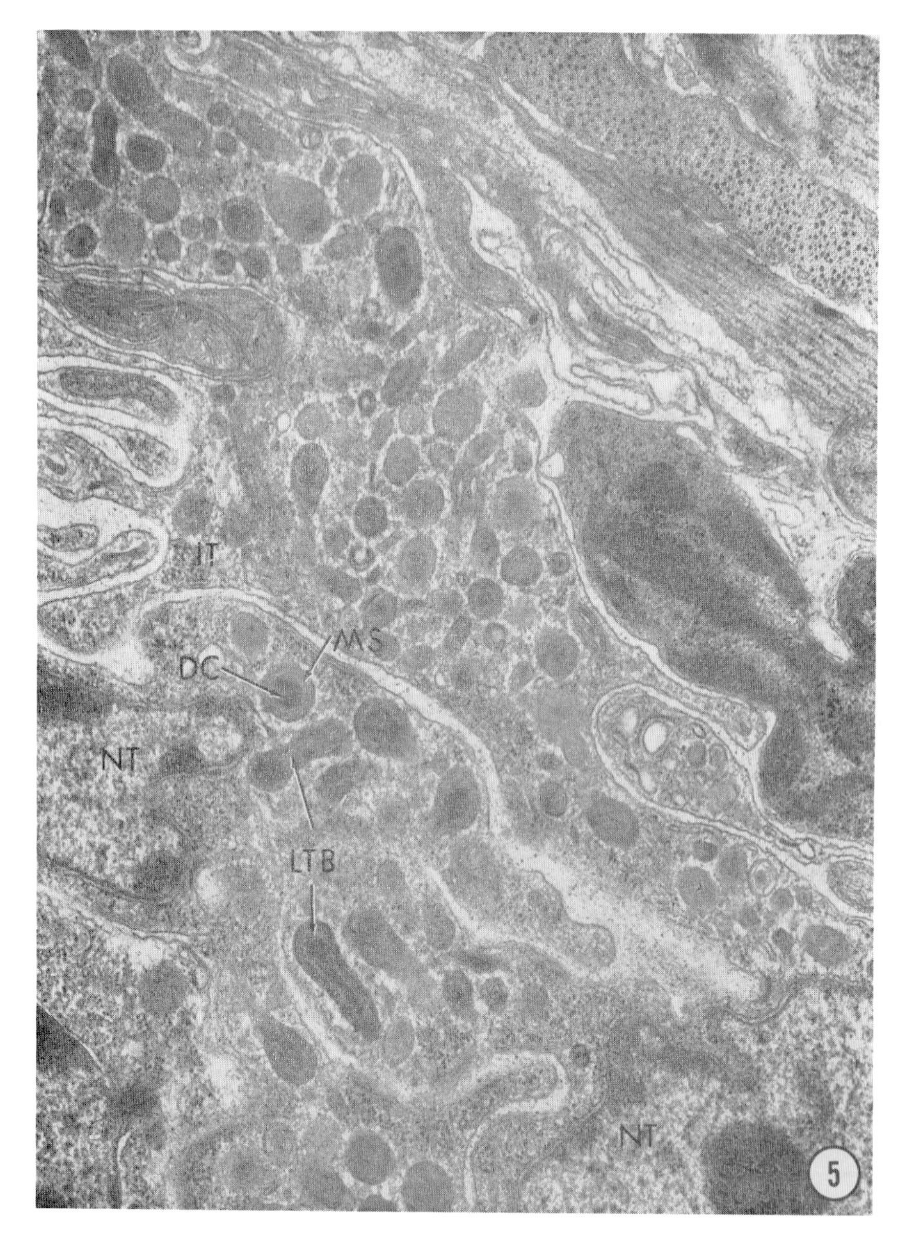

Fig. 5. Electron micrograph showing internal tegumental cyton containing lamellated bodies. Note the electron-dense core and the myelin-like substructure of the body. (×13,500.) *DC*=dense core, *IT*=internal tegument, *LTB*=lamellated tegumental body, *MS*=myelin-like substructure, *NT*=tegument nucleus.

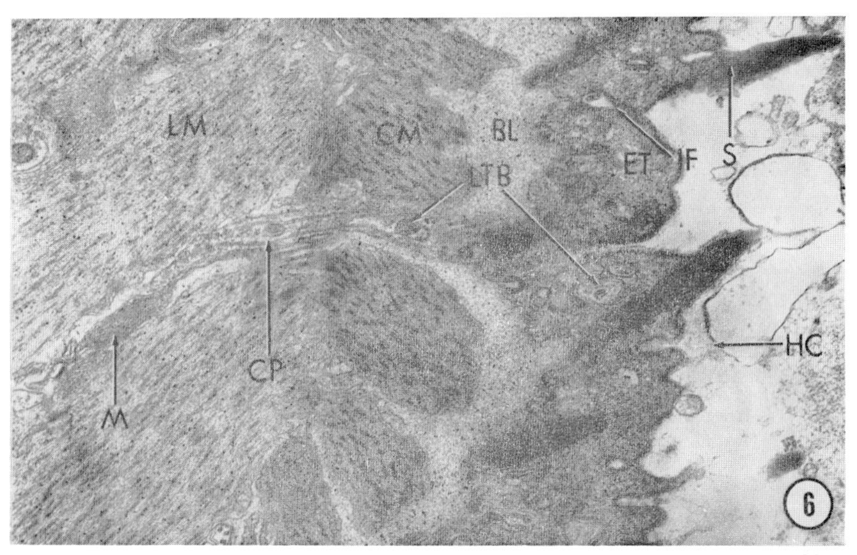

Fig. 6. Electron micrograph showing external tegument and associated structures. Note lamellated bodies within the syncytium (external tegument) and the cytoplasmic bridge. Spines can be seen projecting from the syncytial external tegument, and the hirsute coat can be seen adhering to the outer limiting unit membrane. (×27,500.)

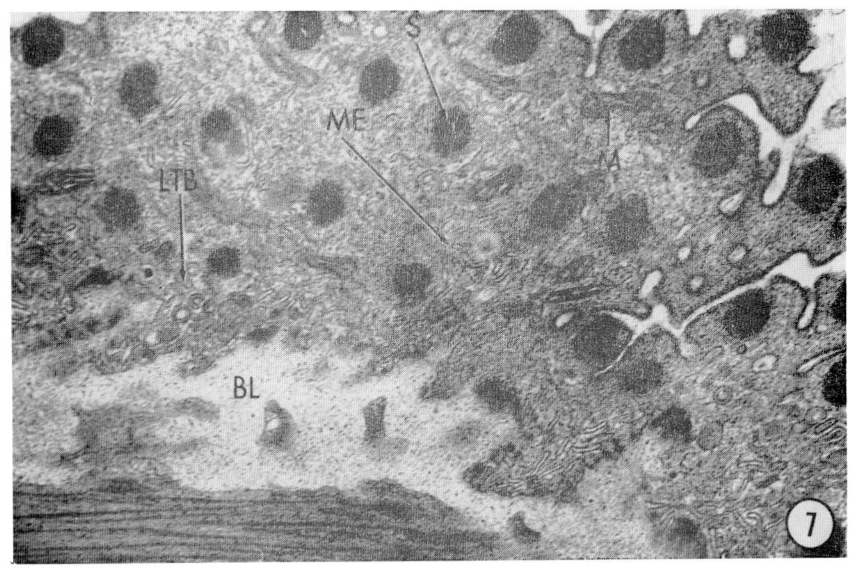

Fig. 7. Electron micrograph showing extensive membranous network within the external tegument. Note the cross-sectional representation of the spines. (×39,000.) *BL*=basal lamina, *CM*=circular muscle, *CP*=connective process (cytoplasmic bridge), *ET*=external tegument, *HC*=hirsute coat, *IF*=infolded tegument, *LM*=longitudinal muscle, *LTB*=lamellated tegumental body, *M*=mitochondria, *S*=spines.

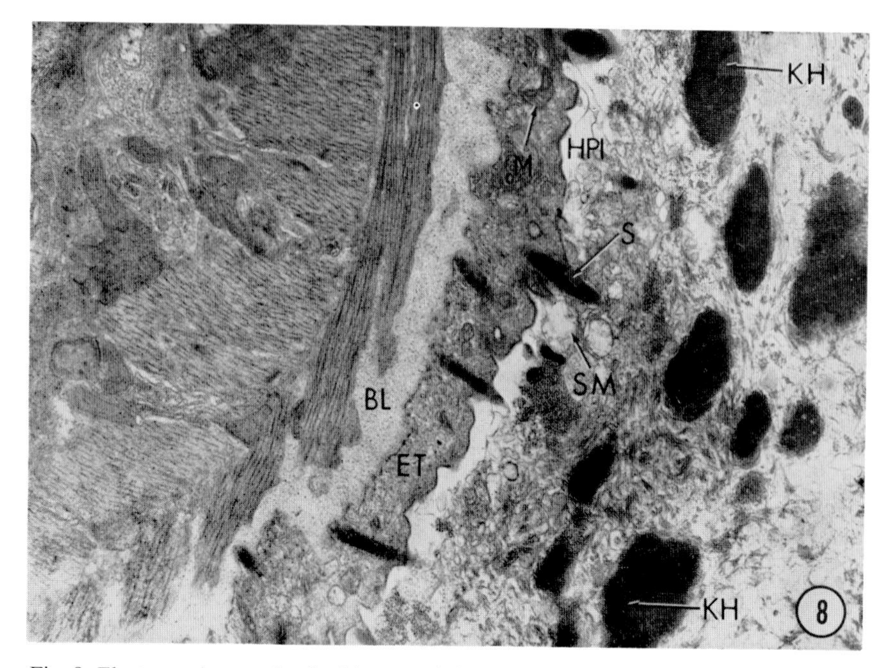

Fig. 8. Electron micrograph of schistosomule-host skin interface showing spinous projections within the host's epidermis. (×29,500.)

physical damage. The cytoplasm of the host cells that have been pierced by the spines of the schistosomule takes on a leached appearance. The host-parasite interface becomes a region of low electron density within which are found the free cytoplasmic organelles of the host epidermis (Figs. 8, 9, 10). Ribosomes, mitochondria, and lamellated bodies, commonly enclosed within their unit membranes, are observed uncontained at this interface.

A hirsute coat is discernible over parts of the membrane bordering the parasite's external tegument. This coat is more densely concentrated over particular regions of the outer unit membrane. In certain areas globular and membranous elements of host origin are juxtaposed against the parasite's external tegument.

The mitochondria of the epidermal cells of the skin in the proximity of schistosomules are swollen and distended (Fig. 10). Cristae are no longer readily apparent, and breaks occur in the outer mitochondrial membrane. It is of interest to note that mitochondria display this morphological appearance in cells which are from 2- to 3-cell layers removed from the tegument as well as in those abutting the schistosomule surface (Fig. 11).

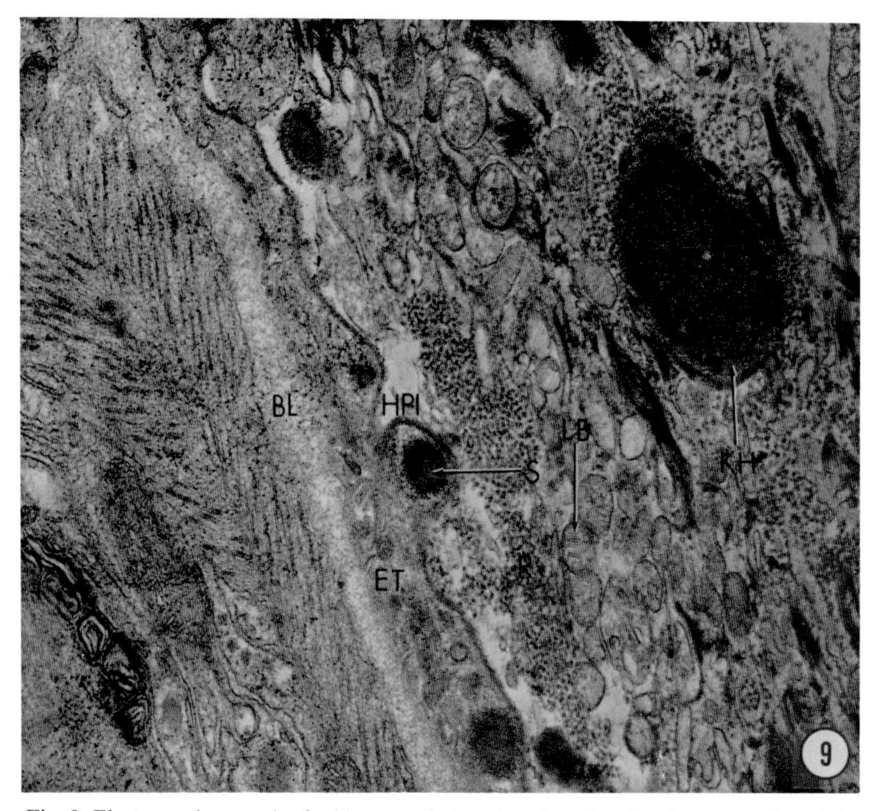

Fig. 9. Electron micrograph of schistosomule-host interface showing free organelles of the host epidermis within the interface. Note the proximity of the parasite's tegument to the host cell. (×34,000.). *BL*=basal lamina, *ET*=external tegument, *HPI*=host-parasite interface, *KH*=keratohylin granule, *LB*=lamellated body (host epidermis), *M*=mitochondria, *R*=ribosomes, *S*=spines, *SM*=swollen mitochondria.

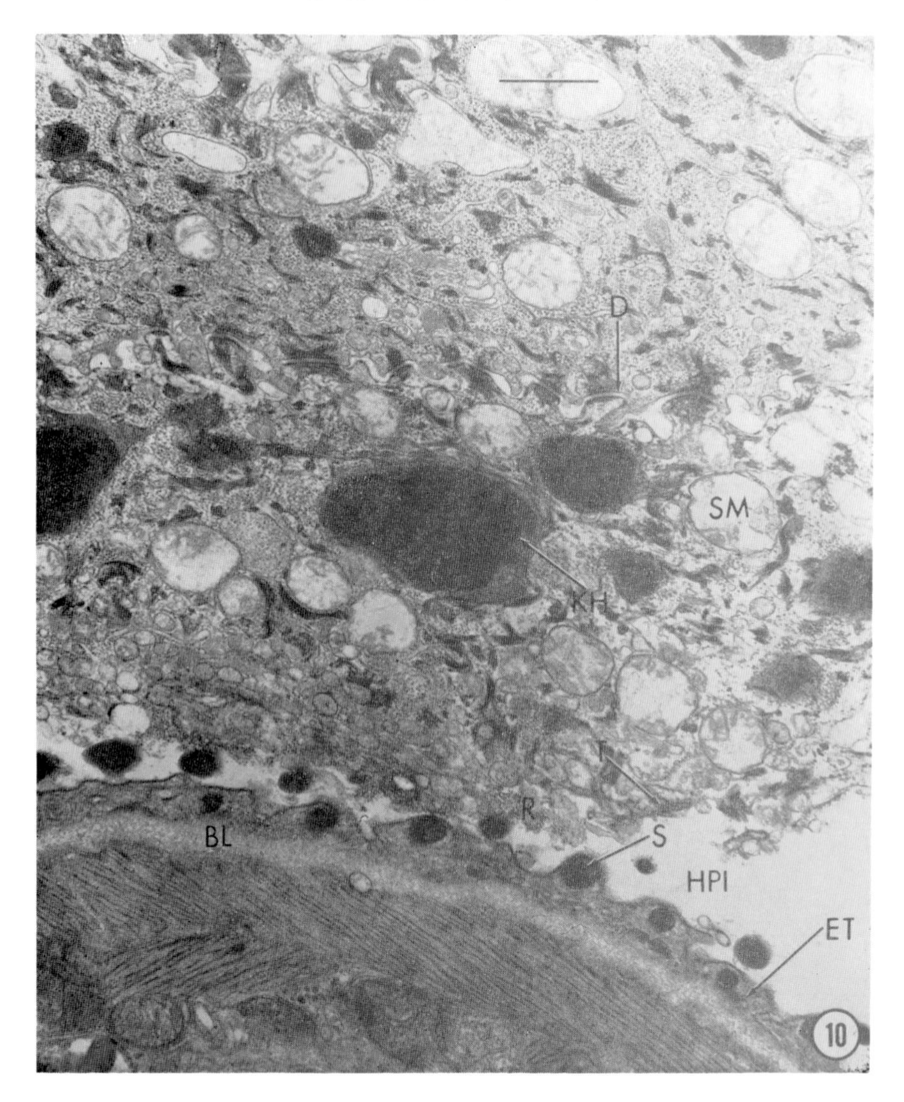

Fig. 10. Electron micrograph of schistosomule-host interface showing distorted and swollen mitochondria in the host epidermal cells. Note cell separating desmosomes and keratohylin granules within the host epidermal cells. (×21,500.) *BL*=basal lamina, *D*=desmosome, *ET*=external tegument, *HPI*=host-parasite interface, *KH*=keratohylin granule, *R*=ribosomes, *S*=spines, *SM*=swollen mitochondria, *T*=tonofilaments.

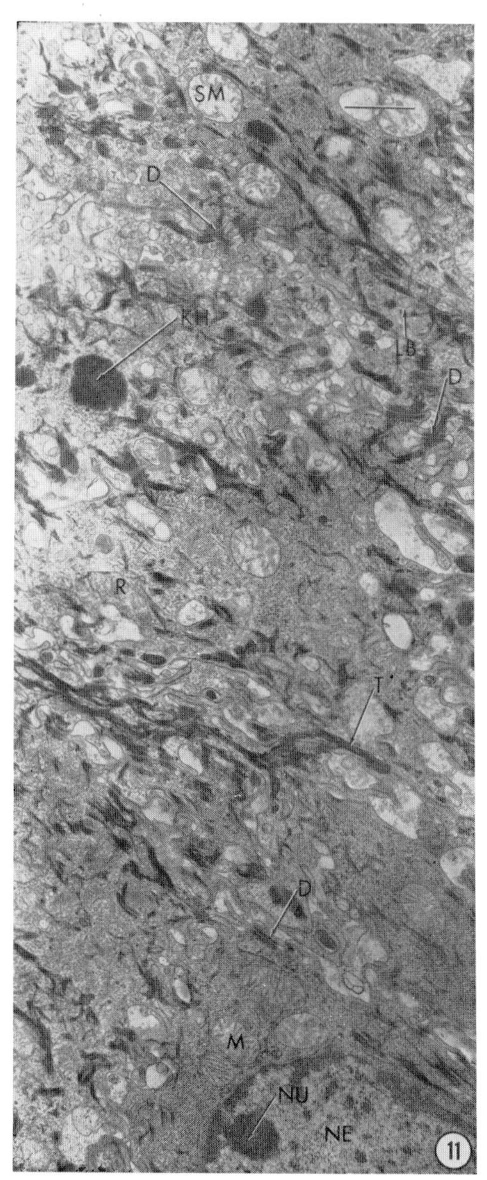

Fig. 11. Electron micrograph of host epidermal cells in the vicinity of, but not adjacent to, the schistosomule tegument. Note the normal mitochondria in the perinuclear region. This nucleated region is furthest from the parasite's tegument. The mitochondria in the cells closest to the parasite are swollen and distorted. The top of Figure 10 corresponds to the top of Figure 11 (note *bar*). (×13,500.) *D*=desmosome, *KH*=keratohylin granule, *LB*=lamellated body (host), *M*=mitochondria, *NE*=epidermal nucleus, *NU*=nucleolus, *R*=ribosomes, *SM*=swollen mitochondria, *T*=tonofilaments.

DISCUSSION

The trematode tegument must maintain its structural and functional integrity in order for the worm to penetrate its host and eventually mature. There is no doubt that the development of the tegument corresponds to its eventual use in the penetration process. Bils and Martin (1966) have described the development of the tegument in six species of trematodes. They have found the tegument of sporocysts, rediae, cercariae, metacercariae, and adults to be composed of a surface membrane usually separated from the basement membrane by a cytoplasmic layer. The surface area of sporocysts, rediae, and developing cercariae is increased by microvilli and irregularities. Cells whose outer cytoplasm formed a syncytium became the tegument of young cercariae. Cytoplasmic bridges connecting this syncytium to the cell bodies remained throughout the life cycle.

If the tegument is to be considered from the standpoint of penetration, then one would necessarily emphasize the role of the external tegument. The absence of Golgi elements, endoplasmic reticulum, and ribosomes suggest that this region is metabolically restricted and probably functions as a protective and absorptive structure. This syncytial layer is in direct contact with the intracellular and extracellular constituents of the host's epidermis. Its ability to penetrate effectively the host's cell layers will determine the future viability of the worm.

The existence of the tegumentary spines has been well established, although their function had been purely speculative. This study would indicate that they are instrumental in causing the physical destruction of the host cells with which they come into contact. Dawes (1963) has shown that the spines of adult *Fasciola hepatica* abrade the superfacial cells of the host's hyperplastic bile duct. He suggests that in this manner the parasite is provided with an abundant source of nutrients. A similar situation may exist at the schistosomule-epidermal interface; that is, the spines might cause the release of nutrients from the host cells for eventual use by the parasite. Pinocytotic and phagocytotic activity might occur in the pockets produced by the convoluted syncytial surface. Spines also appear to serve schistosomules in their progress through host skin.

Although the mucous-like coat found on the cercaria (Kruidenier, 1951a,b, 1953a,b,c,d; Kruidenier and Stirewalt, 1954, 1955; Kruidenier and Mehra, 1957; Stirewalt and Kruidenier, 1961; Stirewalt, 1963, 1966) is not present on the schistosomule surface, amorphous and membranous particles, most probably of host origin, are found there. It is thought that these particles adhering to the external tegument of the schistosomules are host antigens. Smithers *et al.* (1969) have suggested that host antigens incorporated in the tegument of adult *S. mansoni* serve to disguise the worms as host tissue, thus preventing their rejection by the immunologic defenses of the host. It is entirely possible that the adherence of these host antigens begins as the schistosomule penetrates the host epidermis.

The abrasive action of the host cells on the outer limiting unit membrane of the parasite's tegument might present a serious problem to the parasite. It is entirely possible that the membranous elements, which were part of lamellated bodies, supply the phospholipids needed for membrane synthesis, and thus provide for regeneration of a worn or abraded outer membrane. The lamellated bodies may also play a role in the formation of the dense, and possibly protective, stratum below the plasma membrane. A similar stratum was observed in the external tegument of *Gorgoderina* sp. by Burton (1966). Secretory bodies much like the lamellated bodies of the schistosomule were present in the syncytium of *Gorgoderina* sp. These lamellated bodies must be manufactured in the cytons (internal tegument) and transported to the syncytium (external tegument) via the cytoplasmic bridges. The microtubules within the bridges may aid their movement, since microtubules have been implicated as a mechanism for transport. These microtubules may also act as a supporting structure for these thin tortuous processes.

The lysed and distorted mitochondria found in the host epidermal cells may be caused by enzyme-mediated changes in the extracellular spaces. The work of Lewert and Lee (1954, 1957) indicates that in the vicinity of penetrating *S. mansoni* cercariae, enzyme-mediated changes in the ground substance of the host dermis do occur. Enzymes in cercarial extracts have been shown to be capable of degrading a variety of substrates.

The enzyme complex not only demonstrates proteolytic activity (Milleman and Thonard, 1959; Mandlowitz *et al.*, 1960; Gazzinelli and Pellegrino, 1964) but contains mucopolysaccharidase (Levine *et al.*, 1948; Evans, 1953; Lee and Lewert, 1957) and lipase (Mandlowitz *et al.*, 1960).

Wolff and Schreiner (1968) have recently demonstrated a material in epidermal cells that is considered to correspond to the mucopolysaccharide glycokalyx of other cell types. Therefore, the lysing of epidermal mitochondria might be caused by a breakdown in the ground substance, producing a hypotonic matrix, and causing a rapid influx of water into cellular organelles.

SUMMARY

The tegument of *Schistosoma mansoni* schistosomules consists of a cytoplasmic syncytium (external tegument) which is connected to nucleated cytons (internal tegument) by cytoplasmic bridges. The most conspicuous organelle in both levels of the tegument as well as in the bridges is a lamellated body with an electron-dense core and a myelin-like substructure.

The external tegument, which functions as an absorptive and protective structure, comes into direct contact with the host. Extending from this syncytium are spinous projections which tear and destroy the adjacent host epidermal cells during penetration. This results in the release of cytoplasmic organelles (mitochondria, ribosomes, etc.) from the host epidermal cells and thus fills the host-parasite interface with these free constituents.

A hirsute coat found on the surface of the schistosomule differs considerably from the mucin-like layer found on *S. mansoni* cercariae. This coat, which is more discrete over certain regions of the external tegument, contains both membranous and globular elements.

Distorted and swollen mitochondria observed in the strata spinosum and granulosum of the host's epidermis may be caused by enzyme-mediated changes in the extracellular ground substance.

ACKNOWLEDGMENTS

The author gratefully acknowledges the excellent technical assistance of Mrs. Marvelle O'Donnell and Mr. Augusto Uy and wishes to give special thanks to Dr. M. A. Stirewalt for her helpful criticism in preparing the manuscript.

REFERENCES

Bils, R. F. and Martin, W. E. 1966. Fine structure and development of the trematode integument. Trans. Amer. Microscop. Soc. 85:78–88.

Bjorkman, N. and Thorsell, W. 1964. On the fine structure and resorptive function of the cuticle of the liver fluke, *Fasciola hepatica* L. Exp. Cell Res. 33:319–329.

Burton, P. R. 1964. The ultrastructure of the integument of the frog lung-fluke, *Haematoloechus medioplexus* (Trematoda: Plagiorchiidae). J. Morphol. 115:305–317.

Burton, P. R. 1966. The ultrastructure of the integument of the frog bladder fluke, *Gorgoderina* sp. J. Parasitol. 52:926–934.

Dawes, B. 1963. Hyperplasia of the bile duct in fascioliasis and its relation to the problem of nutrition in the liver fluke, *Fasciola hepatica* L. J. Parasitol. 53:123–133.

Evans, A. S. 1953. Quantitative demonstration of hyaluronidase activity in cercariae of *Schistosoma mansoni* by the streptococcal decapsulation test. Exp. Parasitol. 2:417–427.

Gazzinelli, G. and Pellegrino, J. 1964. Elastolytic activity of *Schistosoma mansoni* cercarial extract. J. Parasitol. 50:591–592.

Gordon, R. M. and Griffiths, R. B. 1951. Observations on the means by which the cercariae of *Schistosoma mansoni* penetrate mammalian skin, together with an account of certain morphological changes observed in the newly penetrated larvae. Ann. Trop. Med. Parasitol. 45:227–243.

Hyman, L. H. 1951. The Invertebrates: Platyhelminthes and Rhynchocoela. The Acoelomate Bilateria. McGraw-Hill, New York. 550 p.

Kligman, A. M. 1964. The Biology of the Stratus Corneum. In W. Montagna and W. C. Lobitz, Jr. (eds.) The Epidermis. Academic Press, New York. 649 p.

Kruidenier, F. J. 1951a. Studies in the use of mucoids by *Clinostomum marginatum*. J. Parasitol. 37 (Sect. 2):25–26.

Kruidenier, F. J. 1951b. The formation and function of mucoids in virgulate cercariae, including a study of the virgula organ. Amer. Midland Natur. 46:660–683.

Kruidenier, F. J. 1953a. Studies on the formation and function of mucoids in cercariae: non-virgulate xiphidiocercariae. Amer. Midland Natur. 50:382–396.

Kruidenier, F. J. 1953b. The formation and function of mucoids in cercariae: monostome cercariae. Trans. Amer. Microscop. Soc. 72:57–67.

Kruidenier, F. J. 1953c. Studies on the formation and function of mucoid glands in cercariae: opisthorchoid cercariae. J. Parasitol. 39:385–391.

Kruidenier, F. J. 1953d. Studies on the mucoid secretion and function in the cercariae of *Paragonimus kellicotti* Ward (Trematoda: Troglotrematodae). J. Morphol. 92:531–543.

Kruidenier, F. J. and Mehra, K. N. 1957. Mucosubstances in plagiorchoid and monostomate cercariae (Trematoda: Digenea). Trans. Ill. State Acad. Sci. 50:267–278.

Kruidenier, F. J. and Stirewalt, M. A. 1954. Mucoid secretion by schistosome cercariae. J. Parasitol. 40 (Sec. 2):33.

Kruidenier, F. J. and Stirewalt, M. A. 1955. The structure and source of the pericercarial envelope (CHR) of *Schistosoma mansoni*. J. Parasitol. 41 (Sec. 2):22–23.

Lee, C. L. and Lewert, R. M. 1957. Studies on the presence of mucopolysaccharidase in penetrating helminth larvae. J. Infect. Dis. 101:287–294.

Levine, M. D., Garzoli, R. F., Kuntz, R. E. and Killough, J. H. 1948. On the demonstration of hyaluronidase in cercariae of *Schistosoma mansoni*. J. Parasitol. 34:158–161.

Lewert, R. M. and Lee, C. L. 1954. Studies on the passage of helminth larvae through host tissues. I. Histochemical studies on extracellular changes caused by penetrating larvae. II. Enzymatic activity of larvae *in vitro* and *in vivo*. J. Infect. Dis. 95:13–51.

Lewert, R. M. and Lee, C. L. 1957. The collagenaselike enzymes of skin-penetrating helminths. Am. J. Trop. Med. Hyg. 6:473–479.

Mandlowitz, S., Dusanic, D. and Lewert, R. M. 1960. Peptidase and lipase activity of extracts of *Schistosoma mansoni cercariae*. J. Parasitol. 46:89–90.

Milleman, R. E. and Thonard, J. C. 1959. Protease activity in schistosome cercariae. Exp. Parasitol. 8:129–136.

Morris, G. P. and Threadgold, L. T. 1968. Ultrastructure of the tegument of adult *Schistosoma mansoni*. J. Parasitol. 54:15–27.

Morseth, D. J. and Soulbsy, E. J. L. 1969a. Fine structure of leukocytes adhering to the cuticle of *Ascaris suum* larvae. I. Pyroninophils. J. Parasitol. 55:22–31.

Morseth, D. J. and Soulsby, E. J. L. 1969b. Fine structure of leukocytes adhering to the cuticle of *Ascaris suum* larvae. II. Polymorphonuclear leukocytes. J. Parasitol. 55:1025–1034.

Reynolds, E. S. 1963. The usage of lead citrate at high pH as an electron-opaque stain in electron microscopy. J. Cell Biol. 17:208–212.

Rifkin, E , Cheng, T. C. and Hohl, H. R. 1969. An electron-microscope study of the constituents of encapsulating cysts in the American oyster, *Crassostrea virginica*, formed in response to *Tylocephalum* metacestodes J Invert. Pathol. 14:211–226.

Senft, A. W., Philpott, D. and Pelofsky, A. H. 1961. Electron microscopic observations of the integument, flame cells and gut of *Schistosoma mansoni*. J. Parasitol. 48:217–229.

Smith, J. H., Reynolds, E. S., and von Lichtenberg, F. 1969. The integument of *Schistosoma mansoni*. Amer. J. Trop. Med. Hyg. 18:28–49.

Smithers, S. R., Terry, R. J. and Hockley, D. J. 1969. Host antigens in schistosomiasis. Roy. Soc. (London), Proc., B. 171:483–494.

Stirewalt, M. A. 1959. Isolation and characterization of deposits of secretion from the acetabular gland complex of cercariae of *Schistosoma mansoni*. Exp. Parasitol. 8:199–214.

Stirewalt, M. A. 1963. Chemical biology of secretions of larval helminths. Ann. N. Y. Acad. Sci. 113:36–53.

Stirewalt, M. A. 1966. Skin penetration mechanisms of helminths, p. 41–58. In E. J. L. Soulsby (ed.) *Biology of Parasites*. Academic Press, New York.

Stirewalt, M. A. and Kruidenier, F. J. 1961. Activity of the acetabular secretory apparatus of cercariae of *Schistosoma mansoni* under experimental conditions. Exp. Parasitol. 11:191–211.

Threadgold, L. T. 1963. The tegument and associated structures of *Fasciola hepatica*. Quart. J. Microscop. Sci. 104:505–512.

Wolff, K. and Schreiner, E. 1968. An electron microscope study on the extraneous coat of keratinocytes and the intercellular space of the epidermis. J. Invest. Dermatol. 51:418–430.

Communication Between Fishes in Cleaning Symbiosis

GEORGE S. LOSEY, JR.

Hawaii Institute of Marine Biology
University of Hawaii, Honolulu, Hawaii

INTRODUCTION

Cleaning symbiosis in the marine environment, as defined by Feder (1966), is "a relationship in which certain organisms remove ectoparasites, bacteria, diseased and injured tissue, and unwanted food particles from cooperating hosts and some other organisms that visit them." An idealized summary of cleaning interactions follows: Cleaner fishes are commonly found on topographically conspicuous "cleaning stations." Host fishes approach the cleaner or cleaning station and frequently assume an unusual posture or "pose" in front of the cleaner and may present an erect fin or open mouth to the cleaner in apparent cleaning invitation. The cleaner swims close and "inspects" the host and then may feed on material on the host and even enter the host's mouth or gill cavity. Through this relationship, the cleaner not only gains food material but also is reported to enjoy a relative immunity from predation (see Feder, 1966).

Both vertebrate and invertebrate cleaning organisms have been found in many areas that are frequently visited by divers, particularly in the tropics, but only the cleaner fishes are treated here. Cleaning has been documented for species of various families including the Labridae, Chaetodontidae, Gobiidae, and many others. Some species clean only as juveniles or only occasionally, while others are highly specialized cleaners and rely on this habit for the majority of their food supply. Countless host species have been observed and the list might conceivably be extended to include nearly all fishes found near shore.

Cleaning symbiosis provides a rich field for research in animal communication. It involves a sometimes confoundingly large number of host species, all of which have need to convey approximately the same information to the cleaners. The behavioral, ecological, and evolutionary implications of such studies are abundant.

Contribution No 347 from the Hawaii Institute of Marine Biology, University of Hawaii.

REVIEW OF THE RECENT LITERATURE

The study of cleaning symbiosis has proceeded in the way of most scientific topics. The initial discovery period was marked by a rapid realization of the widespread occurrence of cleaning symbiosis and various hypotheses were formulated as to its ecological and behavioral significance. In general, studies then lapsed into an encyclopedic stage of enumeration of cleaner and host species. Feder (1966) provided a thorough review of these stages in the study of cleaning symbiosis. Few of the existing hypotheses were subjected to detailed scrutiny until the past few years. Students of cleaning symbiosis are now investigating existing hypotheses and making detailed analyses of ecological and behavioral mechanisms which are the primary concern of this paper.

The first and most widely cited experiment was conducted by Limbaugh (1961) in the British West Indies. He removed all cleaning organisms from an isolated patch reef and reported both a subsequent reduction in the number of fish on the reef and a marked increase in parasitism and seeming infections of the remaining fish. It is important to note that while Limbaugh (1961) realized the limited value of his preliminary experiment and criticized his lack of controls, this experiment has been cited and accepted by nearly every subsequent publication dealing with cleaning symbiosis.

In the first detailed ecological and ethological study of cleaning behavior, Youngbluth (1968) was unable to duplicate Limbaugh's results on patch reefs in Kaneohe Bay, Hawaii. He observed no change in the number or the degree of infection of the fish on a reef after he removed all of the *Labroides phthirophagus*, the primary cleaning organism in Hawaii. Neither did he note any change in the frequency of cleaning bouts on a reef when all but one cleaner fish had been removed. As possible reasons for his failure to repeat Limbaugh's results, Youngbluth listed other more occasional cleaners that may have replaced *L. phthirophagus*, as well as the possibility that the configuration of his reefs prevented emigration from the cleanerless reefs. He also cited the theory that some of the parasites of Kaneohe Bay are less noxious than others (Lewis, 1961), thus not producing as profound an effect on the hosts. Another possible basis for the disparity between their results lies in Youngbluth's use of the frequency of cleaning bouts as an indication of amount of cleaning. More sensitive measures such as actual time spent cleaning might yield a more realistic estimation of the amount of cleaning activity, but detailed comparison with Limbaugh (1961) is impossible since no data were presented.

Youngbluth (1968) gave quantitative support to the hypothesis that cleaning stations are stable locations on the reef. While the individuals that occupy any given station may change, the location of the station remains the same.

Hobson (1969) noted one cleaning station in the Gulf of California that had a stable location for six years. Youngbluth (1968) also found that while several individuals may occupy a station at the same time, solitary individuals are more common. These findings are consistent with those of Limbaugh (1961) and with Feder's (1966) statement that colder-water cleaners are "usually highly gregarious or schooling, while the warm-water forms are solitary, paired, or slightly gregarious." Hobson (1969), however, cautioned against overgeneralizing. He presented evidence on the usually gregarious nature of several tropical cleaner fish and pointed out that many temperate cleaners in California are not gregarious.

Hobson (1969) also criticized other generalizations presented by Limbaugh (1961) and Feder (1966) and cautioned against the present tendency to incorporate these generalizations into the literature without critical examination. He cited two other generalizations that he considers unsubstantiated: that "tropical cleaners more nearly approach the condition of full time cleaners," and that "tropical cleaners put on elaborate displays in connection with their cleaning activity that are similar to the mating displays of male fishes" (Feder, 1966). Apparently these result from inadequacy of previous observations. Two other generalizations, that tropical cleaners "are more brightly colored, and more contrastingly marked" and that "tropical cleaning species are more numerous than those in temperate waters, but the number of individuals is less," may perhaps prove to be true, but the basis for these situations may be quite unrelated to cleaning symbiosis (Hobson, 1969).

Youngbluth (1968) showed that while juvenile *Labroides phthirophagus* may be maintained on an artificial diet in the laboratory, the adults are apparently obligate cleaners. He also presented ethograms of the mating and cleaning action patterns of *L. phthirophagus*.

Wickler (1961, 1963, 1967) has studied aspects of the behavior of another cleaner wrasse, *Labroides dimidiatus*, as relating to the cleaner mimic blenny (*Aspidoptus taeniatus*) that feeds on fish scales, mucous, and fin membranes. He concluded that these blennies were preadapted to "parasitize" the cleaning relationship by mimicking the signals of the cleaner fish in order to approach unsuspecting host fish and pursue their predaceous habits.

Fricke (1966a,b) was the first researcher to investigate the reaction of host fishes to a cleaner model by placing a replica of *Labroides dimidiatus* on a reef. He found that host fishes posed in front of a realistically painted model in much the same manner as before a real cleaner.

Davenport (1966) emphasized the interest of symbiotic relationships to behaviorists. They usually involve the exchange of specific information and one frequently finds obvious behavioral indications as to the success of the information exchange. In the study of cleaning behavior, the investigator must consider the behavior of both the cleaner and the host organisms, just as when studying the social behavior of a species one must consider the actions and reactions of both sender and receiver. Youngbluth (1968) was concerned primarily with the behavior of the cleaner and not the host, and this limits the effectiveness of his study in furthering our understanding of cleaning symbiosis.

MATERIALS AND METHODS

In the present study attention is focused on the symbiotic nature of cleaning behavior. By investigating the communication between cleaner and host, the stimuli and responses of both parties must be dealt with simultaneously. Communication, when defined as the exchange of information, is indicated by a change in the probability of a behavioral response in the signal-receiver that is correlated with a stimulus generated by the behavior of the signal sender. This paper presents various features of cleaning behavior as potential sources of information, and offers a model for a portion of the communication system between the Hawaiian cleaner wrasse, *Labroides phthirophagus*, and its various hosts. It is organized to form a sequential development of the communication system in individual sections, each with its own introduction, methods, results and discussion.

Students of the ethology of terrestrial animals have long recognized the need for extended field observations and careful field techniques such as the use of blinds or remote monitoring equipment. Aquatic ethologists have only recently begun to concentrate on field observations and techniques. Most observations are made by free-swimming divers and usually recorded on writing slates. While many behavioral studies may be conducted satisfactorily in this fashion, some studies require more elaborate techniques (Losey, 1968). As a cleaning study will include the greater portion of the reef ichthyofauna, there is obvious need for extensive, carefully executed, field observations. Furthermore, the complexity of the observational data gathered in a study of communication also demands special recording techniques.

My studies include extensive field observations conducted in Kaneohe Bay, Hawaii. Collaborative observations, with less refined techniques, were conducted in several locations in the Hawaiian Islands, Eniwetok Atoll, and Guam.

Most of the observations in Kaneohe Bay were made from an underwater blind (Fig. 1A). Wooden blinds were permanently positioned next to the reef ledge so that a diver could observe the reef through slits in the front of the blind. The diver's "hookah" air was supplied by a large storage cylinder on the surface. The noise caused by exhalation and subsequent bubbles was markedly reduced by having the exhalant chamber of the hookah regulator inside of an air chamber in the blind. Spent air was valved to the surface

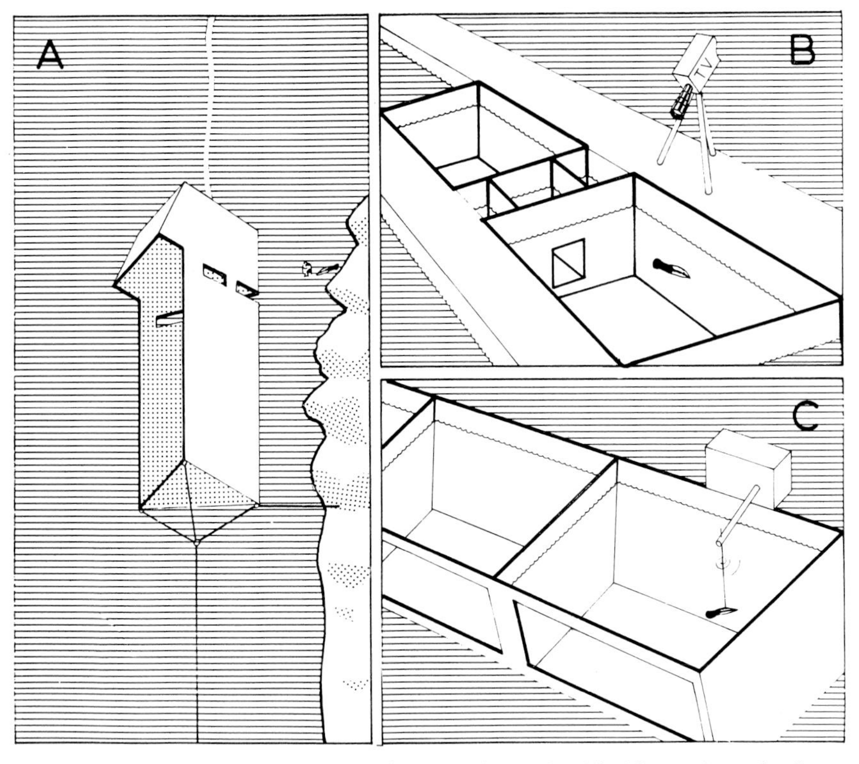

Fig. 1. Schematic drawings of: *A*, an underwater observation blind for use by scuba divers; *B*, the experimental pen with television camera; and *C*, the experimental aquarium with the fish model presentation apparatus.

through an air-filled hose fitted to the top of the blind. During field observations in other areas the diver made use of available cover on the reef for concealment.

Observations were recorded on an underwater tape recorder (Hydro Products, San Diego) and on an event recorder with a 7-channel, underwater keyboard constructed of reed switches imbedded in polyester resin (Fig. 2). Whenever possible, both recording methods were used simultaneously.

Laboratory experiments were conducted at the Hawaii Institute of Marine Biology in a 4 × 1 × 1 m wire pen submerged between two floats in the Coconut Island lagoon as well as in a glass front, 1,000 liter aquarium (Fig. 1B, C). The pen was divided into two major chambers by a trap door compartment used to transfer fish between chambers without handling. The aquarium was divided into three water-tight compartments so that treated water could be used in one without contaminating the others.

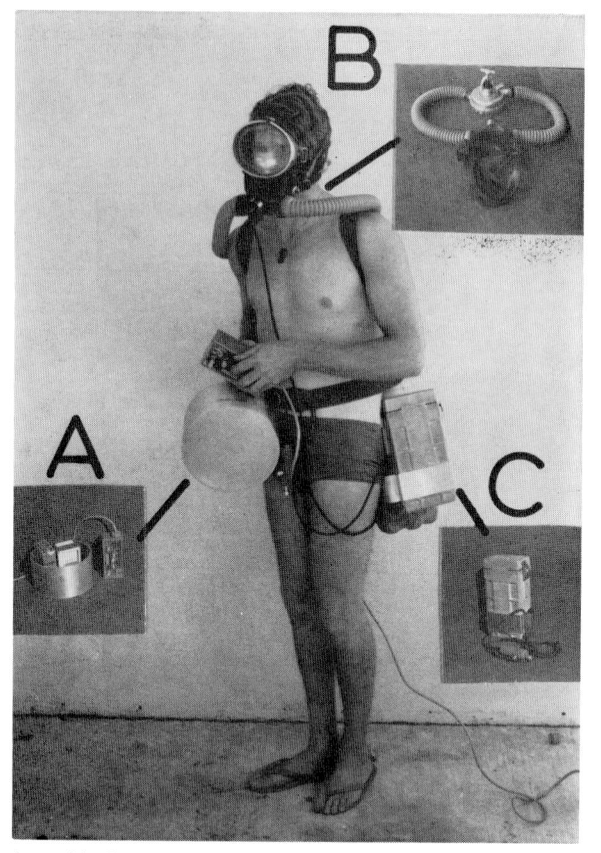

Fig. 2. The author with diving and underwater recording equipment. Shown are: *A*, Event recorder and batteries in custom-made, underwater housing, and underwater keyboard; *B*, Aqualung regulator with microphone-mouthpiece; and *C*, commercial-make, underwater housing and switch for tape recorders.

Fish models were molded of liquid latex in plaster molds made by embedding preserved fish. By this method, series of duplicate models were available. Models were painted with acrylic paints and coated with polyester resin. They were presented by remote control on a thin wire (Fig. 1C). Electric motors and relay circuits caused the model to "swim" back and forth over a distance of about 15 cm. Due to the support of the thin wire and the water's resistance to the model's movement, the "swimming" pattern was rather erratic and quite realistic to the observer.

A closed circuit television system (Ampex, Sony, Telemation, and GBC equipment) was used to obtain detailed sequential observations. Repeated viewing of videotape recordings enabled analyses of the behavior of several individuals over a continuous period of up to one hour.

Assays of ectoparasitism on hosts were accomplished by unaided and light microscopic examination of the body surface, mouth, and gills, and by examining scrapings from the body.

Aquarium water was occasionally treated with copper sulfate to rid the fish of some ectoparasites. Copper levels were maintained from 0.2–0.5 parts per million by analyzing for copper ion concentration with a Taylor Water Analyzer (W. A. Taylor & Co., Baltimore).

The action patterns studied are similar to those presented by Youngbluth (1968). The following terms are used in this paper to connote certain behavior, as indicated:

Cleaner Fish:

(1) *Inspect*: swimming close to the host fish and apparently exploring its body surface.
(2) *Clean*: feeding on matter on the body surface of the host.
(3) *Dance*: a dorso-ventral oscillation of the body as described in detail by Wickler (1963).

Host Fish:

(1) *Pose*: the position or orientation of the host's body and fins and the swimming movements frequently assumed during cleaning interactions. This position varies widely between species (see Table 2).
(2) *Body jerk*: any short, quick movement of the body or head.
(3) *Attack*: darting quickly toward the cleaner.

RESULTS

CLEANER–HOST COMMUNICATION

Stimuli Inherent in the Cleaner and Host in Noncleaning Situations

Introduction.—Fricke (1966a) demonstrated that some hosts *pose* in front of cleaner models and several workers have suggested that the body shape and coloration of cleaners could act as a signal or cleaner guild sign (Wickler, 1963; Eibl-Eibesfeldt, 1955). It is possible that the cleaner recognizes the more frequently rewarding host species; that is, species recognition signs might also serve as stimuli to the cleaner. Thus I tested the hypothesis that some sign or signs inherent in the body shape, coloration, swimming pattern, and so on of both cleaners and hosts serve as signals for cleaning behavior.

Methods.—Natural interactions were observed from an underwater blind and by free-swimming diver, and cleaner models were presented to prospective hosts in the field and in the laboratory. Field observations included continuous recordings of species interacting, the time spent *inspecting* and *cleaning by* the cleaner, and *pose* duration, *pose* position and *attack* by the hosts. Models of a variety of shapes and colorations were used and response to these was recorded as the amount of time the hosts spent *posing*. A standard model of *Labroides phthirophagus* was used as a response control for several species of Hawaiian host fishes. Models were presented for 30 seconds, in random order every 3 minutes for periods of up to 2 hours.

Results.—The field observations provide two pertinent lines of evidence:

(1) Either *pose* or *inspect* may be the initial action in a cleaning sequence. For example, in one typical set of observations, *pose* occurred first 51 times while *inspect* occurred first 42 times. And either of these actions may occur while the other organism is in a noncleaning state; that is, not engaged in cleaning-related movements such as *inspect* or *pose*, or showing no overt awareness of the other animal.

(2) *Labroides* spp. show different responses to different host species. For example, some large fish such as adult parrot fish (Scaridae) or large jacks (Carangidae) may be pursued and *cleaned* by the cleaner as they pass near a cleaning station, while others, such as the wrasse *Cheilinus rhodochrous*, or the damsel fish, *Dascyllus albisella*, may *pose* continually for several minutes and be ignored by the cleaner. Such preferences may be expressed quantitatively by comparing the ratio of time spent posing to time inspected (Table 1). Thus, for example, *Labroides* spp. prefer to clean large scarids to small ones, while chaetodontids must *pose* relatively longer in order to be cleaned.

TABLE 1. *Ratio of time hosts spent* posing *to time* Labroides dimidiatus *spent* inspecting *during 70.9 minutes of observation, where* N = *(number of pose bouts) + (number of inspect bouts). Family comparisons provide a simplified example of species preferences*

Family of hosts	*Pose: inspect ratio*	N
Scaridae <100 mm length	1.6	8
Scaridae <300 mm length	0.7	41
Scaridae >300 mm length	0.3	9
Acanthuridae	1.5	99
Pomacentridae	2.1	23
Labridae	1.0	15
Chaetodontidae	2.3	5

The model experiments show that a graded response is shown to increasingly realistic models of *Labroides phthirophagus*. These results are as yet too incomplete to allow a statement as to relative effectiveness of different elements of the coloration and body shape.

Discussion.—These results show that some stimuli inherent in the noncleaning host (inactive in terms of cleaning relationships) can increase the probability that it will be *inspected* or *cleaned* by the cleaner. The chance that the presence of ectoparasites or infected tissue may affect this interaction is dealt with below. *Labroides* spp. also show a cleaning preference that appears to be related to species recognition, and some preferred species need not *pose* to elicit *inspection* by the cleaner. Hobson (1965a) gave examples of host species preference by the cleaner, *Heniochus nigrirostris*.

The response of the hosts to cleaner models is evidence that stimuli inherent in the noncleaning cleaner increase the probability that the host will pose. The field observations supply additional support for this conclusion since the *host-pose* frequently occurs before any cleaning-related actions by the cleaner.

Host-Pose Stimulus

Introduction.—The poses assumed by many host species are striking features of cleaning symbiosis. Host fish may roll over, do head and tail stands, lapse into prolonged motionless hovering, and change coloration. I tested the hypothesis that *posing* serves as a cleaning invitation signal and that its occurrence alters the probability that the cleaner will *inspect* and *clean* the host.

Methods.—In addition to field observations of natural interactions, laboratory experiments were undertaken to alter the frequency of *posing* in the hosts by depriving them of cleaning. Interactions with an introduced cleaner were then observed. Thirty-two fish of 9 different species (see Appendix I) were placed in the wire pen with an adult *Labroides phthirophagus* which was allowed to clean them for one week. The hosts were then divided into two equal groups; the first group with the cleaner, the second without, for one week. The cleaner was then put with the second group for one week, and then put back with the first. Behavioral observations of species, time spent *posing, inspecting, cleaning,* and in other activities, were made over 5 minute periods, 10 times a day on a systematic schedule. Additional samples were taken on a random schedule for the first 2 days to insure that cyclic processes were not biasing the systematic samples.

Results.—Some hosts such as *Scarus* spp. and *Ctenochaetus* spp. favor one body orientation for *posing*, but others show considerable variability (Table 2). The same may be true for color changes, hovering, and so on.

TABLE 2. *Frequency of pose positions for several host genera in Hawaii based on one series of field observations of cleaning interactions with* Labroides phthirophagus

host	pose positions					N
Scaridae						
Scarus	**89%**	11%	0	0	0	9
Labridae						
Thallasoma	0	38%	62%	0	0	8
Cheilinus	0	14%	25%	54%	7%	28
Acanthuridae						
Ctenochaetus	9%	**91%**	0	0	0	13
Zebrasoma	0	33%	44%	23%	0	9
Naso	0	17%	66%	17%	0	6
Pomacentridae						
Dascyllus	0	40%	40%	20%	0	10
Pomacentrus	0	25%	75%	0	0	4

In the pen experiment, hosts that had been cleaned for a week had low *pose*-levels (Table 3, Fig. 3). After one week of cleaning deprivation, they showed much more *pose* behavior, but this again decreased after about a week of cleaning. *Cleaner-inspect* increased significantly when *pose* increased but *cleaner-clean* showed a significant increase only once. The mean values obtained from the random sampling schedule did not differ from those of the systematic observations ($p > 0.25$ by t test).

The only marked increase in parasitism during the experiment was an infestation of small monogeneid gill trematodes and ectoparasitic protozoans, *Cryptocaryon irritans* (see Wilkie, 1969). A qualitative inspection of relative infestation revealed no differences between "cleaned" fish and fish that had been deprived of cleaning for a week.

Discussion.—Ritualization of communicative displays to a typical intensity is common in social behavior (Morris, 1957; Hinde, 1966) and is thought to increase the signal value of the behavior through a reduction of ambiguity. In some host species, *pose* has a fairly consistent orientation that may be the result of ritualization; but others show considerable variability. More intensive study, including more species and careful control of the motivational state of the host, might reveal more consistency to the *pose* orientations than is now indicated.

The pen experiments show that an increase in the amount of *cleaner-inspect* is correlated with an increase in *host-pose* (Table 3). Cleaning deprivation induced an increase in *host-pose* (this argument will be expanded in the section headed "context of the communication"). This increase may have been caused by increased parasitism but the cleaner did not appear capable of reducing the degree of infestation, probably because of the small size of these parasites and the intimacy of the attachment of the flukes to the gill lamellae. The change in *cleaner-inspect* cannot be correlated to changes in the cleaner's hunger or by the cleaner sensing the parasites since the control observations, in which the hosts were cleaned for one week, both preceded and followed the

TABLE 3. *Mean time in seconds of* host-pose *and* cleaner-inspect *behavior, and mean number of* cleaner-clean *events per five minute period during two pen experiments with the hosts either deprived of cleaning or cleaned for one week. Data are based on all nine species of hosts. Probability of equality was calculated by* "t" *test*

Test conditions	*Variable tested*	$\bar{X}$ *time* ($\bar{X}$ *frequency*)	*Probability of equality*	*Degrees of freedom*
Experiment I				
deprived	*host-pose*	192.1	<0.001	33
cleaned	*host-pose*	34.6		
deprived	*cleaner-inspect*	104.7	<0.01	
cleaned	*cleaner-inspect*	50.9		
deprived	*cleaner-clean*	(18.0)	>0.1	33
cleaned	*cleaner-clean*	(11.2)		
Experiment II				
deprived	*host-pose*	106.3	<0.02	25
cleaned	*host-pose*	53.8		
deprived	*cleaner-inspect*	103.9	<0.01	25
cleaned	*cleaner-inspect*	44.6		
deprived	*cleaner-clean*	(14.2)	<0.02	25
cleaned	*cleaner-clean*	(7.3)		

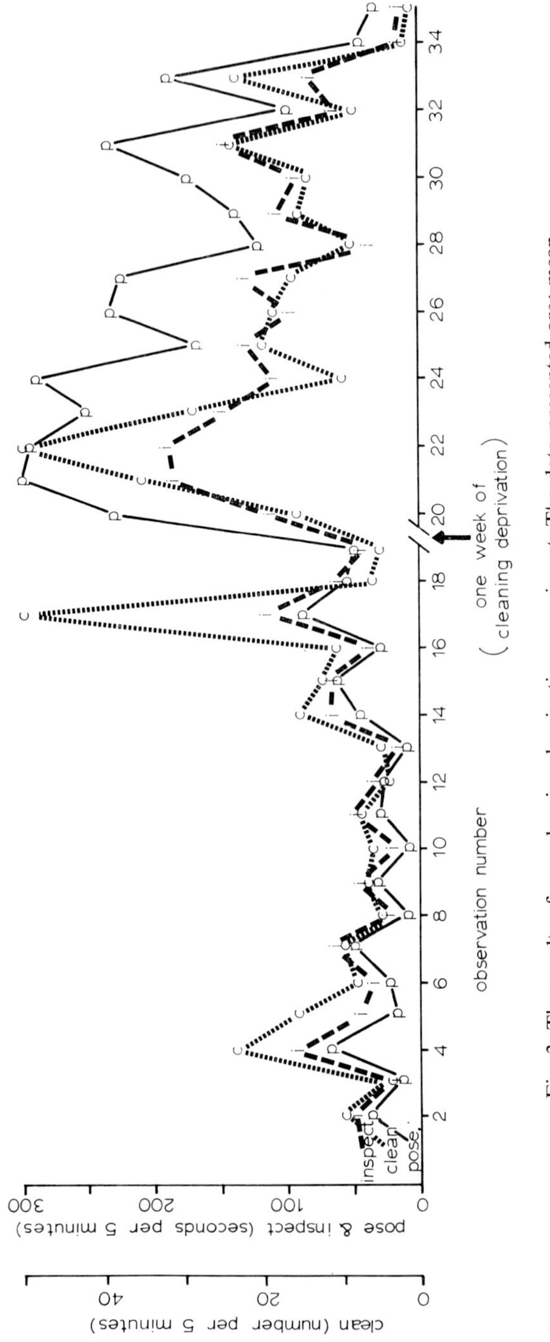

Fig. 3. The results of one cleaning deprivation experiment. The data presented are: mean *pose* durations (*p*) for 22 host fishes, and mean *inspect* duration (*i*) and number of *cleaning* bouts (*c*) for an adult *Labroides phthirophagus* interacting with the above hosts. Statistical relationships are given in Table 3.

deprivation observations, and no visible changes could be found between the level of infestation of cleaned and cleaning deprived hosts.

The change in *cleaner-inspect* must result from the induced change in *host-pose*, so *host-pose* may be interpreted as a communicative behavior. *Host-pose* increases the probability of *cleaner-inspect*.

The number of *cleaner-clean* events increased after deprivation only once. Youngbluth (1968) showed a positive correlation between *inspect* time and number of *cleaning* events. This correlation was also significant in this experiment (Fig. 4; $p<0.05$ by Bartlett regression analysis; Dixon and Massey, 1957) but the number of *cleaning* events did not show a consistent increase

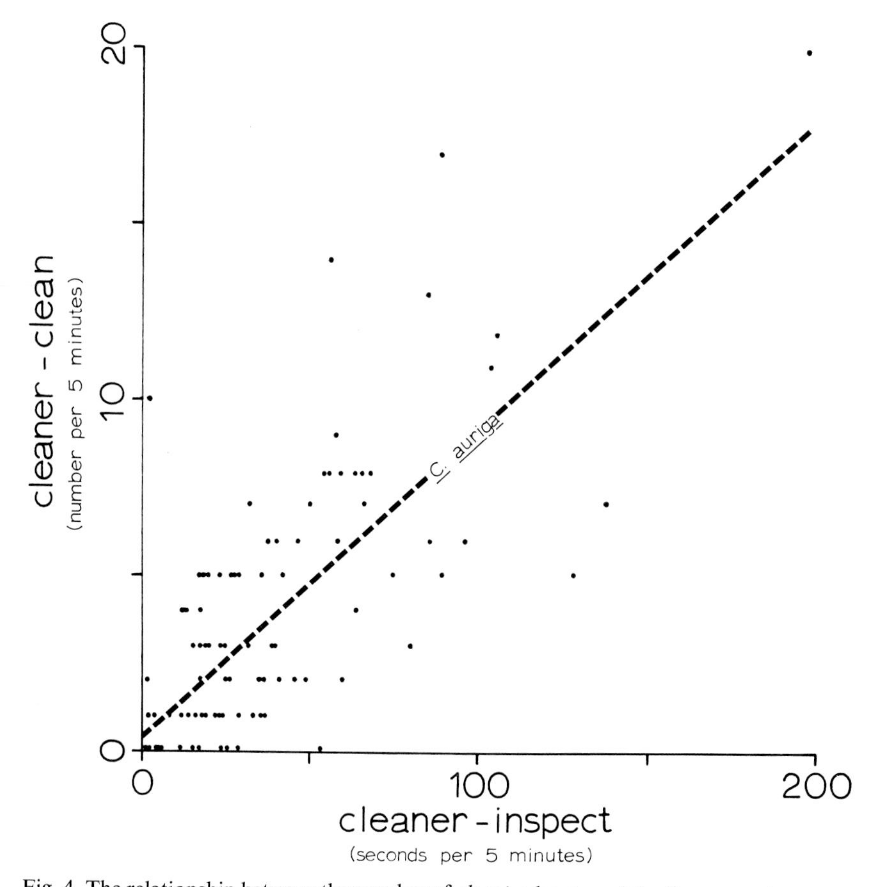

Fig. 4. The relationship between the number of *cleaning* bouts and the time spent *inspecting* by adult *Labroides phthirophagus* in interactions with *Chaetodon auriga* as a representative host. $N=89$, $b=0.09\pm0.02$ at the 95% confidence level.

when *host-pose* increased. Since *clean* is strongly correlated to *inspect*, and considering the present lack of data, I assume that *host-pose* has only an indirect effect on *cleaner-clean*.

Cleaner-Inspect Stimulus

Introduction.—Many cleaning interactions are initiated when the cleaner *inspects* a *nonposing* host (see above), suggesting that the *inspection* itself may serve as a cleaning signal to the host. The present experiment considers the role of *inspection* as a signal.

Methods.—Twelve fish of 8 species (see Appendix 1) were placed in the pen with one adult *Labroides phthirophagus*. Television-videotape recordings were made of continuous activity for one hour at ambient light intensity. Repeated review of these recordings provided extremely accurate sequential data. Only those interactions that began with *host-pose* are considered below.

Results.—For *Chaetodon trifasciatus* hosts, the mean *pose* duration was longer when it was accompanied by *cleaner-inspect* than when the cleaner did not *inspect* (Fig. 5). The mean time lag between the initiation of *pose* and the subsequent *inspect* was, however, much shorter than the mean *pose* duration that was not answered by *inspect*. The other host species tested seemed to give

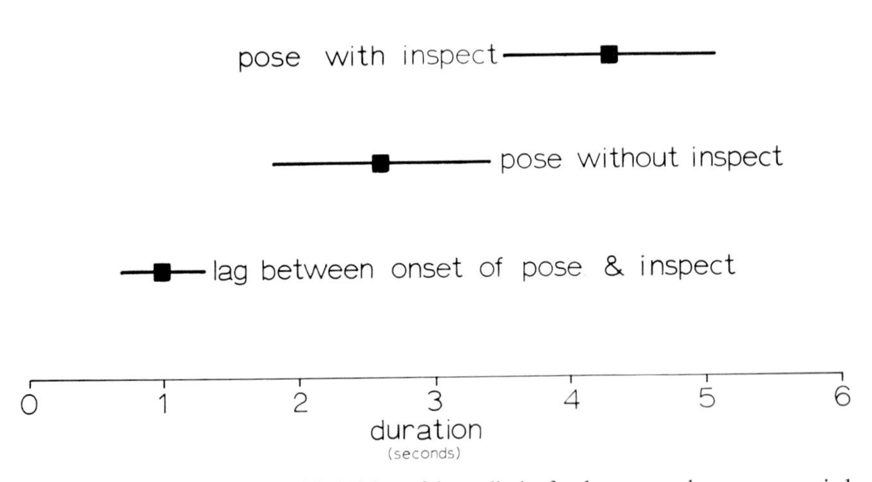

Fig. 5. Mean time durations with 95% confidence limits for *host-pose* when accompanied by subsequent *inspection* by the cleaner, *host-pose* without *inspection* by the cleaner, and the lag between the onset of *pose* and the onset of *inspect* for those *host-pose* bouts that were accompanied by subsequent *cleaner-inspect*. Confidence limits are calculated from the "*t*" relationship.

about the same results but the data available at the moment are not amenable to statistical analysis due to small sample size.

Discussion.—It might be hypothesized that the longer mean *pose* observed when the cleaner *inspects* occurs because the short *poses* are not long enough for the cleaner to respond by *inspecting*. But this hypothesis is not supported by the experiment since the time lag between *pose* and *inspect* is much shorter than the mean *pose* without *inspect*. I conclude that *inspect* is a communicative behavior in that it increases the duration of *posing*.

The communication system as formed thus far depicts *pose* and *inspect* as forming a mutually reinforcing system; each tends to increase the probability of the other's occurrence.

Stimulus Cleaner-Clean

Introduction.—Wickler (1968), while discussing *Labroides dimidiatus,* states that if the "customer does not stay still, the cleaner jabs it violently with its wide-open mouth and then continues with its work." This suggests that contact between the cleaner's mouth and the host's body might serve as a cleaning signal. In pursuit of this problem, I investigated the possibility that the occurrence of *host-pose* might be correlated to *cleaner-clean*.

Methods.—The experimental records from the cleaning deprivation experiments were analyzed in two ways. The first analysis, with *Chaetodon auriga* and *C. lunula* as hosts, utilized cleaning encounters that involved simultaneous *pose* and *inspect*. By means of a three-way analysis of variance, the mean *pose* durations were tested for correlation to length of *inspect*, deprivation state of the amimal, and whether or not they were *cleaned.* In the second analysis, with the above species and *Priacanthus cruentatus* and *Abudefduf abdominalis* as hosts, the observation records were divided into 8.5-second intervals. The frequency of occurrence of behavioral categories was estimated by counting the number of intervals in which they occurred. The observed frequencies of occurrence for individual behavioral acts were then used to calculate theoretical frequencies of co-occurrence (theoretical probability of co-occurrence of acts A and B=observed probability of $A \times$ observed probability of B).

Results.—The three-way analysis of variance reveals that, in *Chaetodon auriga* and *C. lunula*, the mean *pose* duration was correlated to whether or not the host was *cleaned* ($p<0.005$); but inspection of the means reveals that increased *cleaning* was correlated with a shorter *pose* duration.

The second analysis indicates that, in *Chaetodon auriga, host-attack* occurred far more commonly with *cleaner-clean* than it did alone and that this might not be expected if *clean* and *attack* were randomly distributed in time (see Table 4). In the other species tested (*C. lunula, Priacanthus cruentatus*, and *Abudefduf abdominalis*), *host-attack* also occurred with *cleaner-inspect* far more commonly than would be expected if randomly distributed. In *C. auriga*, the frequency of *attack* following *cleaner-inspect* was significantly lower than that of *attack* following *cleaner-clean* ($X^2 = 20$ with one degree of freedom). In the other species the frequencies were not significantly different ($p > 0.1$).

TABLE 4. *Observed and calculated probabilities of occurrence for behavioral acts and pairs of acts in any 8.5 second interval for* Chaetodon auriga. *Number of intervals = 1610*

	Probability of occurrence	
	Observed	*Calculated*
host-attack cleaner without *cleaner-inspect* or *clean*	0.0006	0.012
host-attack cleaner after *cleaner-inspect*	0	0.0007
host-attack cleaner after *cleaner-clean*	0.01	0.0004

In all of the host species, when *attack* occurred with *cleaner-inspect* or *cleaner-clean, attack* always followed the cleaner's behavior ($N = 46$).

Discussion.—The correlation of *cleaner-clean* with shorter mean-*pose* durations indicates that *cleaner-clean* decreases the probability of *host-pose* in the species tested.

In all hosts tested, the comparatively large probability that the host will *attack* the cleaner when it *cleans* or *inspects* as revealed by the second analysis, might be explained in two ways. One hypothesis is that *cleaner-inspect* and *cleaner-clean* might increase the probability of occurrence of *host-attack*. An alternate hypothesis is that the correlations reflect a functional relationship: *inspect, clean,* and *attack* can only occur when the cleaner and host are in close proximity and, during most of the intervals considered, they simply were not close enough for an *attack* to occur. The fact that *attack* never preceded *inspect* or *clean* might support the hypothesis that they are causally related, except that an attack quite probably would eliminate the possibility of subsequent *inspection* or *cleaning* of the host.

There is, however, a strong indication for *Chaetodon auriga* hosts that *cleaner-clean* does increase the probability that the host will *attack* the cleaner. The X^2 value described above represents the significance of correlation between *clean* and *attack* during those intervals when the host and cleaner were known to be in close proximity because the hosts were being *inspected* or *cleaned* by the cleaner. The same does not hold for the other species tested.

This suggests that some forms of *cleaning* may act as negative stimuli for *host-pose* and positive stimuli for *attack* in some hosts, or in anthropomorphic terms, some *cleaning* is painful. But other forms of contact between the cleaner's mouth and the host might serve as positive stimuli to *host-pose*. This is strongly suggested by preliminary model studies in which a moving, realistic model was allowed to gently bump into the hosts. After as few as 20 presentations of the model, each of 30 seconds duration, many hosts learned to *pose* for the bare wire on which the models had been presented. The wire bumped into these hosts similarly to the model. Extension of these experiments may well provide a separation of different visual and tactile stimuli.

Other Communication Signals

Two additional movements have possible communicatory function. *Cleaner-dance* refers to a unique swimming pattern seen in *Labroides* spp. (Wickler, 1963; Feder, 1966; Youngbluth, 1968). These authors hypothesize that this *dance* serves to make the cleaner obvious or as a cleaner recognition signal. My field observations suggest that in adult *Labroides* spp., *dancing* occurs most commonly when the cleaner is approached aggressively (particularly by a diver). The function and causation of *dance* have not been subjected to experimental scrutiny but it is interesting to postulate that *dance* is caused by an aggressive host or predator and that it serves to reduce their tendency to *attack* the cleaner.

Cleaning customers have been noted to jerk or "fidget" near the end of a cleaning sequence and this has been suggested as a host-signal to terminate the cleaning (Feder, 1966; Wickler, 1968). Such movements, collectively regarded here as *body-jerk*, may be signals that decrease the probability of *cleaner-inspect*. I have seen few instances of this behavior. In terminating the cleaning interaction, usually either the cleaner or host just swim away, or the host darts quickly off. Interestingly, *body-jerk* movements are common during model presentation work. Hosts frequently *body-jerk* before, during, and at the end of *posing* and may *body-jerk* after ceasing to *pose*. Its occurrence strongly suggests a conflict between *posing* and fleeing before the model. As such, it could signal the host's relative receptivity to being cleaned to the cleaner. Similar conflict behaviors have been shown to form the basis for signals in many animals (see Hinde, 1966). This hypothesis will be tested in the future.

CONTEXT OF THE COMMUNICATION

The study of animal communication should include the context in which the behavior is observed. Environmental and internal changes may introduce variability into the response of the signal receiver. In considering the response, it is important to differentiate between variability that is due to changes in context, and variability inherent in the communication system itself. For example, a courtship approach by a female may release aggression or only weak response in the males of some animals when they are not sexually ripe, or lead to reciprocal courtship in ripe males (see Hinde, 1966). This variability is probably due to changes in context, or in the sexual maturity of the male. Hinde (1965) and Baerends *et al.* (1955) present series of external and internal stimuli that affect the behavior of the canary and the fish, *Lebistes reticulatus*, respectively.

It is possible to argue that all environmental and internal stimuli probably have some effect on cleaning behavior and that the complexities render the study of context insurmountable. On the other hand, it seems more reasonable to expect some stimuli to have a greater effect than others, and that their study alone will provide a reasonable analysis of the communication system.

The following sections present a few variables that are functionally related to cleaning behavior and appear to have a major effect on the behavior of both host and cleaner. I do not imply that all of the pertinent variables are considered here.

Environmental Stimuli that Affect the Host

The cleaning station is important to cleaning relationships (Feder, 1966). Youngbluth (1968) gave quantitative evidence that cleaning stations are fixed geographic locations and the primary site of cleaning interactions. Even with a lack of experimental evidence, it seems obvious that host fishes are able to recognize the location of cleaning stations. During many of my field observations, hosts have been seen to *pose* on the cleaning station without the presence of a cleaner. It appears that the cleaning station increases the probability that the host will *pose*. To my knowledge, *posing* in the absence of a cleaner has never been seen outside of a cleaning station.

Many other variables, such as the presence of predators (Hobson, 1965b) or prey no doubt have a profound effect on the host's behavior. Of particular interest to future investigation is the possibility of both interspecific and intraspecific social facilitation of *posing* behavior.

Internal Stimuli that Affect the Host

Introduction.—Cleaning deprivation leads to an increase in *host-pose* behavior. Limbaugh (1961) indicated that cleaning deprivation leads to increased parasitism and infections. It may be hypothesized that deprivation results in increased parasitism and infections and that these, or other irritating stimuli, are the causal agents for the observed increase in *host-pose*.

Methods.—Two pertinent experiments have been completed to date, one with *Chaetodon lunula*, the other with *C. auriga*. With each species, a control group and a test group of 4 fish each were placed in adjacent 350 l compartments separated by opaque, watertight partitions and a water filled compartment. The test group was treated for one week with 0.3–0.5 parts per million copper sulfate and seawater solution to remove ectoparasites; the control group was not treated. In each test, a moving model of an adult *Labroides phthirophagus* was presented to the fish for 30 seconds, on a random schedule, with a mean lag time of 5 minutes between presentations. Both groups were tested for duration of *pose* response to the model both before and after copper sulfate treatment. The fish were then assayed for parasites and infections.

Results.—In *Chaetodon lunula*, before treatment there was appreciably more *posing* in the test group than in the control group ($p=0.026$ for equality of medians by the rank sum test: Dixon and Massey, 1957), but after treatment the test group showed far less response than the control group ($p<0.001$; see Fig. 6). Parasite assay revealed that each fish in the control group was infested with about 10–50 gill flukes and 4–10 ectoparasitic flukes on the fin membranes as well as some bacterial infections of "raw" areas. The treated fish had up to about 20 gill flukes, no flukes on the fin membranes, and some bacterial infections.

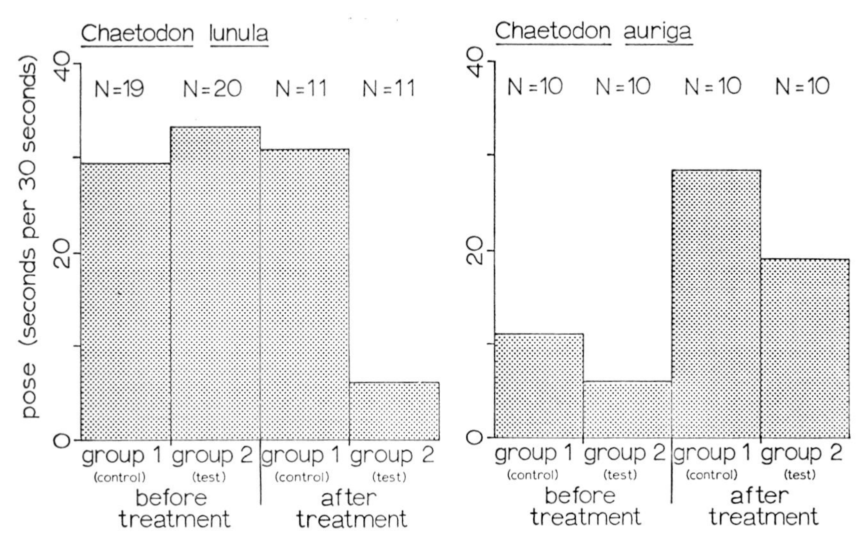

Fig. 6. Mean *pose* durations given in response to 30-second presentations of a realistic *Labroides phthirophagus* model. Control and test groups of four fish each were used for both species. Observations were made both before and after treating the test groups with copper sulfate to remove parasites. N=number of model presentations. Statistical significance is discussed in the text.

In *Chaetodon auriga*, the test group showed less *posing* than the control group both before and after treatment ($p<0.001$) but they responded far more after copper sulfate treatment ($p<0.001$). Parasite-assay revealed that the members of the control group were heavily parasitized with monogenetic trematodes on the fin membranes, up to about 20 per fish, and one had a large ectoparasitic isopod (1 cm long) with associated bacterial infections. I found few trematodes or bacterial infections on the treated group and one individual that had been parasitized by a similar isopod before treatment was no longer parasitized.

No correlation was found between the duration of *pose* response and the time since the last presentation of the model. Also, on four occasions, the model was presented for five minutes continuously. All of the fish continued to *pose*.

Discussion.—These results are somewhat ambiguous. The first experiment suggests that parasitism or infection is the cause of increased *host-pose* after cleaning deprivation. But the second experiment suggests that mere cleaning deprivation or lack of interaction with a cleaner may cause increased *posing*. These differences might reflect species-specific variability or inadequate parasite assays that did not detect some parasites present in *Chaetodon auriga*. Additional experimentation is necessary to resolve this question.

The lack of correlation between *pose* duration and time since last model presentation indicates that *host-pose* is not a decremental behavior and that, at least on the time scale of these experiments, there is little or no habituation to the cleaner model.

Environmental Stimuli that Affect the Cleaner

Cleaner fish frequently *inspect* hosts without *cleaning*. This suggests that the situation in which inspecting is released is not always a sufficient stimulus to release *cleaning*. Many researchers have noted that cleaners tend to pick at conspicuous spots or areas on the hosts (see Feder, 1966). Wickler (1968) notes that one *Labroides* sp. continually attempted to tear off natural spine-like outgrowths from the skin between the boney spines of the puffer, *Diodon* sp. In a personal communication E. S. Hobson noted that the California cleaner wrasse *Oxyjulis californica*, is prone to nibble at conspicuous objects such as a piece of rusty wire protruding from the silver barb of a spear. He attached a fragment of black neoprene to a crude silver fish model and found that this wrasse consistently picked at the neoprene. Some other predators in the kelp bed habitat inspected the model, but only *O. californica* picked at the neoprene.

These observations suggest that conspicuous objects on the hosts (here grouped as "food") may serve as stimuli to increase the probability of *cleaner-inspect* or *cleaner-clean*, but more detailed study is needed.

Internal Stimuli that Affect the Cleaner

Introduction.—Host-pose is probably not a decremental behavior, and hosts *posing* for a cleaner model do not seem to show much habituation to the stimulus (see above). The only behavioral events my study showed that might terminate a cleaning interaction are *cleaner-clean* and *host-attack*. However, many interactions terminate without either of these behavior patterns occurring, so some portion of the cleaner or the host behavior must be vulnerable to other negative processes or stimuli.

Methods.—The data collected in the deprivation experiment, and from the videotape recording, were used to analyze *cleaner-inspect* behavior.

Results.—Host-pose duration bears a positive correlation to the duration of *cleaner-inspect* (Fig. 7). The ratio of the duration of *host-pose* to *cleaner-*

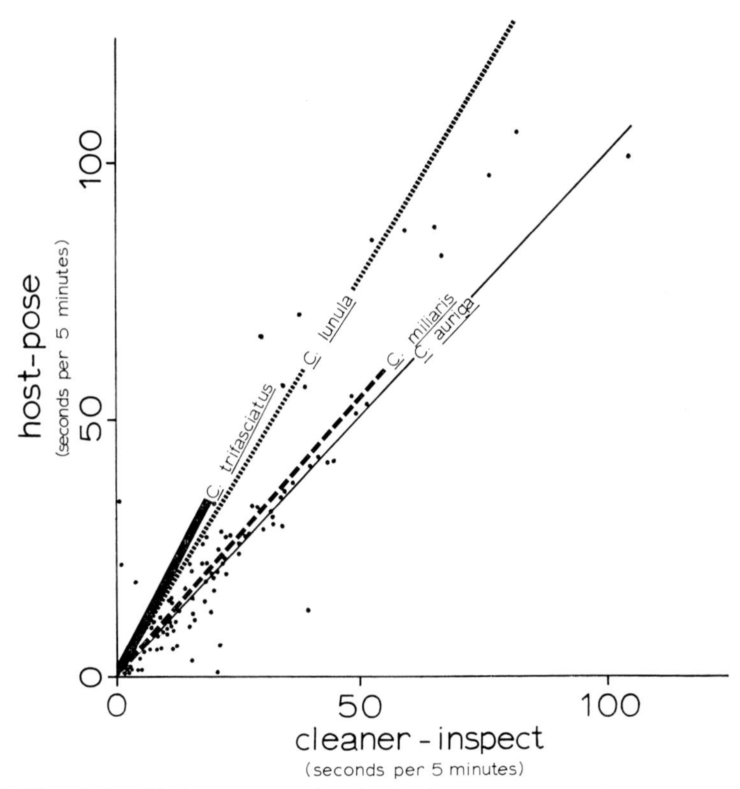

Fig. 7. The relationship between *pose* duration by four host species of *Chaetodon* and the corresponding *inspect* duration by adult *Labroides phthirophagus* cleaners interacting with these hosts. Representative data point spread is presented only for *C. auriga*. The slope of all of the lines are different from 0 ($p<0.001$ by "*t*" test).

inspect is also positively correlated to *cleaner-inspect* (Fig. 8).

The *pose: inspect* ratio for individual cleaning interactions is negatively correlated to the time since the last *pose* by any species that was encountered by the cleaner. There is no such correlation when the interactions of only one species are considered (Fig. 9).

Discussion.—The positive correlation of the *pose:inspect* ratio to *inspect* duration indicates that longer *host-pose* bouts are associated with shorter *cleaner-inspect* durations. This might result from either a more pronounced response by the host to longer *inspection*, or from a decrement in the *inspection* response by the cleaner.

The negative correlation of the *pose:inspect* ratio to time since the last *pose* indicates a response decrement in *cleaner-inspect*. This relationship does not

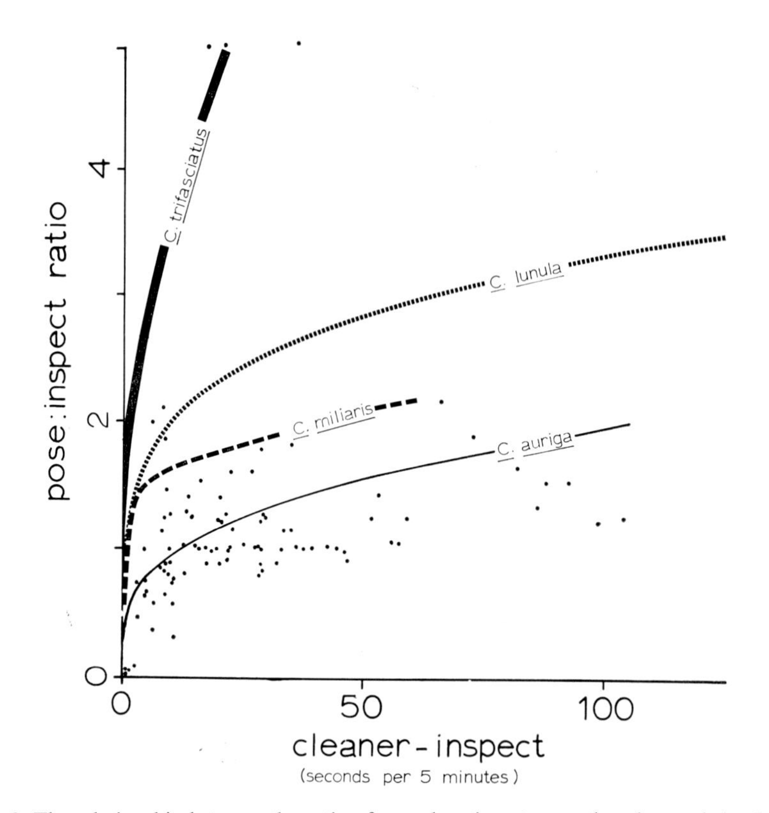

Fig. 8. The relationship between the ratio of *pose* duration: *inspect* duration and the duration of *inspection* by adult *Labroides phthirophagus* for the same data presented in Fig. 7. Representative data point spread is presented only for *Chaetodon auriga*. The slope of all of the lines is different from 0 ($p<0.005$ by "t" test).

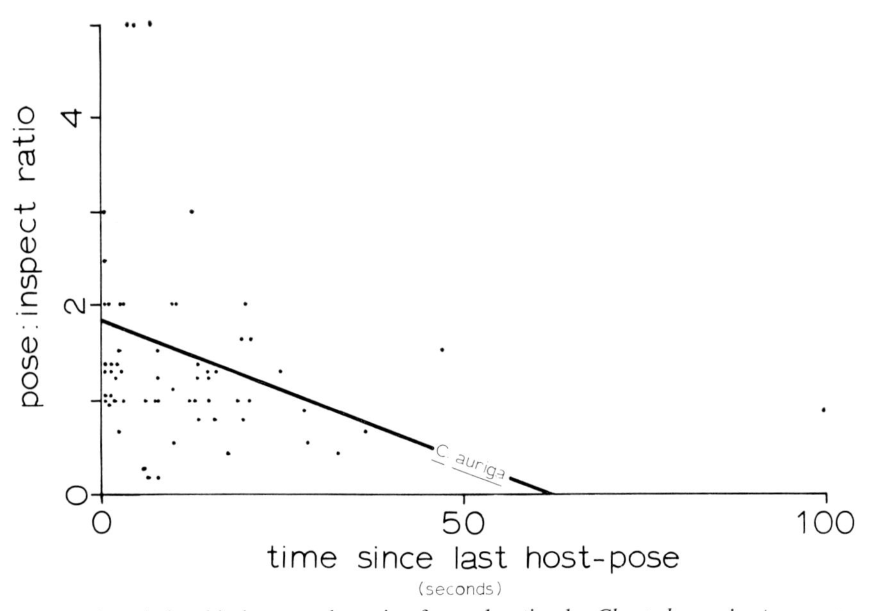

Fig. 9. The relationship between the ratio of *pose* duration by *Chaetodon auriga* to accompanying *inspect* duration by adult *Labroides phthirophagus*, and the time since the last *pose* by any host species. The slope of the line is less than 0 ($p<0.05$ by "*t*" test).

hold when only one of the interacting host species is considered. This multispecific aspect suggests the decrement to be specific to the response. In other words, if the response decrement was specific to the stimulus (habituation) the decrement should be the strongest when only one stimulus type—that is, only one species of host—is considered. These results, however, show that the decrement occurs for many types of *host-pose* stimuli simultaneously. But these results do not exclude the possibility that habituation does occur. The experimental design was not ideally suited for the demonstration of habituation, and additional experimentation is called for.

Hunger is an obvious stimulus that might be hypothesized to increase *inspecting* and *cleaning* behavior, but no experimental work has been conducted on this problem.

GENERAL DISCUSSION

The preceding pages present what appear to be the major elements of a communication system involved in cleaning symbiosis. Some of the features seem to be held in common by many different host species in their interactions with

Labroides spp., but some have been indicated by experiments with only a few host species, primarily of the genus *Chaetodon*. Apparently part of the system also applies to other cleaning organisms (see below). Much research remains to be done on hypotheses that have not been critically analyzed as well as on relationships that are doubtful or seem to show a large species variability. But the work completed to date sheds some light on many aspects of cleaning symbiosis.

Schematic Summarization

A change in the probability of a behavioral response in one animal that is correlated with a stimulus generated by the behavior of another animal is taken as an indication of information exchange or communication. The correlated behavioral changes presented above may be summarized schematically as a communication system (Fig. 10). The *arrows* indicate a change in the probability of the occurrence of the response (at the head of the arrow) that is correlated with the indicated stimulus (at the base of the arrow). The two center columns represent the stimulus-response relationships between

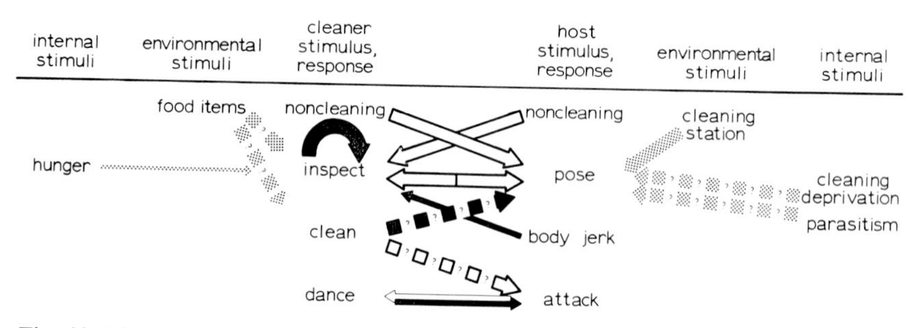

Fig. 10. The correlated behavioral change schematic of stimulus-response relationships for cleaner and host fishes and for contextual variables. *White* or *dotted arrows* indicate an increase in probability that the response at the head of the arrow will occur, given that the stimulus (at the base of the arrow) has occurred. *Solid arrows* indicate a decrease in probability. *Dashed arrows* indicate uncertainty of the proposed relationship due to apparent species variability or inadequate information. *Thin arrows* indicate relationships that are hypothetical and have only scanty observational evidence. For example: *Cleaner-inspect* increases the probability that *host-pose* will occur; *cleaner-clean* may decrease the probability that *host-pose* will occur but this relationship is doubtful: *cleaner-inspect* decreases the probability that *cleaner-inspect* will occur again; and the presence of a cleaning station increases the probability that *host-pose* will occur.

cleaner and host. Solid arrows indicate a correlated decrease in probability and white arrows indicate an increase. Thus, the arrow in Figure 10 from "host-noncleaning" to *"cleaner-inspect"* indicates a positive stimulus-response relationship. Dashed arrows indicate uncertainty due to variable responses and insufficient experimentation, and thin arrows indicate hypothetical relationships that lack firm support. The columns to left and right of center list contextual variables; that is environmental and internal stimuli that affect the cleaner and host respectively. The arrows indicate positive correlations. Thus the arrow from "cleaning station" to *"host-pose"* indicates that cleaning station is an environmental stimulus that increases the probability that the host will *pose*.

Communicative Stimuli

Most of the cleaner and host stimuli also serve noncommunicative functions. *Inspecting* and *cleaning* are both feeding mechanisms of the cleaner and are thus necessary to cleaning symbiosis even had they no communicative function. These actions seemingly do not differ from the noncleaning feeding habits of the species.

Posing is not necessary for cleaning to occur, but it seems to increase the efficiency with which the cleaner can work, and thus is functional even without its communicative role. Most host species vary as to their orientation while *posing*, but some species may *pose* in a characteristic way, and marked differences can be found between species. The *poses* of many species suggest the passive orientation of a motionless fish that results from differences between the centers of mass and buoyancy. Dead fish rising from a reef during a poison station frequently assume *pose*-like orientations and are sometimes cleaned by *Labroides* spp. The presentation of parasitized or infected fins, opening the mouth, and other actions, may also serve two functions in that they allow access to the area to be cleaned and also communicate the location of the infested site.

Stimuli inherent in the host not engaged in cleaning-related activities appear to be related to the species form, shape, coloration, and so on, and as such are functional to the host species in a variety of ways unrelated to cleaning.

Aside from communicative functions, the coloration and body shape of the cleaner may or may not be related to cleaning. A variety of body shapes are seen in cleaning fishes from fusiform wrasses to flattened, high-bodied butterfly fishes. The only functional prerequisite appears to be terminal jaws that are adapted to pick at small objects. The coloration of cleaners is a topic open to wide exploration. While brilliant coloration is quite common in tropical cleaners (see Feder, 1966), Hobson (1969) and many others have pointed out that similar brilliant coloration is also common to many fishes of the tropics, and temperate cleaners tend to be drab (Limbaugh, 1961). Eibl-Eibesfeldt (1955) has suggested that the striped color pattern seen in most *Labroides* spp. and in the Atlantic cleaner goby, *Elacatinus oceanops*, might serve as a guild sign signifying it to be a cleaner. But the fact remains that many cleaners, including other *Labroides* spp. and other gobies, do not have the striped color pattern. My preliminary model studies suggest that guild coloration might be important for initial recognition but that hosts can quickly learn new signals such as the model presentation wire.

My information on the other actions is more limited. The *body-jerk* has been postulated to serve as a signal to terminate a cleaning interaction. Host fish up to 2m long sometimes jerk or dart quickly after a cleaner picks at their skin or fins. This reaction seems to be a response to an aversive or "painful" stimulus. *Labroides* spp. frequently pursue large fish that pass near the cleaning station. On one occasion, a large (1.2–1.6m) carangid was seen to cough vigorously after passing over a cleaning station and spit out an adult *L. phthirophagus* (J. Maciolek, personal communication). Although such actions could hardly be regarded as ritualized displays, *body-jerk* in some species may prove to be a signal that arose from similar predecessors or from movements seen in motivational conflict situations.

The *dance* of *Labroides* spp. does appear to be a ritualized display. Wickler (1963) described the form of the *dance* that probably arose from the already distinctive swimming pattern of the wrasses. The fact that cleaner mimics imitate the *dance* (Wickler, 1963) suggests that it is important to cleaning interactions, but the signal functions of the *dance* are uncertain.

Host-attack needs little mention except that in some cases it is apparently caused by cleaner-clean and probably results from a "painful" stimulus. Some *attacks*, however, may stem from predation or aggressive responses.

Contextual Variables

Behavioral changes that are correlated to contextual stimuli help to explain some of the variability that is observed in cleaning interactions. The contextual variables presented here are all functionally related to cleaning symbiosis, and, as expected, appear to play a major role in determining the animal's responses. The actual mechanisms by which they act are not fully understood.

Stimuli that are inherent in the noncleaning organism such as coloration, body shape, and so on, might be more accurately grouped as contextual variables. But some of these stimuli, such as the coloration of the cleaners, may well have evolved as signals for cleaning symbiosis. As brought out by Smith (1969), many behavioral signals sometimes act as contextual variables to alter the "message" of another action.

The Communication System

The stimulus-response scheme for cleaner-host interactions functions as a communication system; that is, there is an exchange of information, as evidenced by behavioral changes correlated to the appropriate stimuli. For many of the relationships the possibility remains that the correlations reflect only a functional relationship and that the causal agent has not yet been found. But experiments such as the model studies presented here provide a means for the firm establishment of causal relationships.

The communication model presented here provides an explanation for some of the variability seen in cleaning interactions. Areas such as Kaneohe Bay, Hawaii, are characterized by large numbers of a primary cleaner, for example, *Labroides phthirophagus*. Interactions in these areas may be loosely characterized as having a low ratio of *host-pose* to *cleaner-inspect* duration, and little if any cleaning is performed by species other than the primary cleaner. Some of the deeper areas in Hawaii, such as Pokai Bay, have fewer *L. phthirophagus* as compared to the number of hosts. Here the *pose:inspect* ratio is usually much higher and other species of more occasional cleaners are active. And finally, in the deprivation experiments, the *pose:inspect* ratios are very high and species of genera that are not known to clean in the field, such as *Scarus*, sometimes show cleaning behavior. The communication system offers an explanation as follows: The host fishes in the three situations presented above represent increasing levels of cleaning deprivation as indicated by the *pose:inspect* ratio. This change in context renders *host-pose* more likely and thus it may occur when the other stimuli such as the shape, coloration, and behavior of the cleaner are less optimal. The hosts would be expected to show the observed increase in disposition to *pose* for, or allow *cleaning* by, increasingly dissimilar species of prospective cleaners. It remains only that a nearby organism be disposed to pick at other fishes or react to the host's signals, and cleaning behavior occurs. Thus the more occasional cleaners clean only when "called upon" to do so by the hosts.

The communication system postulated for cleaning behavior may be compared with that of other behavioral interactions. In comparison to the social behavior of fishes, one of the most striking features is the lack of ritualized signals and behavior that serves only a signal function. Many communicative signals are thought to evolve from behavior that previously had other functional and causal relationships through processes referred to as ritualization (see Hinde, 1966). The resulting displays frequently differ from their predecessors in both form and causal structure.

A tempting hypothesis is that there is a lesser amount of information to be transferred in cleaning symbiosis than in social behavior. Smith (1969) lists twelve message types found in vertebrate communication. The messages transmitted during cleaning symbiosis might well be added to this list as a new category; that is, promotion or termination of close proximity between dissimilar species for purposes of cleaning. These needs should not demand as specialized a communication system as does much vertebrate social behavior. It will be interesting to apply information theory to this hypothesis to provide a comparative estimate of both the potential information value of the behavior and the minimum information transfer.

The almost panspecific nature of cleaning symbiosis is strikingly different from other communicative behavior that usually involves at most only a few species. The necessary inclusion of many species into the evolutionary scheme may hinder or even reduce the adaptive value of signal specialization. Such a scheme would probably favor behavior with less phylogenetic programming and a greater dependency on learning mechanisms.

Even if specialization were of adaptive value, lower selective importance might favor slower evolutionary specialization. Cleaning symbiosis must be less vital to a species than reproductive or aggressive demands that are closely linked to the success and survival of the species. And cleaning is probably more recently evolved than agonistic, sexual, or other forms of social behavior which are our best known sources of communicative displays. As such, cleaning would have had less chance to evolve and might exist today in a more primordial state.

When comparing cleaning behavior with other symbioses, a difference is found in that most symbionts communicate only recognition and attraction signals or other signals whose original purposes were other than interspecific communication. Parasites frequently use some existing host information for recognition and attraction, such as odor or temperature. Some relationships involve interspecific recognition of intraspecific signals, such as the alarm calls in mixed bird flocks or interspecific agonistic behavior. Although mimicry involves the evolution of special signals for interspecific communication, these signals are usually the recognition, attraction, or repulsion signals of the model species. A few mutualistic relationships do utilize what seem to be specialized interspecific signals: the hermit crab *Dardanus venosus* taps the anemone attached to its shell as a signal to release its grip and be transferred to a new shell (Cutress and Ross, 1969). Several other crabs do not show this signal and forcibly remove the anemone.

The Probabilistic Approach to Communication Analysis

The schematic presentation of the stimulus-response complex in terms of correlated behavioral changes bears a superficial resemblance to the early schematization of the stimulus-response nature of courtship in the stickleback, *Gasterosteus aculeatus* (Tinbergen, 1951; see Fig. 11). But the differences between the methods are marked.

The early stimulus-response scheme of Tinbergen was based on the hypothesis that certain stimuli are releasers for definite responses in such a manner that a characteristic sequence is formed. The schematic representation of the stimulus-response system served to illustrate the sequential nature of the idealized pattern. But the ideal pattern is not the most realistic presentation of the events. Barlow (1962) and Morris (1958) present the actual amount of overlap in similar schemes for courtship behavior (Fig. 11). While the idealized stimulus-response schemes are not obligatory, they do emphasize the fact that the observed sequences are by no means random (Baerends *et al.*, 1955; Morris, 1958).

Basing stimulus-response schemes on probabilistic relationships alleviates many of the complications common to these earlier methods and yet serves equally well to present the sequential nature of the observations. The system may be presented as correlated behavioral changes that show communication in terms of both information transmission and dependency upon contextual variables. Since only significant correlations are applicable, rare sequences

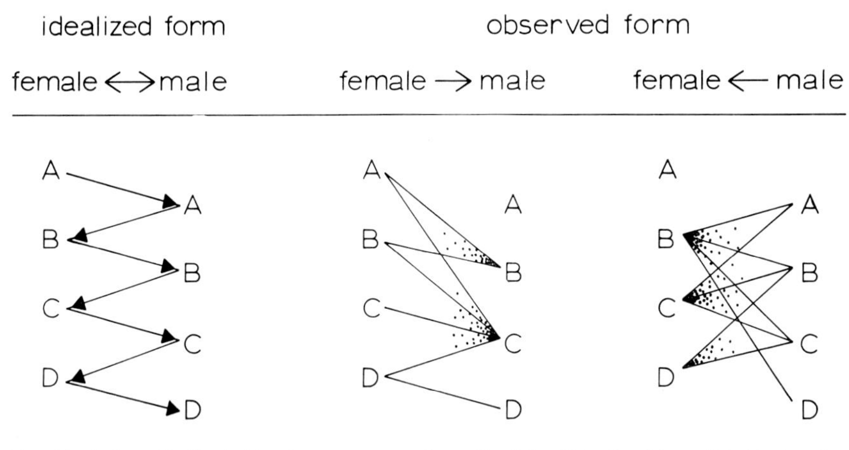

Fig. 11. A hypothetical stimulus-response schematic. The "idealized form" indicates the chain reaction nature of much courtship behavior; the *arrows* indicate that the action at the head of the arrow is released by the stimulus at the base. The "observed form" schematics present the actual stimulus-response relationships and overlap based upon observations of sequential relationships. (Adapted freely from Tinbergen, 1951 and Morris, 1958.)

have little effect on the formation of the hypothesis. The resultant schematic (Fig. 10) presents both a pictorial representation of the results and new, testable hypotheses. While the diagram presented here includes only qualitative relationships, work is underway to establish quantitative probability relationships through the use of standardized response to models. The results will produce a communication model that is easily tested for adequacy in terms of the postulated relationships as well as the interaction between signals whereby one acts as a contextual variable for another.

Even at this stage of investigation it is possible to postulate situations that can be used to test the adequacy of the hypothesized communication system. For example, the only negative stimuli that are indicated for *host-pose* result from *cleaner-clean*. As a result, it holds that once a host begins to *pose*, and if the cleaner does not *clean*, then the host should usually continue to *pose* until the positive stimuli (presence of cleaner, *cleaner-inspect*, and/or cleaning deprivation) are removed. In some cases (perhaps with *Chaetodon auriga*) mere interaction with the cleaner might be sufficient to terminate *posing* behavior. Inspection of Figure 10 will probably reveal other testable hypotheses to the reader. Note that situations can be chosen to test the hypothesized system that bear little or no relation to the arguments that were used in its formulation.

Cleaning as Related to Other Symbiotic Behavior

Cleaning is a unique type of symbiosis. Many symbiotic interactions might be characterized as relatively constant, intimate, and between dissimilar species. Cleaning, in one case or another, violates all of these rules. Some cleaning relationships may be quite constant and specific. For example, remoras (Echeneidae) may remain attached to one host for long periods of time and have been shown to act as cleaners (Strasburg, 1959; Randall, 1962). But most cleaners service a broad cross section of reef fishes, including both bottom and pelagic forms. The degree of intimacy ranges from long-term attachment with a specific group of hosts, as in the case of remoras and specialized cleaners entering the oral or gill cavity of the hosts, to forms that normally pick at the substratum for food and only occasionally pick at the skin of other fishes. Some cleaners, such as *Labroides phthirophagus*, usually clean other species but may show identical intraspecific cleaning behavior. And some species (*Abudefduf abdominalis*) that do not normally clean may begin to clean their species mates after having been isolated with them in an aquarium for long periods (P. Helfrich, personal communication). In these latter cases there appears to be a continuum between what must be termed intraspecific social behavior, and interspecific symbiotic behavior. But while some cleaning relationships might be classed as social behavior and some as phoresis, as defined by Cheng (1967), the highly developed cooperation and communication seen in most other cleaning relationships suggests that they have evolved far beyond casual or phoretic relationships.

ACKNOWLEDGMENTS

I thank the staff of the Hawaii Institute of Marine Biology, University of Hawaii, and the Atomic Energy Commision, Eniwetok Marine Biological Laboratory, for their facilities and generous support during the course of this research. I specifically thank Dr. Ernst S. Reese for his advice and assistance throughout this project, Dr. Edmund S. Hobson for reading the manuscript, and Mr. Dennis R. Kull for his assistance with illustration and animation of the motion picture that accompanied the presentation. This work was supported by Research Fellowship 1 FO2 GM38866-01, -02 the National Institutes of Health.

SUMMARY

Cleaning symbiosis provides a rich field for research and is now being subjected to critical examination by several workers. In the interactions between some cleaner and host fishes, the following list of stimuli act as communicative signals and increase the probability of the response shown in parentheses; morphology, coloration, and general swimming movements of the cleaner (*host-pose*), morphology, coloration, and general swimming movements of the host (*cleaner-inspect*), *host-pose* (*cleaner-inspect*), and *cleaner-inspect* (*host-pose*).

Cleaner-clean may be a communicative signal in that it lowers the probability that some host fishes will *pose*. *Cleaner-inspect* behavior shows a response-specific decrement of performance in that its performance lowers the probability that it will occur again.

The following contextual stimuli increase the probability of the behavior indicated in parentheses: cleaning station area (*host-pose*), deprivation of cleaning or increased parasitism (*host-pose*), and prospective food items (*cleaner-inspect* and *clean*).

The correlated behavioral changes presented above form a hypothetical communication system for cleaning symbiosis. This system helps to explain some observed variability in cleaning interactions and suggests additional hypotheses that may be used to test the adequacy of the system as a communication model.

The communicative signals involved in cleaning symbiosis may be less specialized or ritualized than many seen in vertebrate social behavior.

Cleaning is a distinct type of symbiotic behavior and yet the highly developed communication system suggests that it is much more than just a casual relationship.

APPENDIX

I. Species, and in parentheses, the number of specimens, used in the deprivation experiments.
Chaetodon auriga (6), *C. lunula* (4), *C. trifasciatus* (2), *C. miliaris* (4), *Scarus* sp. (12), *Apogon menesemus* (2), *Abudefduf abdominalis* (4), *Dascyllus albisella* (2), *Priacanthus cruentatus* (8).

II. Species, and in parentheses, the number of specimens, used to make the videotape recordings.
Chaetodon auriga (6), *C. lunula* (4), *C. trifasciatus* (2), *Scarus* sp. (2), *Apogon menesemus* (1), *Abudefduf abdominalis* (5), *Dascyllus albisella* (2).

REFERENCES

Baerends, G P , Brouwer, R, and Waterbolk, H. T. 1955. Ethological studies on *Lebistes reticulatus* (Peters). I. An analysis of the male courtship pattern. Behaviour 8:249–334.

Barlow, G. W. 1962. Ethology of the Asian Teleost, *Badis badis*: IV. Sexual behavior, Copeia 1962:346–360.

Cheng, T. C. 1967. Marine molluscs as hosts for symbioses (Adv. Marine Biol., Vol. 5). Academic Press, London. 424 p.

Cutress, C. E. and Ross, D. M. 1969. The sea anemone, *Calliactis tricolor*, and its association with the hermit crab *Dardanus venosus*. J. Zool. London 158:225–241.

Davenport, D. 1966. The experimental analysis of behavior in symbiosis, p. 381–429. In S. M. Henry (ed.) Symbiosis, Vol. 1. Academic Press, New York.

Dixon, W. J. and Massey, F. J., Jr. 1957. Introduction to Statistical Analysis. McGraw-Hill Book Co., New York. 488 p.

Eibl-Eibesfeldt, I. 1955. Über Symbiosen, Parasitismus, und andere besondere zwischenartliche Beziehungen tropischer Meeresfische. Z. Tierpsychol. 12:203–219.

Feder, H. M. 1966. Cleaning symbiosis in the marine environment, p. 327–380. In S. M. Henry (ed.) Symbiosis, Vol. 1. Academic Press, New York.

Fricke, H. 1966a. Zum Verhalten des Putzerfisches, *Laborides dimidiatus*. Z. Tierpsychol. 23:1–3.

Fricke, H. 1966b. Attrappenversuche mit einigen plakatfarbigen Korallenfischen im Roten Meer. Z. Tierpsychol. 23:4–7.

Hinde, R. A. 1965. The integration of the reproductive behavior of female canaries. In F. A. Beach (ed.) Sex and Behavior. John Wiley & Sons, New York.

Hinde, R. A. 1966. Animal Behaviour. McGraw-Hill Book Co., New York. 534 p.

Hobson, E. S. 1965a. A visit with el Barbero. Underwater Natur. 3:5–10.

Hobson, E. S. 1965b. Diurnal-nocturnal activity of some inshore fishes in the Gulf of California. Copeia 3:291–302.

Hobson, E. S. 1969. Comments on generalizations about cleaning symbiosis in fishes. Pacif. Sci. 23:35–39.

Lewis, A. G. 1961. A contribution to the biology of caligoid copepods parasitic on acanthurid fishes of the Hawaiian Islands. Ph.D. Thesis, Univ. Hawaii. 471 p.

Limbaugh, C. 1961. Cleaning symbiosis. Sci. Amer. 205:42–49.

Losey, G. S. 1968. The comparative behavior of some fishes of the genus *Hypsoblennius* Gill. Ph.D. Thesis, Univ. Calif., San Diego. 276 p.

Morris, D. 1957. "Typical intensity" and its relation to the problem of ritualisation. Behaviour 11:1–12.

Morris, D. 1958. The reproductive behaviour of the ten-spined stickleback (*Pygosteus pungitius* L.). Behaviour, Suppl. No. 6:1–154.

Randall, J. E. 1962. Fish service stations. Sea Frontiers 8:40–47.

Smith, W. J. 1969. Messages of vertebrate communication. Science 165:145–150.

Strasburg, D. W. 1959. Notes on the diet and correlated structures of some central Pacific echenied fishes. Copeia 3:244–248.

Tinbergen, N. 1951. The Study of Instinct. Clarendon Press, Oxford. 228 p.

Wickler, W. 1961. Über das Verhalten der Blenniiden *Runula* und *Aspidontus* (Pisces, Blenniidae). Z. Tierpsychol. 18:421–440.

Wickler, W. 1963. Zum Problem der Signalbildung, am Beispiel der Verhaltensmimikry zwischen *Aspidontus* und *Labroides* (Pisces, Acanthopterygii). Z. Tierpsychol. 20: 657–679.

Wickler, W. 1967. Specialization of organs having a signal function in some marine fish. Stud. Trop. Oceanogr. Miami 5:539–548.

Wickler, W. 1968. Mimicry. World University Library, New York. 249 p.

Wilkie, D. W. 1969. Outbreak of Cryptocaryoniasis in marine aquaria at Scripps Institution of Oceanography. Calif. Fish and Game 55:227–236.

Youngbluth, M. J. 1968. Aspects of the ecology and ethology of the cleaning fish *Labroides phthirophagus* Randall. Z. Tierpsychol. 25:915–932.

Nudibranch Associations as Symbioses

LARRY G. HARRIS
Department of Zoology
University of New Hampshire
Durham, New Hampshire

INTRODUCTION

During the past decade there has been an increased interest in the marine environment and, paralleling this, a greater interest in the study of symbiotic associations involving marine organisms. Cheng (1967) and Smyth (1969) have suggested that the study of different categories of symbiotic associations are not only useful in understanding symbioses, but can be valuable as models for the elucidation of basic biological problems. They have advocated strongly the study of symbiotic associations as a unit rather than looking exclusively at the symbiont as has usually been the case.

The purpose of this report is to discuss a group of opisthobranch molluscs, the nudibranchs, as symbionts. They are well known to marine researchers as grazers on sessile organisms including sponges, bryozoans, and coelenterates; however, many species of nudibranchs are so closely associated with their food species that they easily fit the criteria for a symbiotic association. Cheng (1967) has described a symbiont as "any animal or plant that spends a portion or all of its life cycle intimately associated with a different and larger species of animal or plant, during which there is a physiological exchange." It is the belief of the author that nudibranch associations have a high potential as models for the study of basic biological problems and may provide a new perspective into the understanding of the evolution of parasitism from predator-prey relationships. To illustrate this point, the current state of knowledge in three aspects of the biology of nudibranch associations will be described: (1) the induction of metamorphosis in veligers, (2) chemorecognition and chemotaxis, and (3) the utilization of host pigments and nematocysts by nudibranchs. Sources for these topics include the combined information from the literature and five years of personal observations by the author on nudibranch associations, particularly the association between aeolid nudibranchs of the genus *Phestilla*, and their respective scleractinian coral hosts (see Harris, 1970a).

INDUCTION OF METAMORPHOSIS

A good laboratory research subject must submit to serial laboratory cultivation. In a symbiotic association this problem is doubled because two species are involved; therefore the induction of metamorphosis of the larval stage of a marine symbiont is a relevant problem. Another consideration is that a symbiont species should have a short planktonic stage; in nudibranchs, a species that has a short, nonfeeding (lecithotrophic) larval stage is ideal.

To date, few species of nudibranchs have been followed through the complete life cycle, but those that have, suggest great potential for study. Nudibranch species with planktonic larval stages have been successfully raised by Rasmussen (1944), Thompson (1958, 1962), Swennen (1961), Tardy (1962), and Harris (1970a). Hadfield (1963) has reviewed the literature on larval cultivation attempts. All species with a planktonic stage that have been raised will metamorphose only in the presence of the living host, though contact with the host species is not a consistent requirement.

Thompson (1958) has found that *Adalaria proxima* veligers would settle only if the bryozoan, *Electra pilosa*, is present and alive, although contact is not necessary. Thompson (1962) also has found that veligers of *Tritonia hombergi* will settle only in the presence of *Alcyonium digitatum*. Tardy (1962) has reported that the small aeolid, *Capellina exigua*, will metamorphose only on contact with the hydroid *Kirchenpaueria pinnata*, and has stated that a broth made from this hydroid will induce settling, but that contact with the hydroid is necessary for complete metamorphosis. In contrast, broths made from other coelenterates and *Kirchenpaueria pinnata* broth that had been boiled or treated with alcohol do not induce any response in *C. exigua* veligers.

Phestilla melanobrachia veligers must feed before they are able to settle. They will only settle and undergo metamorphosis in close proximity to living coral tissue of the family Dendrophylliidae on which the adults feed. This is the only species of nudibranch with an obligatory feeding veliger stage that has been successfully raised through metamorphosis. *Phestilla sibogae* veligers, which do not feed, will metamorphose only in the presence of living *Porites* coral tissue, but contact is not necessary. They will complete metamorphosis any place in the container, including directly on the coral tissue, within 24 hours after the coral has been introduced.

Nudibranch veligers must have the prey (or host) species present at a specific time for successful metamorphosis, for there is a limited period during which metamorphosis can occur. If the association is made too early, the veligers are not ready; if it is made too late, the molluscs will have lost their ability to respond. Thompson (1958, 1962) has found that this period is about a week for *Adalaria proxima* and *Tritonia hombergi*. Under laboratory conditions the two *Phestilla* species are capable of settling for only two or three days.

The factors responsible for inducing settling and metamorphosis in nudibranch veligers are water-soluble. According to Tardy (1962), the factor from *Kirchenpaueria pinnata* is deactivated by heat and alcohol treatment, which suggests that it may be proteinaceous. Work is being conducted by M. G. Hadfield at the University of Hawaii to isolate and identify the active agents from species of *Porites* that induce metamorphosis in *Phestilla sibogae*. Hadfield and Karlson (1969) have found that the site of activity is located in the mucus from the coral and have isolated a fraction which is heat labile and will pass through a dialysis membrane and millipore filter, but its exact identity has not yet been established. Once the active factor has been identified, studies into the actual mechanism of metamorphosis can be pursued. This is a problem basic to many groups of marine invertebrates.

CHEMORECOGNITION AND ORIENTATION

Due to the relative simplicity of their nervous systems, there has been considerable interest in invertebrates as tools for neurophysiological and behavioral studies. Of particular emphasis has been their ability of chemorecognition and chemotaxis. Symbiotic associations are good models for such studies because host recognition is often a necessary requirement in at least some stage of the symbiont. In some nudibranch-coelenterate associations it would be possible to study and compare chemorecognition in both members of the association, as well as chemotaxis in the nudibranch.

Sea anemones have proved to be particularly useful animals for study because of their participation in numerous symbiotic associations (see Davenport, 1966; Mariscal, 1970) and for their ability to recognize and respond to specific substrates (see Robson, 1966).

Certain anemone species are attacked by aeolid nudibranchs of which *Aeolidia papillosa* is the best known. The sea anemones *Stomphia coccinea* and *Actinostola* sp. respond to *Aeolidia papillosa* mucus by swimming (Robson, 1966; and personal observations). Robson (1966) has reviewed swimming in *Boloceroides* sp. from Hawaii as described by March and Josephson (1964). The author has collected a species of aeolid, as yet unidentified, which actively attacks *Boloceroides* and elicits a swimming response in the anemone. Rosin (1969) has found that the anemone *Anthopleura nigrescens* detaches from the substrate when attacked by the aeolid *Herviella* species.

The intertidal anemone *Anthopleura elegantissima* shows chemorecognition of the nudibranch *Aeolidia papillosa* with which it forms an association. If an *A. papillosa*, or a piece of noble agar coated with mucus from *Aeolidia*, is dropped on the tentacles of an *Anthopleura* collected from an area free of *A. papillosa*, the anemone tries without success to eat the nudibranch and does eat the agar. However, if an *A. papillosa* is first allowed to attack the anemone for a short period, then the anemone, after resuming an open position, rejects *A. papillosa* and mucus-coated noble agar. The anemone's reaction lasts for several hours and it is specific for *A. papillosa*, for the anemone will readily take other nudibranchs and pieces of shrimp or squid. *Anthopleura* does not react to fresh noble agar controls.

The specificity of nudibranchs regarding food preferences suggested to some of the early workers that these animals must have the ability of distance chemoreception and concluded that the rhinophores were the most likely site for this sense (see Kohn, 1961). Nudibranchs have been shown to have chemosensory ability including chemotaxis; this would be an important sense for larger, longer-lived nudibranchs that may need to find new prey should they deplete their immediate food supply.

Arey (1918) has found, in a series of experiments on nudibranchs in Bermuda, that the rhinophores are the most sensitive areas of the body to tactile stimuli, that they are used for rheotaxis, and that they are sensitive to concentrations of certain salts, acids, bases, and essential oils. The nudibranchs only respond to direct application, and Arey has concluded that the rhinophores are not olfactory in function; however, he did not test a water current from a food source. Agersborg (1922) has found that the rhinophores of *Hermissenda crassicornis* are sensitive to tactile stimuli and to acids, but the animal could not find food a few millimeters away except by touch. He reported that applying broth of a substance that *H. crassicornis* would eat to the rhinophores will cause disorientation in the nudibranch. *Hermissenda* would move up a current, but Agersborg has concluded that the rhinophores are not for distance olfactory orientation. Agersborg (1925) has reported that the oral tentacles of *H. crassicornis* are gustatory and tactile in function; he has described two types of sensory cells which he believed to be responsible for these abilities.

The classic experiments of Davenport (1950) and Davenport and Hickock (1951) clearly have demonstrated the phenomenon of distance chemotaxis in marine invertebrates using polynoid commensals of certain echinoderms. Davenport (1950) has reported that polynoid worms of the genus *Arctonoe*, which are symbiotic on various asteroids, will move up a stream of water coming from a source containing a suitable host. These polynoid worms are able to descriminate between a stream of water from a suitable host and a stream of water from either no host or from an unsuitable host at the junction of a Y-tube. Davenport and Hickock (1951) have found that the factor that attracted the worms is unstable, heat labile, and will not diffuse through a dialysis membrane. They also have discovered that injury to the host will produce substances that overcome the attraction and cause repulsion by the worms instead.

Stehouwer (1952), using a setup similar to that reported by Davenport (1950), has shown that *Aeolidia papillosa* is attracted to the anemone *Metridium senile.* Instead of a Y-tube, Stehouwer used two separate glass siphons; this enabled the nudibranchs to reach the prey being tested. Stehouwer has found that *A. papillosa* demonstrates a preference for *M. senile* over other anemones in chemosensory and predation tests; she also has found that *A. papillosa* prefers anemones that are open rather than closed, young specimens over old, and damaged over undamaged specimens. *A. papillosa* is also attracted to containers with *Metridium* and feeding nudibranchs in preference to aquaria with *Metridium* and no nudibranchs. Braams and Geelen (1953) have extended Stehouwer's work by showing that *A. papillosa* is attracted to *M. senile* over several species of hydroids, while another aeolid, *Cratena* (= *Trinchesia*) *aurantia,* is attracted to its hydroid food but not to *M. senile.* Haafton and Verwey (1960) have shown that *A. papillosa* and *Trinchesia aurantia* will orient into a current of low velocity if water from a source containing a suitable prey (or host) is introduced into the aquarium. If no water from the prey is added, the animals tend to move in random directions. Therefore, according to Haafton and Verwey, "An animal scenting a certain favorable substance could then find its food simply by facing the current."

Cook (1962) has shown that the small dorid *Rostanga pulchra* is attracted to its food sponge, *Ophlitaspongia pennata,* while another dorid, *Archidoris montereyensis,* shows no attraction for its normal prey, *Halichondria panicea,* even though it ate only that sponge.

Phestilla melanobrachia has been tested for chemotaxis using a Y-tube setup similar to that reported by Davenport (1950) with the modification that the nudibranch had to crawl into the Y-tube of its own volition and then all the way to the top of the correct arm of the Y before the test was considered positive. This gave the nudibranch the possibility of only one positive choice and three negative choices; besides the choice of arms of the Y, the animal could refuse to enter the Y-tube, or start up and then turn around and go out again. The results summarized in Table 1 show that *Phestilla melanobrachia* is attracted to corals that it will eat, but not to ones that it will not eat, even when starving. *Phestilla sibogae* has not been tested, except for one small experiment in which all runs were positive for its normal food species, *Porites compressa.* The positive chemotaxis demonstrated in both species of *Phestilla* is combined with a positive rheotaxis similar to that reported by Haaften and Verwey (1960). The attractive factor(s) is heat labile and relatively stable, for the nonheated fraction remained active after even 18 hours.

Stehouwer (1952) has shown that *Aeolidia papillosa* is preferentially attracted to injured anemones over healthy ones. This is directly opposite the results of Davenport and Hickock (1952), which have shown that polynoid symbionts are repelled by injured hosts; however, this should be expected since an injured anemone is likely to be an easier prey than a healthy one. There was not sufficient time to test the attractiveness of injured prey to *P. melanobrachia*.

Collected data for *P. melanobrachia* in Hawaii, Japan, Singapore, and Australia suggest that the veliger stage is preferentially attracted to coral colonies already occupied by other *P. melanobrachia*; it would, therefore, be interesting to test whether the injured host or the combination of host and another symbiont is the key to this preferential selection. This would be an important adaptation to insure successful reproduction in short-lived species and seems that it may be a common phenomenon for ectosymbionts in general.

The information available on induction of metamorphosis and chemotaxis in nudibranch associations may indicate that in some closely associated nudibranch-coelenterate relationships, the factor which attracts the veliger to the host may also induce metamorphosis, attract the adult nudibranch, and elicit feeding behavior. Most behavior in nudibranchs is probably genetically programmed and one could envisage that, at a certain stage in the development of a veliger, low concentrations of a particular molecule could elicit positive chemotactic behavior based on a positive rheotaxis; but when the concentration of the molecule reaches a certain threshold, metamorphosis would be triggered. After metamorphosis the same molecule could then elicit

TABLE 1. *Results of Y-tube experiments on the chemosensory ability of* Phestilla melanobrachia. Phestilla melanobrachia *feeds on the first five corals listed, but will not eat any of the other corals listed.*

Corals tested	*Positive*	*Negative*	*No result*
Tubastraea aurea (orange)	20[a] (63)[b]	11 (34)	1 (3)
T. aurea (yellow-pink)	42 (87.5)	5 (10.4)	1 (2.1)
T. aurea (pink)	29 (90.6)	3 (9.4)	0
T. diaphana	29 (90.6)	3 (9.4)	0
Dendrophyllia elegans	30 (93.8)	1 (3.1)	1 (3.1)
Fungia scutaria	0	0	8 (100)
Montipora sp.	0	0	8 (100)
Porites compressa	1 (9.1)	0	11 (90.9)
P. compressa[c]	5 (100)	0	0
Control	12 (37.5)	0	20 (62.5)
Heat test[d] (heated)	1 (10)	0	9 (90)
Heat test (non-heated)	7 (70)	2 (20)	1 (10)

[a]=number of tests
[b]=percentages in parentheses
[c]=*Phestilla sibogae* used in this experiment
[d]=*Tubastraea aurea* (yellow-pink) was the coral used

rheotaxis-linked chemotaxis, and, above a certain threshold, feeding behavior. Such a mechanism would be efficient because only one type of chemoreceptor cell would be necessary for the veliger and the adult animal.

UTILIZATION OF PIGMENTS

Nudibranchs have long attracted attention because of the striking coloration of many species. Thompson (1960) and Edmunds (1966) have stated that most nudibranch species, including many of the brightly colored ones, are actually cryptically colored to blend with their background which is usually their food species. Carotenoids and concentrations of pterine or purine granules that reflect colors ranging from white to blue are the two most prominent sources of coloration reported for nudibranchs (Fox and Vevers, 1960; Bürgin, 1965), while carotenoids, melanin, flavones, and pigments contained in symbiotic zooxanthellae are important pigments in sponges and coelenterates (Fox, 1953; Fox and Pantin, 1944).

It is generally accepted that animals are not able to produce their own carotenoid pigments, but must obtain them from a plant source or through a food chain from animals that have received them from plants originally (Fox and Vevers, 1960). Animals are able to modify existing carotenoids, as is evidenced by the fact that the carotenoid pigment, astaxanthin, which is common in animals, occurs in only a few plants. It is, therefore, to be expected that many highly pigmented and brightly colored nudibranchs receive their pigments directly from their prey. What is surprising is that this phenomenon has been documented so few times in the literature and that nothing is known about the mechanisms of pigment uptake and deposition.

Elmhirst and Sharpe (1920) have reported that the red color of the anemone *Actinia equina* is due to a fat-soluble pigment that is most likely a carotenoid, while the brown coloration of *Anemonia sulcata* is due mainly to pigments contained in zooxanthellae in the tissues of the anemone. Payne (1931) has found that carotenoids are responsible for the coloration of hydroids in the families Haliciidae and Antennulariidae and that flavone-like pigments are the coloring agents in the hydroid family Sertulariidae. Kropp (1931) has stated that the blue pigment in the siphonophore *Velella spirans* is a carotenoid-protein complex which frees a red carotenoid in alcohol. Fox and Pantin (1941) have found that varying concentrations of melanin and carotenoids are responsible for the many color varieties in the anemone *Metridium senile*. The most common carotenoid has been found to be astacene, an oxidized form of astaxanthin, which does not occur in nature, but which is produced during extraction. Kawaguti and Tokohama (1966) have reported that carotenoids are at least partially responsible for the coloration of the red dendrophylliid coral *Dendrophyllia cribrosa*.

The first report of a nudibranch obtaining pigmentation from its coelenterate prey (or host) was made by Kropp (1931); however, the pigment is not permanent and is limited to the digestive gland. He has stated that the aeolid nudibranch *Fiona pinnata* contains the same blue pigment found in *Velella spirans* when the nudibranch has been feeding on the siphonophore. The pigment is localized in the digestive gland in the cerata and fades considerably when the nudibranchs are fed fish flesh which does not contain the blue pigment.

Abeloos (1932) has found that the dorid nudibranch *Doris* (=*Archidoris*) *tuberculata*, which feeds on the sponge *Halichondria panicea*, contains a yellow carotenoid that is identical to a yellow carotenoid extracted from the sponge. Both the sponge and the nudibranch also contain a blue, water-soluble pigment. The blue pigment is more highly concentrated in the digestive gland of the nudibranch than in the sponge, thereby indicating that it is being accumulated. The yellow carotenoid in *D. tuberculata* is not restricted to the digestive gland, but is present also in the body tissues. Abeloos has stated that he is of the opinion that *Doris tuberculata* is obtaining its pigments from *Halichondria panicea*.

Strain (1949) has described a pink xanthophyll from the bright pink dorid *Hopkinsia rosacea*, and designated it hopkinsiaxanthin. *Hopkinsia rosacea* feeds on a pink encrusting bryozoan of the genus *Eurystomella* (see Harris, 1970b) and this would seem to be the source of the xanthophyll.

Goodwin and Fox (1955) have reported carotenoids in two species of nudibranchs, a dorid, *Archidoris montereyensis*, and an aeolid, *Flabellina iodinea*. *Archidoris montereyensis* yields a yellow carotenoid similar to β-carotene, while *Flabellina iodinea* contains a carotenoid similar to astaxanthin. The pigment in *Archidoris* is free, while the one in *Flabellina* is in the form of a protein complex in the purple body and in a free state in the orange cerata. *Archidoris* feeds on the yellow sponge *Halichondria panicea*, which is the same species of sponge that Abeloos (1932) reported as supplying a yellow carotenoid to *Doris tuberculata*. *Flabellina* feeds on an orange hydroid, *Eudendrium* sp. (Harris, 1970b), which could be responsible for its orange carotenoid. This is the only case known to the author where the nudibranch has possibly modified a carotenoid obtained from its prey; in this case, a protein may have been added, assuming the orange carotenoid comes from the hydroid.

Bürgin-Wyss (1961) has found an orange pigment in the aeolid *Trinchesia coerulea*, which is probably a carotenoid. He also has described structural colors due to concentrations of granules in epithelial cells that behave like purines or pterines. Bürgin (1965) has reported a similar situation in the aeolid *Hermissenda crassicornis*. He has stated that the digestive gland takes on the color of certain highly pigmented prey organisms such as red hydroids. The author (unpublished) also has observed that the digestive gland in the cerata of *Hermissenda crassicornis* turns blue when specimens are fed on *Velella* sp.

Cullon (1966, and per. com.) has found that the small red dorid, *Rostanga pulchra*, has red carotenoids with absorption maxima identical to those of the red sponges on which it feeds. Furthermore, the pigments of the sponges vary specifically according to species, and *R. pulchra* taken from a particular species of sponge contains only the pigments that occur in that species. Cullon has found that it is possible to identify the species of sponge on which an individual *R. pulchra*, collected away from sponge, has been feeding by matching the absorption maxima of the pigments from the nudibranchs with those of one of the red sponge species.

The many shades of red, orange, yellow, and pink found in the dendrophylliid corals suggest that carotenoids are important pigments, while zooxanthellae appear to be the most important source of coloration in *Porites* spp. Field observations, laboratory feeding experiments, and spectrophotometric studies all confirm that *Phestilla melanobrachia* obtains its pigments from its dendrophylliid coral prey (or host). There are five separate pigment types found in *P. melanobrachia*, depending on the coral species on which the nudibranch has been feeding. All specimens of *P. melanobrachia* contain concentrations of reflecting granules in the epithelium of the dorsal surface similar to those described by Bürgin (1965). Nudibranchs feeding on *Turbinaria* spp. have zooxanthellae in the digestive gland which impart a gray color to the cerata. Individuals eating red-colored species of coral accumulate a red epiphasic carotenoid with a single absorption band at 470 mμ permanently in the epidermis and salivary glands; this corresponds closely to astaxanthin. Black-colored corals contribute a black, granular pigment that accumulates permanently in the digestive gland cells. This pigment is insoluble in all solvents tested including KOH and glacial acetic acid. Red, orange, pink, yellow, and black dendrophylliid corals contain one or more yellow, water-soluble pigments, which are similar to flavones and have absorption peaks around 380 mμ and 395 mμ. These pigments are limited to the digestive gland and are replaced slowly. This can be illustrated by changing a nudibranch from one color of coral to another. The color of the cerata varies according to the coral being eaten, although the black pigment, once ingested, remains, and the body does not lose the red pigmentation once acquired. Predation experiments suggest that cryptic coloration is relatively more important to young animals than it is to older, larger individuals, which are able to rely more effectively on other mechanisms such as defensive secretions. Therefore the ability to take on the color of the coral species being eaten, by utilization of its pigments, gives newly metamorphosed nudibranchs flexibility as to the color of coral on which they may feed in relative safety.

Phestilla sibogae has a mottled appearance due to concentrations of reflecting granules in the epithelium and zooxanthellae in the digestive gland. No separate pigments, such as carotenoids, have been observed in *P. sibogae*. The study of pigmentation in nudibranchs, particularly in relation to their host species, is a potentially rewarding subject for future studies.

UTILIZATION OF NEMATOCYSTS

Aeolid nudibranchs have long been known to store nematocysts, but almost nothing is known about the mechanisms involved or how a member of the Gastropoda can ingest, transport, and store, in a functional state, organelles from members of the Coelenterata, and then in turn utilize them for defense.

The types of nematocysts found in the cnidosacs of aeolids have been discussed by a few researchers, notably Graham (1938), Kepner (1943), and Edmunds (1966). Graham and Edmunds have described the types of nematocysts found in a number of aeolid species, and Edmunds has categorized aeolids into two groups: those that select only one or two types of nematocysts, and those that show essentially no selection but store all types occurring in the prey species. Kepner (1943) has given the only description of the fate of nematocysts within an aeolid. He has found that *Aeolis* (=*Cratena*) *pilata* selected only one type of nematocyst, the microbasic mastigophore, from its hydroid prey, *Pennaria tiarella*; he has also reported that unfired nematocysts travel through the digestive gland into the cnidosac where they are ingested by the cnidosac cells. Of the four types of nematocysts found in *P. tiarella*, only the microbasic mastigophores are maintained in the cells of the cnidosac. The other three types are digested by the same cells that maintain the microbasic mastigophores. Kepner has suggested that this type of nematocyst would be the most effective deterrent to predators because it is a penetrating nematocyst which functions by injecting toxins and digestive enzymes into prey organisms.

Thompson and Bennett (1969) have found that the pelagic aeolids *Glaucus atlanticus* and *Glaucilla marginata* selectively store only the largest penetrating nematocysts of their prey, the portuguese man-o-war, *Physalia utriaulus*.

Personal observations on a number of coelenterate-eating nudibranchs have revealed that many nematocysts are fired during feeding. Examination of the stomach contents of nudibranchs immediately after feeding has indicated that often as many as 50 % of all nematocysts are fired prior to or during ingestion. There are some nudibranch species that feed by penetrating the perisarc of hydroid stalks or stolons and suck out the contents, thereby obtaining a high proportion of immature, and therefore nonfuctional, nematocysts, and yet, the nudibranch's cnidosacs contain functional nematocysts. These observations suggest the hypothesis that at least some species of aeolid nudibranchs may be able to select immature, nonfunctional nematocysts and develop them to maturity in the cnidosac cells.

The two species of *Phestilla* are atypical of most aeolid nudibranchs for they do not store nematocysts. At least 50 % of the nematocysts seen in the stomach contents of *P. melanobrachia* immediately after feeding had been fired; the rest are unfired and remain intact as they move through the digestive tract and are egested in feces. The high percentage of fired nematocysts in the stomach of *P. melanobrachia* following feeding would seem to indicate that a high percentage of those nematocysts that are intact are nonfuctional, though this has not been tested.

The utilization of functional cell organelles from animals in one phylum by animals in quite a distant phylum promises to be a fertile area for study. Whether nudibranchs select mature or immature nematocysts is a limited portion of a complex mechanism that has possible implications for the general understanding of the phenomenon of tissue acceptance and rejection.

DISCUSSION

There is a vast number of problems that can be studied, depending on the nudibranch association, of which the induction of metamorphosis, chemotaxis, and utilization of host materials are only a sampling.

The aeolid nudibranch *Phestilla sibogae* and corals of the genus *Porites* form an association which has almost unlimited possibilities for research as the following facts should indicate. Corals of the genus *Porites* are important members of the tropical reef community and, therefore, are of interest by themselves for studies of the role of zooxanthellae and reef ecology. *Phestilla sibogae* thrives under laboratory conditions and large numbers can be raised in a relatively short period; the generation time from egg to egg is about 40 days. Nudibranchs are hermaphroditic and, once sexual maturity is reached, each animal lays approximately 1.5 egg masses per day for almost 100 days, each egg mass containing 2,000–3,000 eggs. The veligers hatch in 5–6 days after laying and they are ready to settle within 4 additional days. If a small piece of living *Porites* is added on the third day, 50–60 % of the veligers present will metamorphose in about 24 hours. Two hundred veligers can be raised in 200 ml of clean sea water. *Phestilla sibogae* grows to a length of 40 mm. It is large enough to manipulate easily and since adult nudibranchs lack a shell, all body areas are readily accessible. As has been mentioned in the section on the "Utilization of Pigments," *P. sibogae* has zooxanthellae in the digestive gland which are digested, but prior to digestion, they are still functional; therefore, a soft-bodied nudibranch may be an excellent site to study some aspects of the biology of zooxanthellae.

Not all nudibranch associations have the same research potential as the one involving *Phestilla sibogae*, but this association does indicate some of the advantages of working with a nudibranch association including a short generation time, soft body, prolific reproduction, and hosts of research interest.

There are also nudibranchs associated with most other groups of sessile organisms ranging from sponges to barnacles, and these associations cover the spectrum from generalized predators, like the aeolid *Hermissenda crassicornis* which feeds on various coelenterates, small crustaceans, and other nudibranchs, to the highly specialized aeolid *Cratena aurantia* which lives among the stolons of *Tubularia* colonies and feeds by drilling through the perisarc of a stolon and sucking out small amounts of coenosarc. It is surprising that a group with such promising research potential has been so neglected. It is likely, however, that as more researchers become interested in marine symbioses, nudibranch associations will receive the attention that their potential seems to merit.

SUMMARY

Many nudibranch gastropods form highly specific associations with food species that fulfill the criteria for symbiotic relationships as defined by Cheng (1964). To illustrate this and to emphasize the research potential of nudibranchs as members of symbiotic associations, three aspects of the biology of nudibranch associations are discussed in this paper, including (1) the induction of metamorphosis of veligers, (2) chemorecognition and chemotaxis, and (3) utilization of host pigments and nematocysts. This information is derived from the literature and from the author's studies on aeolid nudibranchs of the genus *Phestilla* and their associations with scleractinian coral hosts.

REFERENCES

Abeloos, M. et R. 1932. Surs les pigments hepatiques de *Doris tuberculata* Cuv. (Mollusque Nudibranche) et leurs relations avec les pigments de L' Eponge *Halichondria panicea* (Pallas). C. R. Soc. Biol. Paris 109:1238–1240.

Agersborg, H. P. K. 1922. Some observations on qualitative chemical and physical stimulations in nudibranchiate mollusks with special reference to the role of the "rhinophores." J. Exp. Zool. 36:423–444.

Agersborg, H. P. K. 1925. The sensory receptors and their structure in the nudibranchiate mollusk *Hermissenda crassicornis* (Eschscholtz, 1831) syn. *Hermissenda opalescens* Cooper, 1862. Acta Zool. 6:167–182.

Arey, L. B. 1918. The multiple sensory activities of the so-called rhinophores of nudibranchs. Amer. J. Physiol. 46:526–532.

Braams, W. G. and Geelen, H. F. M. 1953. The preference of some nudibranchs for certain coelenterates. Arch. Neerl. Zool. 10:241–264.

Bürgin, U. F. 1965. The color pattern of *Hermissenda crassicornis* (Eschscholtz, 1831). The Veliger 7:205–215.
Bürgin-Wyss, U. 1961. Die Ruckenanhange von *Trinchesia coerulea* (Montagu). Rev. Suisse Zool. 68:461–582.
Cheesman, D. F., Lee, W. L. and Zagalsky, P. F. 1967. Carotenoproteins in invertebrates. Biol. Rev. 42:132–160.
Cheng, T. C. 1964. The biology of animal parasites. W. B. Saunders Co., Philadelphia. 727 p.
Cheng, T. C. 1967. Marine molluscs as hosts for symbioses. (Adv. Marine Biol., Vol. 5) Academic Press, London. 424 p.
Cook, E. F. 1962. A study of food choices in two opisthobranchs, *Rostanga pulchra* McFarland and *Archidoris montereyensis* (Cooper). The Veliger 4:194–196.
Cullon, C. 1966. Senior Honor's Report, Department of Zoology, Univ. California, Berkeley.
Davenport, D. 1950. Studies in the physiology of commensalism. 1. The polynoid genus *Arctonoe*. Biol. Bull. 98:81–93.
Davenport, D. 1966. The experimental analysis of behavior in symbiosis, p. 381–429. *In* Henry, S. M. (ed.) Symbiosis, Vol. 1. Academic Press, New York.
Davenport, D. and Hickock, J. F. 1951. Studies in the physiology of commensalism. 2. The polynoid genera *Arctonoe* and *Halosydna*. Biol. Bull. 100:71–83.
Edmunds, M. 1966. Protective mechanisms in the Eolidacea (Mollusca Nudibranchia). J. Linn. Soc. (Zool.) 47:25–71.
Elmhirst, R. and Sharpe, J. S. 1920. IV. On the colours of two sea anemones, *Actinia equina* and *Anemonia sulcata*. I. Environmental. II. Chemical. Biochem. J. 17:48–57.
Fox, D. L. 1953. Animal biochromes and structural colours. Cambridge Univ. Press, Cambridge. 379 p.
Fox, D. L. and Pantin, C. F. A. 1941. The colours of the plumose anemone, *Metridium senile* (L.). Roy. Soc. London Philos. Trans. 230:415–450.
Fox, D. L. and Pantin, C. F. A. 1944. Pigments in the Coelenterata. Biol. Rev. 19:121–134.
Fox, H. M. and Vevers, G. 1960. The nature of animal colours. Sidgwick and Jackson Ltd., London. 246 p.
Goodwin, T. W. and Fox, D. L. 1955. Some unusual carotenoids from two nudibranch slugs and a lamprid fish. Nature 75:1086–1087.
Graham, A. 1938. The structure and function of the alimentary canal of aeolid molluscs, with a discussion on their nematocysts. Trans Roy. Soc. Edinburgh 59:267–307.
Haaften, J. L. v. and Verwey, J. 1960. The role of water currents in the orientation of marine animals. Arch. Neerl. Zool. 13:493–499.
Hadfield, M. G., 1963. The biology of nudibranch larvae. Oikos 14:85–95.
Hadfield, M. G. and Karlson, R. H. 1969. Externally induced metamorphosis in a marine gastropod. Amer. Zool. 9:1122.
Harris, L. G. 1970a. Studies on the aeolid nudibranch, *Phestilla melanobrachia* Bergh, 1874. Ph.D. Thesis. Univ. California, Berkeley.
Harris, L. G. 1970b. Observations on the distribution and food preferences of some West Coast nudibranch mollusks. (MS in preparation.)
Kawaguti, S. and Yokohama, T. 1966. Electron microscopy on the polyp and pigment granules of an ahermatypic coral. Biol. J. Okayama Univ. 12:69–80.
Kepner, W. A. 1943. The manipulation of the nematocysts of *Pennaria tiarella* by *Aeolis pilata*. J. Morphol. 73:297–311.
Kohn, A. J. 1961. Chemoreception in gastropod molluscs. Amer. Zool. 1:291–308.
Kropp, B. 1931. The pigment of *Velella spirans* and *Fiona marina*. Biol. Bull. 60:120–123.
March, S. C. and Josephson, R. K. 1964. The swimming performance of a sea-anemone. Amer. Zool. 4:384.
Mariscal, R. 1970. Experimental studies on the protection of anemone fishes from sea anemones. In this volume.
Payne, N. M. 1931. Hydroid pigments. I. Feneral discussion and pigments of the Sertulariidae. J. Mar. Biol. Ass. U. K. 17:739–749.

Rasmussen, E. 1944. Faunistic and biological notes on marine invertebrates. I. The eggs and larvae of *Brachystomia rissoides* (Hanl.), *Eulimella nitidissima* (Mont.), *Retusa truncatula* (Brug.) and *Embletonia pallida* (Alder & Hancock) (Gastropoda marina). Vidensk. Medd. Dansk. Naturh. Foren. 107:207–233.

Robson, E. A. 1966. Swimming in Actiniaria, p. 333–360. *In* Rees, W. J. (ed.) The Cnidaria and their evolution (Symp. Zool. Soc. Lond. No. 16). Academic Press, London. 449 p.

Rosin, R. 1969. Escape response of the anemone, *Anthopleura nigrescens* (Verrill) to its predatory eolid nudibranch *Herviella* Baba spec. nov. The Veliger 12:74–77.

Smyth, J. D. 1969. Parasites as biological models. Parasitology 59:73–91.

Stehouwer, H. 1952. The preference of the slug *Aeolidia papillosa* (L.) for the sea anemone *Metridium senile* (L.). Arch. Neerl. Zool. 10:161–170.

Strain, H. H. 1949. Hopkinsiaxanthin, a xanthophyll of the sea slug, *Hopkinsia rosacea*. Biol. Bull. 97:206–209.

Swennen, C. 1961. Data on distribution, reproduction and ecology of the nudibranchiate, molluscs occurring in the Netherlands. Netherlands J. Sea Res. 1:191–240.

Tardy, J. 1962. Observations et experiences sur la metamorphose et la croiseance de *Capellinia exigua* (Ald. & H.) (Mollusque Nudibranche). C. R. Acad. Sci. Paris 254:2242–2244.

Thompson, T. E. 1958. The natural history, embryology, larval biology and post-larval development of *Adalaria proxima* (Alder and Hancock) (Gastropoda, Opisthobranchia). Phil. Trans. Roy. Soc. London, B 242:1–58.

Thompson, T. E. 1960. Defensive adaptations in opisthobranchs. J. Mar. Biol. Ass. U. K. 39:123–134.

Thompson, T. E. 1962. Studies on the ontogeny of *Tritonia hombergi* Cuvier (Gastropoda, Opisthobranchia). Phil. Trans. Roy. Soc. London, B. 245:171–218.

Thompson, T. E. and Bennett, I. 1969. *Physalia* nematocysts: utilized by mollusks for defence. Science 166:1532–1533.

Respiratory Metabolism of a Trematode Metacercaria and Its Host

WINONA B. VERNBERG
and
F. JOHN VERNBERG
Belle W. Baruch Coast Research Institute
and
Department of Biology
University of South Carolina
Columbia, South Carolina

INTRODUCTION

Host-parasite relationships represent a dynamic adjustment on the part of both partners. Not only must they adjust to each other, but also this interspecific interplay affects the ability of each to meet the stresses of their physical environments. Digenetic trematodes are an excellent group of animals for host-parasite studies since they have complex life histories with the various stages occupying diverse habitats. The present study, involving the metacercarial stage of *Zoogonius lasius* and its second intermediate host, the nereid worm *Leonereis culveri*, has two specific purposes: (1) to determine oxygen uptake rates of encysted metacercariae during development; and, (2) to ascertain the effects of the developing cysts on the metabolism and enzymatic activity of the host animal acclimated to different thermal levels.

Development of metacercariae, the final larval stage of digenetic trematodes, varies among different species. In many species encystment occurs as soon as the free-living cercariae make contact with suitable vegetation or the proper host. In others, such as some of the gymnophallid trematodes, the cercaria attaches to the mantle of the body wall of bivalve molluscs and grows unencysted almost to adult size (Stunkard and Uzmann, 1958). In contrast, in schistosomes this larval stage is completely bypassed.

Literature on the length of development of encysting metacercariae has been reviewed recently by Donges (1969), and he has divided these metacercariae into three groups: (1) those that encyst in the open — in these species

This research, supported in part by Grant R01 A107174–03 from the National Institutes of Health, represents Part V of a series concerned with "Interrelationships Between Parasites and Their Hosts."

the metacercariae are infective to the definitive host immediately after encystment; (2) those in which the metacercaria is inside the body but does not grow in the second intermediate host — such metacercariae are not infective to the final host for several days; and (3) species that grow and develop within the second intermediate host — these metacercariae are infective only after a developmental period of several weeks.

To survive, all organisms must be able to generate energy. Although the ultimate source of energy for animals is derived from its external environment in the form of food, energy liberation depends on oxidation, and in most, but not all animals, oxidation eventually involves oxygen. In those metacercariae in which little development occurs, energy needs and oxidative demand appear to be low. But in species where there are great developmental changes, it is expected that energy needs would be large. In unencysted gymnophallid metacercariae, which have been seen actively feeding on host tissue, larvae increase five-fold in size during metacercarial development (Stunkard and Uzmann, 1958). In these metacercariae there is no cyst wall to act as a barrier to the passage of food or of molecular oxygen needed for oxidative processes. In metacercariae with well developed cyst walls, on the other hand, both food and molecular oxygen probably would be considerably less available.

Some studies indicate that development of encysted metacercariae may not involve molecular oxygen in oxidative processes. The metacercaria of *Echinostoma revolutum* is one that undergoes morphological changes following encystment. In this species, glycogen deposited in the giant cells of its digestive tract is used up within six weeks after encystment, and there is as gradual decrease of fat in the excretory bladder accompanying the decrease in glycogen, thus indicating anoxidative glycolysis (Zdarska, 1964). There is evidence, however, that cyst walls of metacercariae are not completely impermeable barriers. Metacercariae of *Clinostomum campanulatum*, which encyst in the subcutaneous tissue and muscles of fish, absorb ^{14}C-glucose *in vivo* and incorporate glucose carbon into glycogen, amino acids, and some intermediates of the TCA cycle (Thomas and Gallicchio, 1967). The fish hosts were injected intraperitoneally with labeled glucose solution, and within one hour detectable quantities had been absorbed by the worms within the cyst. Such findings indicate that the developing worms are obtaining energy from the host animal.

When the intensity of an environmental factor fluctuates, many organisms show compensatory changes in metabolic rate. In a recent review, W. B. Vernberg (1969) has concluded that larval trematodes (including *Z. lasius*) influence the thermal acclimatory ability of their first intermediate host (the gastropod, *Nassarius obsoleta*), but the metabolic response pattern to temperature of the larvae and its host may not be similar. However, the influence of metacercariae of *Z. lasius* on the metabolic-temperature response of their host is not known.

MATERIALS AND METHODS

Zoögonus lasius cercariae used to infect experimentally the second intermediate host were obtained from naturally infected marine mud-flat snails, *Nassarius obsoleta*, the first intermediate host. The second intermediate host, a polychaete worm, *Leonereis culveri*, was collected from an area in Beaufort, North Carolina, where naturally occurring infections with metacercariae of *Z. lasius* were very low. Infections of the polychaetes are easily detected since cysts occur in the parapodia and are readily observed. In the laboratory, uninfected polychaete worms, which were of similar size, were isolated singly in compartmentalized plastic boxes, then placed in constant temperature cabinets at either 10°C (cold acclimated) or 25°C (warm acclimated) under a 12 hour light/12 hour dark photoperiod for periods of not less than 10 days nor more than 2 weeks. Each polychaete was then exposed to $\pm$50 freshly shed cercariae for a period of approximately 2 hours and then placed at the appropriate temperature in plastic boxes.

Cartesian diver respirometers having a volume of 10–12μl were used to determine the metabolic rate of metacercariae at different stages of development. Polychaetes infected with metacercariae that were used in this phase of the study were maintained only at 25°C. Cysts of a known age were removed from the polychaete host and rinsed thoroughly in filtered seawater with the salinity adjusted to 30 ‰. From 6 to 8 cysts were placed in each diver in filtered seawater and readings were made hourly over a 3–4 hour period. All measurements were made at 25°C. Metabolic rates were determined 1, 2, 3, 4, 7, and 11 days following encystment; 8–19 determinations were made at each stage of development. To determine the metabolic rate of an 11-day excysted metacercaria, the cyst wall was carefully removed by teasing it away with a pair of fine needles. From 2 to 3 excysted metacercariae were used per diver. To determine the weight of the encysted and excysted metacercariae, 10–12 cysts were placed on a pre-weighed piece of plastic wrap (Cut-rite brand), dried to a constant weight at 110°C, and weighed on a microtorsion balance having a sensitivity of $\pm 2\mu$g.

Metabolism of infected and noninfected *L. culveri* was determined in Warburg respirometers. From 3 to 4 worms were used in each 3–4 ml flask which contained both filtered seawater (salinity of 30 ‰) and 20 % KOH in the center well. The manometers were not shaken. Each set of worms was used only once. Determinations on infected worms were made 48 hours following exposure to cercariae. Determinations were made on warm-acclimated and cold-acclimated worms at 20, 25, and 30°C. From 9 to 25 determinations were made at each temperature. At the end of each determination the worms were placed on a small piece of aluminum foil, dried to a constant weight at 110°C, and weighed. Results are expressed as $\mu l\ O_2$ per hour per mg dry weight.

Spectrophotometric assays of cytochrome *c* oxidase activity were made and calculated using the method of Smith (1955). Details of the analyses are given in a previous paper (Vernberg and Vernberg, 1968). Assays were made on tissues of uninfected *L. culveri*, and on tissues of polychaetes infected 48 hours previous to the assay. Metacercariae were removed, and, as far as it was possible to determine, only polychaete tissue was used. Results are expressed as mμ—moles cytochrome *c* oxidized/mg protein per second. Protein determinations were made on an aliquot sample of the tissue homogenate by the method of Lowry, *et al.* (1951).

Significance of difference of means was calculated by the method given in Simpson, *et al.* (1960) for small samples.

RESULTS

The metabolic rate of the metacercariae is relatively high one day after encystment, and, although the rate drops slightly during the following 2 days, it is not until day 4 that there is a precipitous drop in oxygen uptake. There is a further sharp decrease at day 7, but between days 7 and 11 the rate does not change significantly (Fig. 1). While the metabolic rate decreases, the weight of the cysts is increasing. However, the decrease in metabolic rate is much greater than the size increase of the cysts: the weight increased approximately 2.5 times between day 1 and day 11, but the metabolic rate decreased approximately 8 times during this same period (Figs. 1, 2).

When 11-day-old metacercariae were excysted, they became very active, and the metabolic rate increased sharply over that of encysted metacercariae of the same age. When the metabolic rate of 11-day-old encysted metacercariae was calculated on the same weight basis as for excysted worms, the rate of the excysted worms was still approximately 3 times that of encysted worms. The heightened activity of excysted worms undoubtedly accounts for much of the increased oxygen uptake (Fig. 3).

A comparison of the response of infected and noninfected polychaetes to cold acclimation indicates that the metacercariae do affect the response of their host. While noninfected polychaetes have a typical metabolic response to cold at 25°C (that is, the metabolic rate of cold-acclimated worms is translated to the left of warm-acclimated ones), infected worms show no acclimation response at this temperature (Fig. 4).

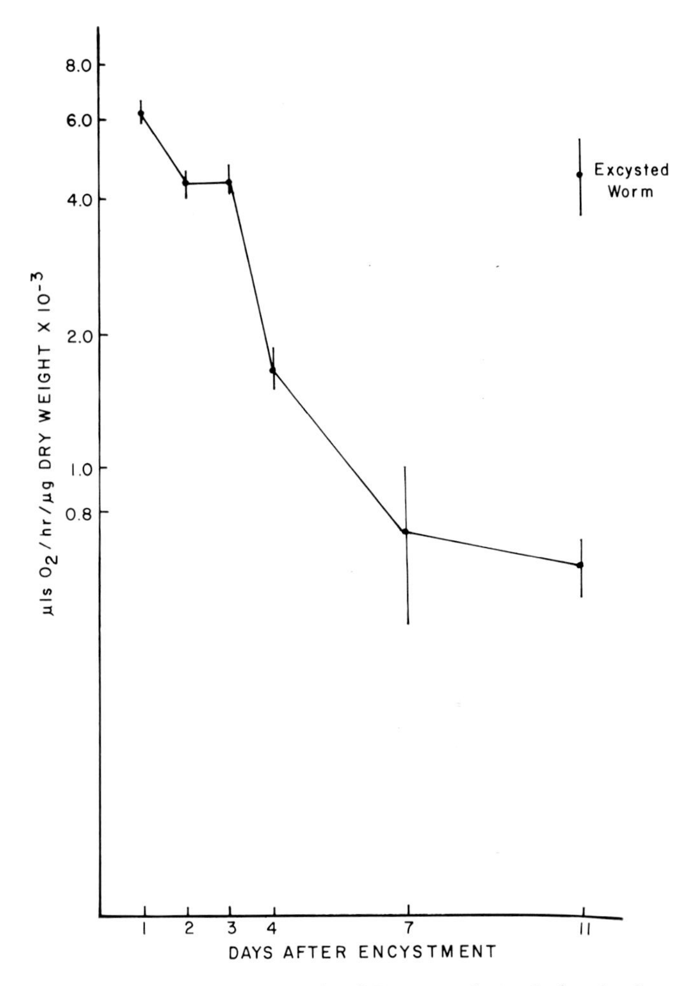

Fig. 1. Metabolic rate of the metacercaria of *Zoogonius lasius* during development.

Forty-eight hours following exposure to cercariae of *Z. lasius*, the metabolic rate of cold-acclimated polychaetes was significantly lower than that of cold-acclimated, noninfected worms at both 20° and 25°C. There was no statistically significant difference between warm-acclimated infected and noninfected worms at 20, 25, and 30°C, although there was a tendency for higher oxygen consumption rates in infected worms at 30°C.

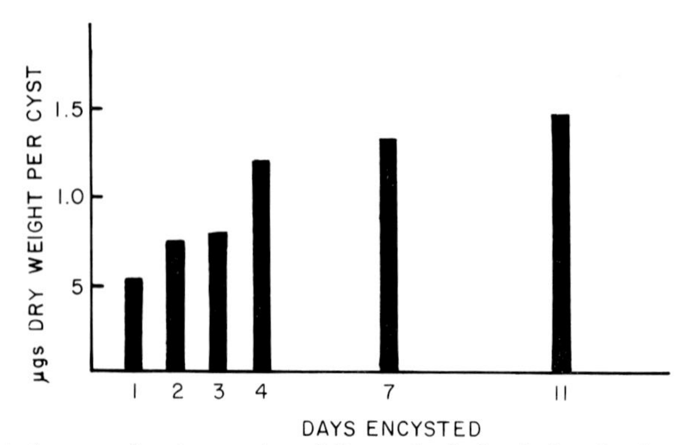

Fig. 2. Weight changes of metacercariae of *Zoogonius lasius* during development.

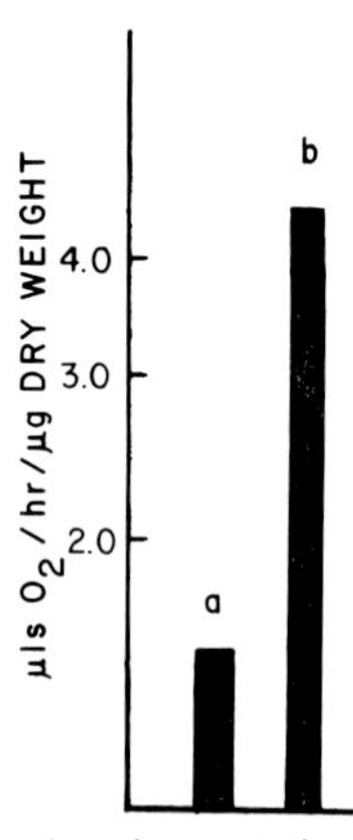

Fig. 3. Metabolic rates of encysted and excysted metacercariae of *Zoogonius lasius* calculated on the excysted weight basis. *a*, the metabolic rate of the encysted larvae; *b*, the metabolic rate of the excysted larvae.

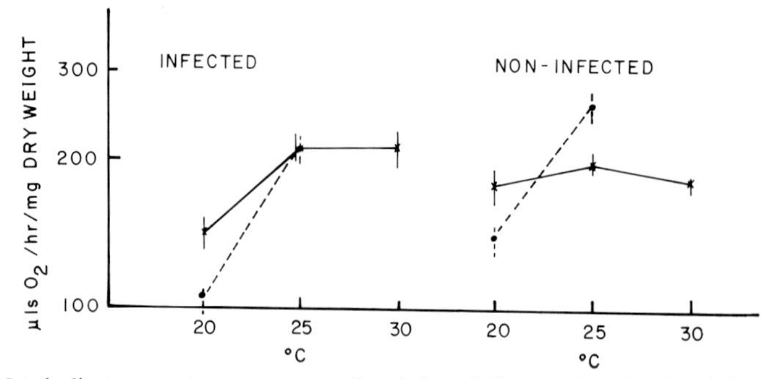

Fig. 4. Metabolic-temperature response of uninfected *Leonereis culveri* and *L. culveri* infected 48 hours previously with metacercariae of *Zoogonius lasius*.

Thermal acclimation patterns of cytochrome *c* oxidase activity in tissues of the host animal are also modified by parasitism (Fig. 5). The response of infected cold-acclimated worms was nearly the reverse of that of noninfected ones. In cold-acclimated, infected polychaetes cytochrome *c* oxidase activity was shifted to the right of warm-acclimated infected worms at 10, 25, and 35°C, while enzymatic activity in tissue of cold-acclimated, noninfected polychaetes was shifted to the left of tissues of warm-acclimated, noninfected worms at 10 and 20°C. This type of acclimation pattern is one of the most common reported for tissues of other marine invertebrates (Vernberg and Vernberg, 1968; W. B. Vernberg, 1969).

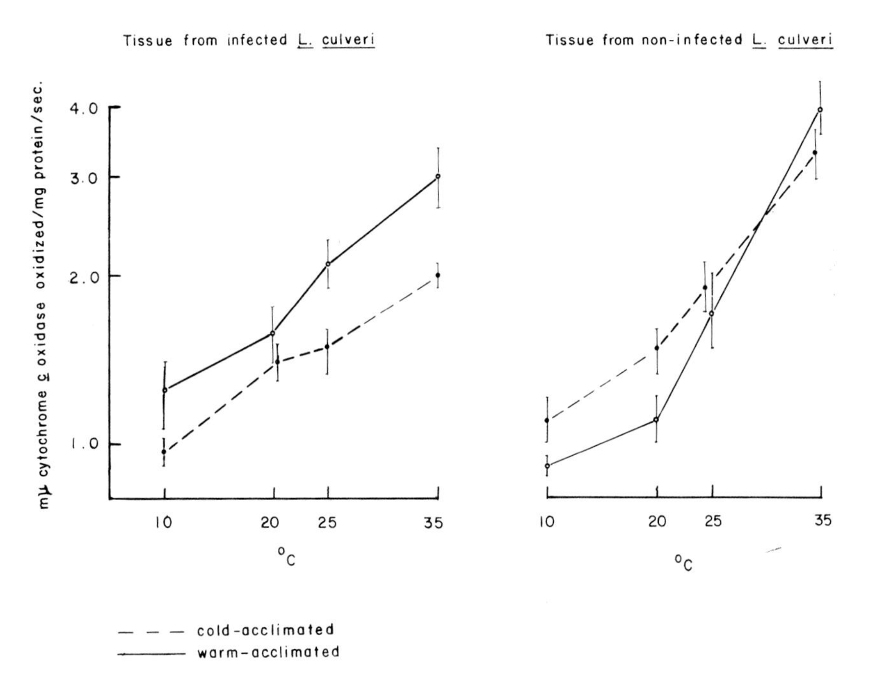

Fig. 5. Thermal acclimation patterns of cytochrome *c* oxidase activity in tissue from non-infected *Leonereis culveri* and in tissue of *L. culveri* infected 48 hours previously with *Zoogonius lasius*.

The rate of activity of cytochrome *c* oxidase tended to be higher in tissues of warm-acclimated, infected worms than in tissues of noninfected ones, and lower in tissue of cold-acclimated, infected polychaetes when compared with activity rates in tissues from cold-acclimated, noninfected ones (Fig. 6, Table 1).

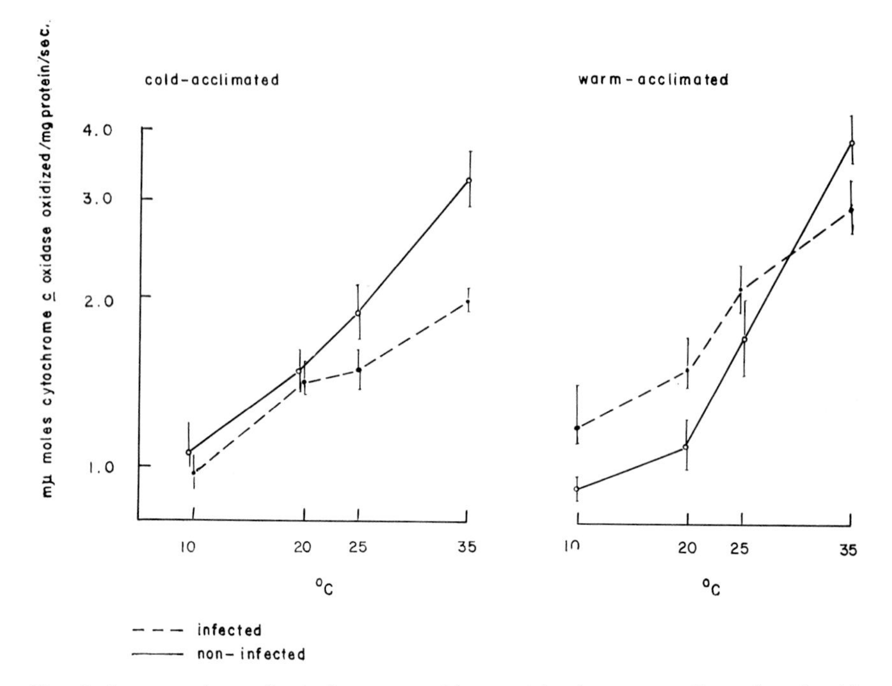

Fig. 6. A comparison of cytochrome *c* oxidase activity in warm-acclimated and cold-acclimated, infected and noninfected, *Leonereis culveri*.

TABLE 1. *Comparison of the rate of activity of cytochrome* c *oxidase in tissue of* Leonereis culveri *infected with* Zoogonius lasius *metacercariae and in tissue of noninfected* L. culveri.

Temperature of determination	*Warm-acclimated*	*Cold-acclimated*
10	+[a]	
20	+	
25		−[b]
30	+	−

[a] (+) indicates that the rate of activity of tissue from infected *L. culveri* is significantly higher than in tissue from noninfected *L. culveri*

[b] (−) indicates that the rate of activity from infected *L. culveri* is significantly lower than in tissue from noninfected *L. culveri*

DISCUSSION

The morphological changes accompanying the development of metacercariae of *Zoogonius lasius* in *Nereis virens* have been described in detail by Shaw (1933). Although she does not state the temperature at which infected worms were maintained, since the work was done during the summer months at Woods Hole, Massachusetts, temperatures were most likely several degrees cooler than the 25°C used in the present study, and development probably was somewhat slower in the Woods Hole worms. Other minor differences in the developmental pattern of *Z. lasius* might exist because a different host was used in our study.

The cercariae accumulate in the parapodial tissues of the polychaete host. Initially the larvae lay down a thin, clear cyst wall. After a few days a thick coat of connective tissue produced by the host develops around the metacercariae. Shaw (1933) reported a well developed outer cyst wall by day 10; in worms used in this study the outer wall was well developed 7 days following encystment. During development the larval structures are resorbed and the organs of the adult differentiate. In the Woods Hole larvae, the cysts had attained 80 % of their maximum size by day 13, many of the larval structures had been resorbed or greatly reduced in size, and adult structures were beginning to develop. Measurements given by Shaw would indicate that maximum development is completed between 27 and 45 days. Metacercariae develop to the point where, according to Stunkard (1938), the only change a maximally developed metacercaria undergoes in the definitive host is the maturing of the reproductive organs.

With such marked developmental changes occurring during the metacercarial stage it seems unlikely that the larvae rely solely on stored energy, and, considering the relatively high metabolic rates of the developing cysts, it also seems unlikely that all energy is liberated by anoxidative processes. Shaw, noting that the metacercariae are always found in the parapodia-region, stated "no doubt the rich blood supply in this region affords especially favorable conditions for development of the metacercariae." The rich blood supply, in addition to offering the cysts access to food, would also insure access to oxygen.

The metabolic pattern of developing metacercariae is very similar to the U-shaped pattern reported for insect pupae (Vernberg and Meriney, 1957). At the beginning of pupation, during the initial reorganization of the larvae, the metabolic rate is relatively high; following this period the rate drops to a very low level. Then, at the end of the pupal period, the metabolic rate again increases sharply. Metacercariae have a high metabolic rate at the beginning of their development, then there is a gradually decreasing rate which drops to a very low level as they become more developed; excystment triggers a sharp increase in metabolic rate.

What effect does metacercarial development have on its host? As is true for many second intermediate hosts, a massive invasion of cercariae will kill *L. culveri*. Lighter infections, while not lethal, do modify capacity adaptation responses of their hosts at least in their earlier stages of development.

The presence of metacercariae influences the ability of the polychaete to metabolically acclimate to temperature, especially low temperature. For example: (1) noninfected, warm-acclimated hosts maintained a relatively constant metabolic rate over the thermal range of 20–30°C while the rate for infected, warm-acclimated animals dropped when going from 25° to 20°C (Q_{10} of 2.2); (2) at 25°C, the metabolic rates of cold-and warm-acclimated infected animals were similar, but, in contrast, the rates of cold-acclimated noninfected animals were statistically higher than those of warm-acclimated animals; and (3) the metabolic rate of cold-acclimated, infected polychaetes was lower than that of cold-acclimated, noninfected animals. Although slight differences in experimental design do not permit unqualified comparison, there is a similarity of response between the metacercariae and its host and between the sporocyst stage of *Z. lasius* and its host, *N. obsoleta* (Vernberg and Vernberg, 1965, 1967).

It has been demonstrated that temperature acclimation influences cytochrome *c* oxidase activity in tissues of a number of invertebrates (Vernberg and Vernberg, 1968). The response found most often is enhancement of cytochrome *c* oxidase activity following exposure to cold temperature. Parasitism, however, can modify this response. In the gastropod *Nassarius obsoleta*, infections with different species of sporocysts or rediae of larval trematodes differentially affect the thermal acclimation curves of rate of activity of cytochrome *c* oxidase in digestive gland tissue. Generally, the rate in tissue from cold-acclimated, infected snails tends to be lower than that in tissue from cold-acclimated, noninfected snails, and the rate in tissue from warm-acclimated, infected snails is higher than that in tissue from warm-acclimated, noninfected snails. This is the same response found in noninfected and infected *L. culveri* (Table 1).

TABLE 2. *Type of thermal acclimation pattern of cytochrome* c *oxidase activity in tissue of* L. culveri *infected with metacercariae of* Z. lasius, *in comparison with tissue from noninfected* L. culveri. *(Based on the classification of Precht, 1958)*

Temperature of acclimation (°C)	*Tissue from infected* L. culveri	*Tissue from noninfected* L culveri
10	5	4
25	4	3

— In Type 3, there is partial acclimation: that is, cold enhances the rate of activity. Type 4 represents no acclimation; and Type 5 is inverse acclimation, where cold depresses rate of activity.

Using Precht's (1958) scheme of classification of acclimation patterns, a comparison of thermal acclimation patterns of cytochrome *c* oxidase activity of tissue of infected and noninfected *N. obsoleta* showed that only two patterns emerged in tissue of parasitized snails; either type 4 (no acclimation) or type 5 (inverse acclimation) (W. B. Vernberg, 1969). These are the same patterns that are found in tissues of *L. culveri* infected with *Z. lasius* (Table 2). Tissue from uninfected polychaetes shows acclimation when going from warm to cold, but no acclimation going from cold to warm.

SUMMARY

Metabolic rates have been determined for metacercariae of the trematode *Zoogonius lasius*, which utilize the polychaete *Leonereis culveri* as their host. During metacercarial development, metabolic rates are relatively high during the first few days, then the rate drops sharply to a low level. Following excystment, the metabolic rate again rises to a level approaching that of the early larval development.

Thermal metabolic response in the second intermediate host, *Leonereis culveri*, is modified by metacercarial infection. Uninfected *L. culveri* metabolically acclimate to cold temperatures, but infected *L. culveri* do not.

Cytochrome *c* oxidase activity is also modified by metacercarial infection. In noninfected *L. culveri* where there are significant differences in activity between cold acclimated and warm-acclimated worms, enzyme activity is higher in worms that are cold acclimated. In infected worms, this pattern is reversed. Not only are acclimation patterns different in infected and noninfected polychaetes, but actual enzyme activity rates of tissue of warm-acclimated, infected worms tend to be significantly higher than tissue from warm-acclimated, noninfected ones, and following cold acclimation, the rate in tissue from infected polychaetes is lower than in noninfected ones.

ACKNOWLEDGMENTS

We are greatly indebted to Mrs. Marilyn B. Barnes for technical assistance. We thank Dr. Marian Pettibone for identification of the polychaetes used in this study.

REFERENCES

Donges, J. 1969. Entwicklungs- und Lebensdauer von Metacercarien. Zeit Parasitenk. 31: 340–366.

Lowry, O. H., Rosenbrough, N. J., Farr, A. L., and Randall, R. J. 1951. Protein measurement with the Folin phenol reagent. J. Biol. Chem. 193:265–275.

Precht, H. 1958. Concepts of the temperature adaptations of unchanging reactions of cold-blooded animals, p. 50–78. *In* Prosser, C. L. (ed.) Physiological adaptation. Amer. Physiol. Soc., Washington, D. C.

Shaw, C. R. 1933. Observations on *Cercariaeum lintoni* Miller and Northup and its metacercarial development. Biol. Bull., 64:262–275.

Simpson, G. G., Roe, A., and Lewontin, R. C. 1960. Quantitative zoology. Harcourt, Brace & World, New York. 450 pp.

Smith, L. 1955. Spectrophotometric assay of cytochrome *c* oxidase, p. 427–434. *In* Glick, D. (ed.) Methods of Biochemical Analysis, Vol. 2. Interscience, New York.

Stunkard, H. W. 1938. *Distomum lasium* Leidy, 1891 (Syn. *Cercariaeum lintoni* Miller and Northup, 1926). The larval stage of *Zoögonus rubellus* (Olsson, 1868) (Syn. *Z. mirus* Loss, 1901). Biol. Bull. 75:308–334.

Stunkard, H. W. and Uzmann, J. R. 1958. Studies on digenetic trematodes of the genera *Gymnophallus* and *Parvatrema*. Biol. Bull. 115:276–302.

Thomas, R. E. and Gallicchio, V. 1967. Metabolism of ^{14}C-glucose by metacercariae of *Clinostomum campanulatum* (Trematoda). J. Parasitol. 53:981–984.

Vernberg, F. J. and Meriney, D. K. 1957. The influence of temperature and humidity on the metabolism of melanistic strains of *Drosophila melanogaster*. J. Elisha Mitchell Sci. Soc., 73:351–362.

Vernberg, W. B., 1969. Adaptations of host and symbionts in the intertidal zone. Amer. Zool. 9:357–365.

Vernberg, W. B., and Vernberg, F. J. 1965. Interrelationships between parasites and their hosts. I. Comparative metabolic patterns of thermal acclimation of larval trematodes with that of their host. Comp. Biochem. Physiol. 14:557–566.

Vernberg, W. B., and Vernberg, F. J. 1967. Interrelationships between parasites and their hosts. III. Effect of larval trematodes on the thermal metabolic response of their molluscan host. Exp. Parasitol. 20:225–231.

Vernberg, W. B. and Vernberg, F. J. 1968. Physiological diversity in metabolism in marine and terrestrial crustacea, Am. Zool. 8:449–458.

Zdarska, Z. 1964. Contribution to the knowledge of metabolic and morphological changes in the metacercariae *Echinostoma revolutum* (Frölich, 1802) Dietz, 1909 (Trematoda) Vestnik Cesk Spol. Zool. 28:285–289.

Enhanced Growth as a Manifestation of Parasitism
and
Shell Deposition in Parasitized Mollusks

THOMAS C. CHENG

Institute for Pathobiology
and
Department of Biology
Lehigh University
Bethlehem, Pennsylvania

INTRODUCTION

Although parasites are generally considered to be deleterious to their hosts as the result of living at their expense, recent evaluations of the nature of the host-parasite relationship have intentionally avoided employing "the infliction of harm" as a criterion in distinguishing parasitism from other categories of symbiosis (Smyth, 1962; Lincicome, 1963; Cheng, 1967, 1970; and others). Although this modern concept has evolved primarily because the evaluation of what constitutes injury or harm to the host is difficult to recognize without engaging in anthropomorphism, there is yet another reason. Instances are known wherein true parasites enhance, rather than retard, aspects of the growth of their hosts. It is to this latter phenomenon that this paper is directed. Specifically, the phenomenon commonly termed "gigantism" is under consideration.

ENHANCED GROWTH IN INVERTEBRATES

Our knowledge of increased growth of parasitized invertebrates, commonly designated as "gigantism," is almost exclusively limited to mollusks at the present. It was first recognized by Wesenberg-Lund (1934) during his extensive study of the parasites of the fresh water molluscan fauna in Denmark. He reported that specimens of *Radix auricularis* (=*Lymnaea auriculata*) which had been naturally parasitized by germinal sacs of digenetic trematodes were larger than nonparasitized specimens. He offered the explanation that the gigantism had resulted from increased growth induced by the presence of the parasites, suggesting an excessive consumption of food by the snails to meet the demands of the trematodes. Further investigation of this interesting phenomenon occurred a few years later when Rothschild (1936, 1938), while examining a population of the estuarine gastropod *Hydrobia* (=*Peringia*) *ulvae* at Plymouth, England, found that those snails parasitized by the asexually reproducing precursor stages of *Cercaria oocysta* and *C. ubiquita* were larger than nonparasitized specimens. She offered the suggestion that the assumed abnormal increase in size was not due to greater food intake but "it seems more likely that this [condition] is brought about by the destruction of the gonads and other glands." By this, it would appear that Rothschild was of the opinion that gigantism in mollusks is a manifestation of parasitic castration. In her 1938 paper she reported additional observations on *Hydrobia ulvae* and was convinced that gigantism of parasitized specimens was real. Furthermore, after studying the correlation between host size and number of cercariae produced, she concluded that: "The gigantism of the host, which involves an increase in the soft parts of the body as well as the shell, is thus of great advantage to the parasite. The faculty of producing this increase in size is a character which is presumably most susceptible to selection and its widespread occurrence among the Trematoda is therefore not surprising."

Rothchild's investigations on gigantism in parasitized snails resulted in three additional publications. Specifically, Rothschild and Rothschild (1939), besides reporting their observations on the growth rate of *H. ulvae* under various laboratory conditions, pointed out that parasitized snails grew faster than nonparasitized ones. Rothschild (1941a) further confirmed the fact that parasitized *H. ulvae* grew larger than did nonparasitized ones. She reported that the former grew to lengths of 9–10mm while their uninfected counterparts averaged 6.7mm in length. Finally, Rothschild (1941b) reported that the incidence of parasitism was greater among naturally parasitized specimens of *Littorina neritoides* than among nonparasitized ones. She offered three possible explanations: (1) infections may be lethal to young snails and hence these were killed off; (2) young snails were "unattractive" to miracidia; and (3) the growth of snails may be so greatly diminished after attaining a certain size that the time factor alone could account for the greatly increased percentage of infection in the larger size groups. She considered all three of these to be possible auxiliary factors, but, nevertheless, considered parasite-induced gigantism to be real. By plotting the infection rate against body size in *L. neritoides* which served as the first intermediate host (that is, it harbored the germinal sacs and emitted cercariae) and in specimens which served as the second intermediate host (that is, they harbored encysted metacercariae), Rothschild (1941b) demonstrated two different curves (Fig. 1). The first revealed a relatively low rate of infection in the small-size groups but a steep upward slope rising to 91 % in the larger-size groups. The second curve revealed a fairly uniform increase up to 87 % infection. From these data she concluded that the difference in the two curves was probably because of the fact that those which served as first intermediate hosts had been accelerated in their growths. Rothschild's thesis and supporting data were summarized by Rothschild and Clay (1952).

Lysaght (1941) reexamined Rothschild's findings and also reported that the larger specimens of *Littorina neritoides* were more frequently parasitized in nature by trematode germinal sacs. However, she was more cautious in her interpretation. Specifically, she wrote: "The data as they stand do not, on first inspection, suggest that the parasites stimulate growth; but no conclusion can be drawn, since the real information required, namely comparison of growth-rates in infected and uninfected specimens of the same initial size, is not available."

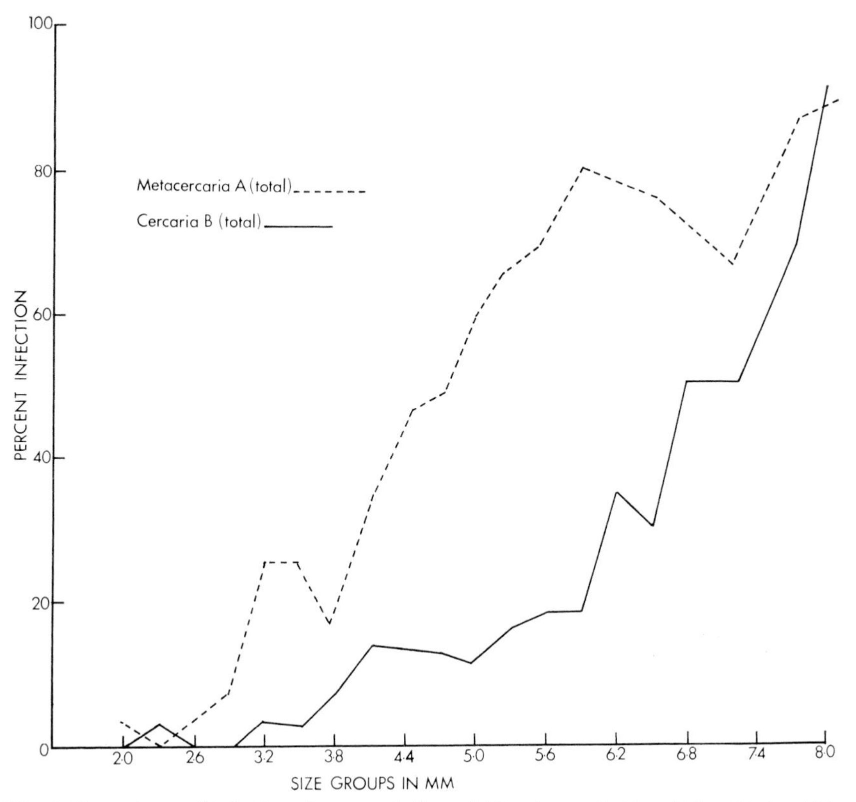

Fig. 1. Percentages of infections in a population of *Littorina neritoides*. (After Rothschild, 1941b.)

A decade lapsed before further investigation of variations in the growth of parasitized mollusks was resumed. This time, Hoshina and Ogino (1951) reported that oysters, *Crassostrea gigas*, which served as second intermediate hosts for the trematode *Gymnophalloides tokiensis*, did not grow as fast as uninfected ones. Upon close scrutiny, it would appear inappropriate to compare the report by these Japanese investigators with the earlier reports since the metacercariae of *G. tokiensis* are encysted in the oysters' mantles rather than deep within the mollusks' soft tissues.

A further suggestion of gigantism in a mollusk harboring trematode germinal sacs was contributed by Menzel and Hopkins (1955a,b). Basing their conclusion on extremely limited observations — specifically on a single specimen of the American oyster, *Crassostrea virginica*, parasitized by the dendritic sporocysts of *Bucephalus cuculus* — they reported that the parasitized oyster grew unusually quickly during the first year of its life but this was followed by a lag in growth rate, relative to nonparasitized oysters. Furthermore, the specimen failed to increase in length or height during the last two months of the observational period.

It should be apparent that in spite of the general claim in the literature that the studies summarized up to this point, especially those by Rothschild, have revealed the occurrence of gigantism in mollusks parasitized by larval trematodes, the evidences are inconclusive, especially in view of Lysaght's criticism, since this general assumption has been extrapolated almost exclusively from observations on naturally-infected mollusks of undetermined age. It is of interest to note at this point that Croll (1966) has suggested that the correlation between higher percentages of parasitism and the size of snails may not be caused by parasite-induced increased growth at all, but is the result of the attraction of miracidia to the mucous trail of the snail. It thus follows that as the larger mollusks produce more mucus, they are more attractive to invading miracidia.

A single report of gigantism in parasitized mollusks appeared in the 1950's when Abdel-Malek (1952) reported that specimens of *Helisoma corpulentum*, naturally parasitized by the sporocysts of *Uvulifer ambloplites* or the rediae of *Petasiger chandleri* or *Clinostomum camplanatum*, were heavier than their nonparasitized counterparts.

A renewed interest in the phenomenon of gigantism in parasitized mollusks occurred in the 1960's. Andrade (1962), in Brazil, reported that in natural populations of *Biomphalaria glabrata*, the larger specimens were more commonly parasitized by *Schistosoma mansoni*. In addition to this report, experimental studies carried out, especially by the Harvard school, shed new light on the situation. Chernin (1960) reported that axenically reared specimens of *Biomphalaria glabrata* infected with *Schistosoma mansoni* tended to grow faster than did uninfected control snails during the initial 36 days after infection, but that they were approximately the same size (6.2 mm shell diameter in infected snails, and 5.8 mm shell diameter in uninfected ones) by the 43rd day after infection. Similarly, Pan (1962) reported that starting with specimens of *B. glabrata* averaging 7.3 mm in shell diameter, those parasitized with *S. mansoni* grew faster, as determined by shell size, than did the uninfected controls during the first six weeks after infection and reached a maximum size of 13.8 mm by the 7th week. The growth of the controls caught up with that of the infected group by the 8th week and reached a maximum size of 14.5 mm by the end of the 15th week.

In a more detailed report, Pan (1965) presented evidence that when the growth rates of adolescent *B. glabrata* (7.3 mm in shell diameter) mass infected with *S. mansoni* were compared with those of uninfected control snails of the same initial size, the former were larger between the 2nd and 6th weeks after infection, reaching 13.7 ± 1.2 mm in shell diameter by the end of the 6th week, while the controls only averaged 13.0 ± 1.0 mm by this time. But the infected snails reached their maximum size (13.8 ± 1.1 mm) by the 7th week after infection after which the uninfected specimens gradually surpassed them in size and reached a maximum diameter of 14.6 mm in week 21.

It is apparent from the adequately controlled studies by Chernin and Pan that although there is an increase in the growth rate of *Biomphalaria glabrata* parasitized by *Schistosoma mansoni*, as determined by shell measurements, the slight acceleration in the growth of these snails (Fig. 2) is definitely limited and is not manifested in fully grown specimens. In fact, Pan (1965) reported that the temporary increased growth of *B. glabrata* associated with parasitism by *S. mansoni* did not occur in snails that had reached maturity at the time of infection. It is of interest to note that a number of investigators (Pesigan *et al.*, 1958; Moose, 1963; Zischke and Zischke, 1965) have reported that the growth of snails harboring trematode sporocysts and rediae is retarded.

From these reports it must be concluded that if gigantism does indeed occur among parasitized mollusks, it is either restricted to the juvenile period or does not occur in all cases. Furthermore, as the criterion employed has been shell dimensions in the controlled studies, the question may be raised as to whether the so-called increased growth was of the entire mollusk or only of the shell. The greater weight of parasitized mollusks as reported by Rothschild (1938) and Abdel-Malek (1952) need not of course be attributable solely to enhanced growth of the whole organism. It could have also been due to thicker shells and/or additional total weight resulting from the parasites embedded in the host's tissues.

It should be noted at this point that Etges (1961a,b), while examining specimens of *Helisoma anceps* parasitized by the progenitor stages of the echinostome *Cercaria reynoldsi* from Mountain Lake, Virginia, noted that their shells were thicker than normal, as were those of the same species of

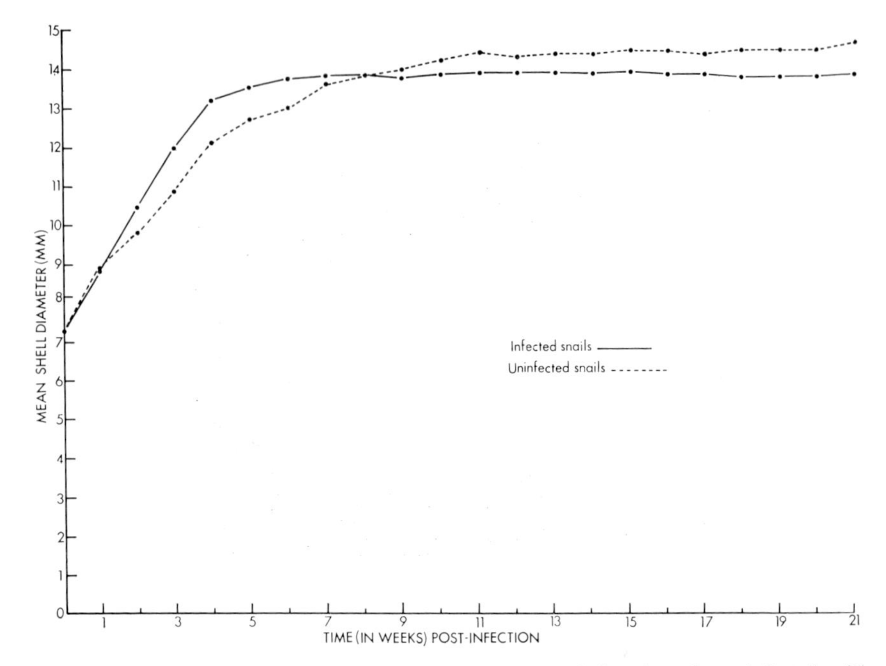

Fig. 2. Comparative growth of *Biomphalaria glabrata* infected and noninfected with *Schistosoma mansoni*. (Data from Pan, 1965. See Pan for statistical data.)

snails parasitized by the sporocysts of *Diplostomulum scheuringi*. This interesting fact is considered further later in this paper.

The question should be raised whether the term "gigantism" is really appropriate in describing the comparatively larger sizes of growing parasitized mollusks. It should be obvious that the enhanced growth of these animals, if it is indeed true growth involving soft tissues, is rather subtle. For the time being, perhaps "enhanced growth" is a more appropriate designation. To the cursory reader the term "gigantism" tends to conjure visions of some type of monstrosity.

The only other known example of parasite-induced growth in an invertebrate host is that associated with the *Nosema-Tribolium* relationship. West (1960) and Fisher and Sanborn (1962a,b) reported that larvae of the beetle *Tribolium* attained larger sizes when parasitized by the sporozoan *Nosema* sp. Infected beetle larvae underwent as many as six supernumerary molts and commonly died as giant larvae which weighed twice as much as nonparasitized controls. The enhanced growth in parasitized *Tribolium*, however, was somewhat different from that in parasitized mollusks in that the weights of both parasitized and nonparasitized beetles increased at the same rate until the 12th day. It was only after that time that a more rapid increase in body weight occurred among members of the infected group. This enhanced growth lasted until day 24, at which time the weights of infected beetles diminished and death ensued after about 30 days of larval life. In the case of the controls, the weight increase occurred until the 19th day, followed by pupation on the 20th–25th day. An explanation for the enhanced growth in parasitized *Tribolium* larvae may be extrapolated from the studies by Fisher and Sanborn (1962b, 1964). In brief, they demonstrated that implanted *Nosema* into allatectomized nymphs of the roach *Blaberus* replaced the corpora allata. If the situation in *Tribolium* is similar to that in *Blaberus*, and there is no reason to believe that it is not, then the enhanced growth in parasitized *Tribolium* could be attributed, directly or indirectly, to the contribution of a juvenile hormone-like material by *Nosema*.

ENHANCED GROWTH IN VERTEBRATES

Parasite-induced growth in vertebrates is much better substantiated than in invertebrates. The earliest report appears to be that by Neumann (1905) who reported that animals parasitized with *Fasciola hepatica* "have a tendency to fatten more readily." He offered the explanation that this may have been because of "the stimulus the young flukes impart at first to the liver [of the host], and consequently there is more abundant biliary secretion and better assimilation." As far as I have been able to determine, the next account of this nature in a vertebrate occurred 44 years later when Whitlock (1949) reported that there is an initial period of accelerated growth in lambs parasitized by trichostrongylid nematodes.

Modern studies on parasite-induced enhanced growth in vertebrates have been contributed primarily by two investigators, David R. Lincicome and his associates at Howard University,* and Justus F. Mueller at the State University of New York's Downstate Medical Center at Syracuse. Lincicome *et al.* (1960, 1963), as the result of following the growth of albino rats parasitized by the hemoflagellate *Trypanosoma lewisi*, reported that those rats intraperitoneally challenged with an initial small inoculum of *T. lewisi* (0.2×10^6) consistently revealed increased growth, manifested as greater body weights (Fig. 3) and a trend in this direction was also revealed by rats challenged with larger inocula (4.0×10^6) of trypanosomes, although the difference

* In addition to the stimulatory effects of *Trypanosoma lewisi* and *T. duttoni* on the growth of rodents summarized herein, Lincicome (in this volume) has reported that the nematode *Trichinella spiralis* also enhances the growth, as determined by weight, of laboratory rats.

was not consistently significant. These investigators offered five possible explanations for the enhanced growth of parasitized rats; namely:

(1) Animals under parasitic stress may have consumed more food than nonparasitized, control animals;
(2) the trypanosomes, in some yet undetermined way, accelerated the host's cell reproduction and/or metabolic rates;
(3) the increased body weight may have been due to a pathological inadequacy in water balance of the body, resulting in edema;
(4) the pituitary glands of parasitized rats may have been stimulated by some metabolic substance elaborated by the trypanosomes, and hyperactivity of this endocrine gland, in turn, could have accounted for the excessive growth;
(5) the thyroid glands of parasitized rats may have been stimulated by some product of parasitic origin and hyperthyroidism, in turn, caused the increased growth.

Similar results were obtained by Lincicome and Shepperson (1961, 1963) by employing the *Trypanosoma duttoni* – laboratory mouse model. In brief, 82 female albino mice were inoculated intraperitoneally with either $3.8–5.3 \times 10^4$ or $3.8–5.3 \times 10^5$ trypanosomes. Their subsequent growth rates, expressed as body weights, were compared to those of 50 uninfected control mice. During the ensuing 30–58 days, the parasitized mice revealed enhanced growth. These studies by Lincicome and his associates, which have been reviewed by Lincicome (1963), have provided quantitative evidence of an example of parasite-stimulated increased growth. However, as far as I have been able to determine, the causal relationship, as expressed in physiological and biochemical terms, has not yet been established.

Mueller's studies on this subject have involved primarily the mouse – *Spirometra mansonoides* relationship. *S. mansonoides* is a pseudophyllidean tapeworm, the adult of which is an intestinal parasite of the domestic cat, bobcat, dog, and raccoon. The vertebrate definitive host acquires this parasite by ingesting spargana (larvae) encysted in a number of poikilothermic and homeothermic vertebrates, except fishes (see Mueller, 1966). Mueller (1962, 1963a) has established a technique whereby solices clipped from spargana reared in other mice can be injected subcutaneously into experimental mice. By employing this technique, he has examined the growth rates of various rodents harboring spargana regenerating from implanted scolices. It was reported (Mueller, 1962, 1963a) that beginning with carefully matched mice (in pairs ranging from 11.8–21.37g), those harboring small numbers (6–8) of spargana gained weight faster than did the controls (Figs. 4, 5). This effect occurred in both male and female mice (Mueller, 1963b) and at starting weights ranging from 10 to 40g. On the other hand, if mice with starting

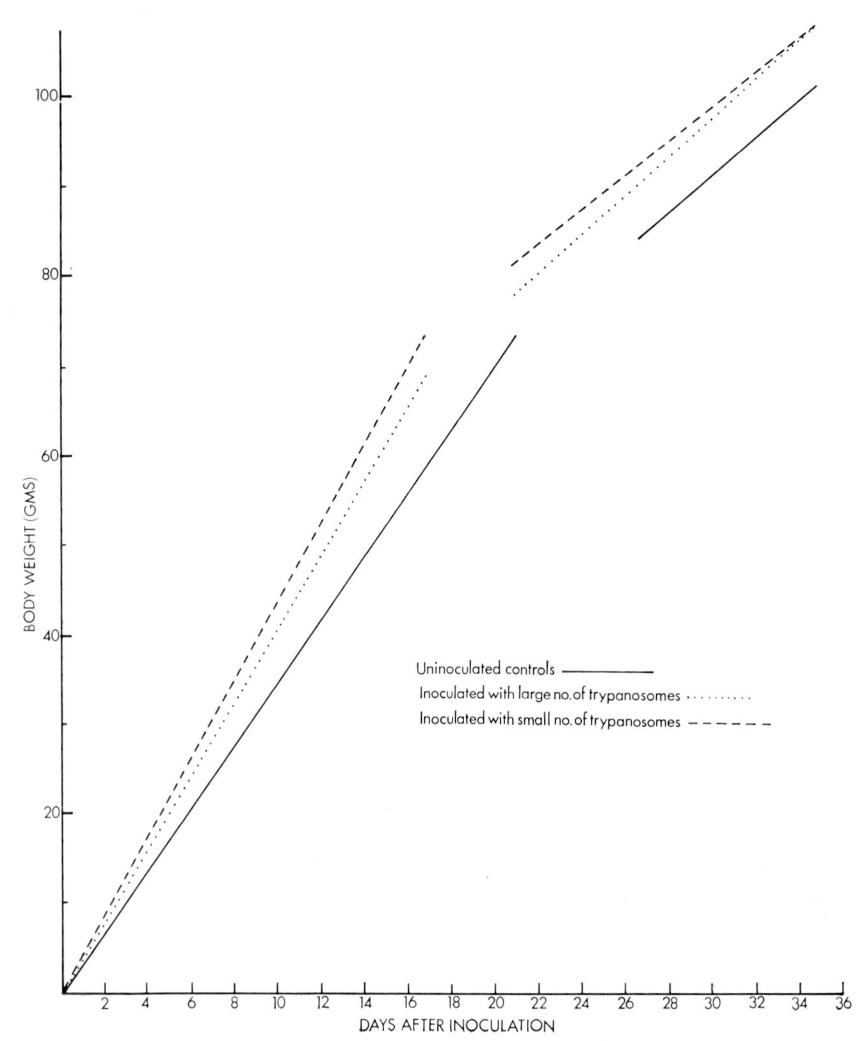

Fig. 3. Comparative increases in body weights of laboratory rats inoculated and uninoculated with *Trypanosoma lewisi*. The graphs are plotted from data obtained from rats inoculated with 12×10^4 trypanosomes (small number of trypanosomes) and 12×10^5 trypanosomes (large number of trypanosomes). (Slightly modified after Lincicome *et al.*, 1963.)

weights of 9.5–10 gm were each implanted with seven scolices, there was an initial period of retarded growth of several weeks and thereafter there was a slight gain over the matched controls. According to Mueller (1966), who has reviewed his own work, this deviation was probably the result of the parasite load being disproportionate to the weight of the host.

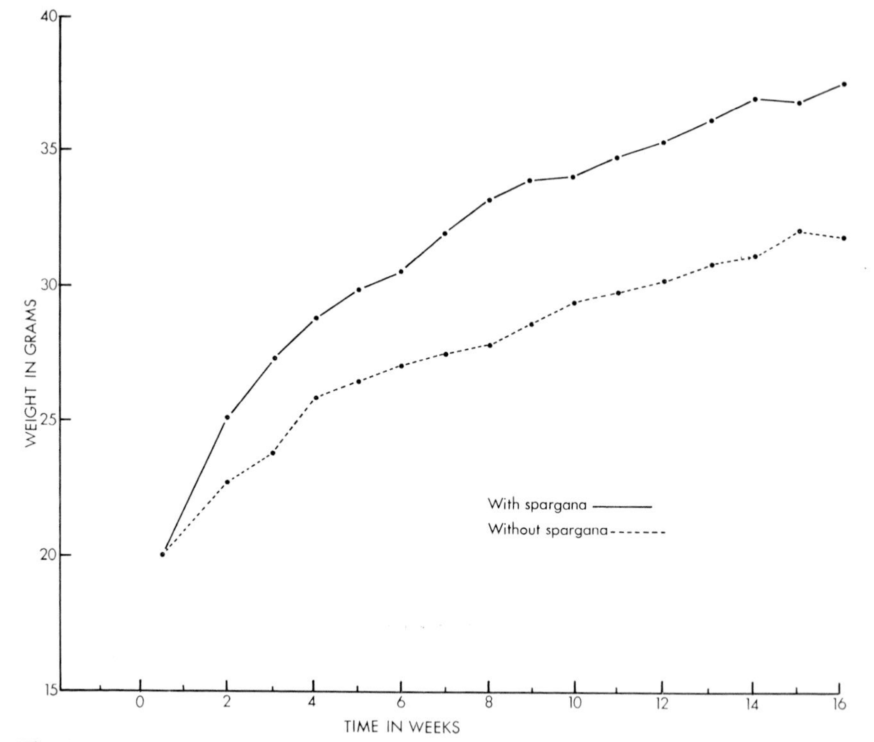

Fig. 4. Comparative body weight increases in laboratory mice implanted with *Spirometra mansonoides* spargana. Each of the infected mice received seven worms. Each point on the graph represents the mean weight of eight animals. (After Mueller, 1963a.)

This parasite-induced weight gain occurs not only in mice but also in the deer mouse, *Peromyscus* (Fig. 6), and in the golden hamster (Mueller, 1965a). In the case of the hamster, it was shown that both true growth (as revealed by X-ray examination) and obesity occurred. In the same paper (Mueller, 1965a) it was reported that the weight gain was additive, increasing with the number of spargana introduced, but reached a plateau at about 12 worms per mouse of 20g. Beyond this point there was either no weight gain or increased mortality as the result of parasite invasion of the hosts' pleural cavities or some other vital area. The increased weight, according to Meyer *et al.* (1965), was distributed among the internal organs, liver, skin, and carcass, in that order. Furthermore, of the 61 % increase in dry weight, 46 % was attributable to lipids, 14 % to proteins, and 1 % to nucleic acids, glycogen, and ash.

It is noted in passing that Mueller (1965a) reported that the Oriental species of *Spirometra, S. ranarum*, has an even greater stimulatory effect than *S. mansonoides*.

Fig. 5. Photographic comparison of laboratory mice with and without *Spirometra mansonoides* spargana. The mouse on the left was infected with eight spargana under the dorsal skin and weighed 68.6 g, while the one on the right was uninfected and weighed 38.7 g. Each mouse weighed 21.35 g at the start and the photograph was taken 69 weeks later. (After Mueller, 1963a: with permission of the New York Academy of Sciences.)

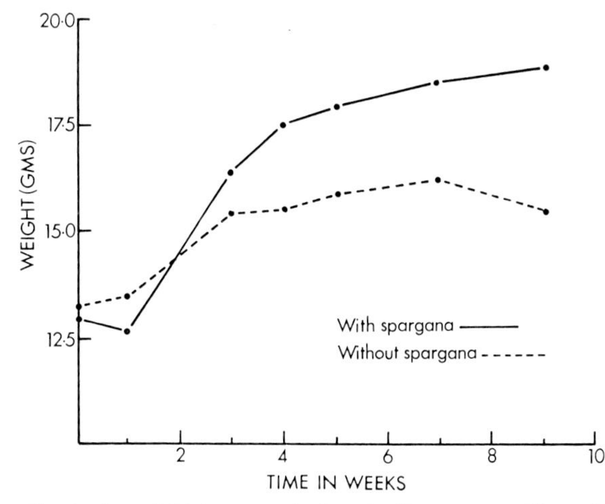

Fig. 6. Comparative body weight increases in female *Peromyscus* with and without *Spirometra mansonoides* spargana. The point of the graph represents the mean increase of three animals. Each of the infected *Peromyscus* harbored seven spargana. (Slightly modified after Mueller, 1965a.)

The possible cause of the enhanced growth due to the spargana of *Spirometra* spp. in rodents has been the subject of some study. Doisy (in Mueller, 1966) and Harlow *et al.* (1964) reported that spargana contained an insulin-like substance as determined by the rat epididymal fat-pad bioassay. This finding, naturally, tempted Mueller and his associates to speculate that the insulin-like substance may have leaked into the host, caused hypoglycemia, and consequently increased the host's appetite. However, Sadun *et al.* (1965), upon analysis of blood samples from experimentally infected mice, demonstrated that hyperglycemia, rather than hypoglycemia, occurred. Furthermore, as the result of comparative quantitative food intake studies, Mueller (1965b) reported that the enhanced growth, including weight gain, could not be explained on the basis of any appreciable difference in relative food intake.

More recently, Mueller (1968) and Mueller and Reed (1968) reported that the implantation of *S. mansonoides* spargana into hypophysectomized and hypothyroid rats stimulated growth. They concluded that the secretion of a growth hormone-like substance by the parasites into the altered host is strongly suggested. These results and conclusions have been verified by Steelman *et al.* (1970) who demonstrated significant increases in the weights of the kidney, thymus, and liver of hypophysectomized rats implanted with spargana. The increase in the weights of these organs, together with a doubling of the width of the epiphyseal cartilage, strongly suggested the action of a growth hormone-like substance. However, Steelman *et al.* are of the opinion that: "Growth hormone secretion *per se* by SMS [*S. mansonoides* spargana] seems unlikely but cannot be excluded. Another possible explanation is that a substance is released which causes sulfation factor to be produced. Daughaday and Kipnis . . . showed that growth hormone has this action. Alternatively, the growth factor could be sulfation factor like."

Before departing from the studies by Mueller and his associates, it appears appropriate to point out that certain aspects of immunology are involved. Specifically, Mueller (1966) reported that the enhanced growth in rodents was dependent upon the tolerance of the spargana by the host. The difference between the growth rates of parasitized and nonparasitized hosts was correlated with the host's cellular reactions, with maximum difference occurring when the subdermally situated parasites were not encapsulated. If encapsulation occurred, the weight gain was less apparent. Also, a secondary pyrogenic manifestation also interfered with the effect and usually killed the parasites. Whether these deaths were caused by the fever or by a heightened cellular reaction in the host is not known.

SHELL DEPOSITION IN PARASITIZED MOLLUSKS

There are suggestions that the growth of mollusks parasitized by larval trematodes may be enhanced, although the available quantitative, expermental data only permit the conclusion that the enhanced growth, as determined by shell measurements, is temporary, being limited to young, growing specimens. However, the observations by Etges (1961a,b) on the thicker shells of *Helisoma anceps* parasitized by the rediae of *Cercaria reynoldsi* and the sporocysts of *Diplostomulum scheuringi*, are of considerable interest. It is noted, however, that the thick-shelled condition described by Etges (1961a) was not consistent. He reported that among 107 specimens of *H. anceps* examined, 8 possessed thick shells and all of these were parasitized. However, at least one of the thin-shelled snails was also parasitized. Nevertheless, these observations, although cursory, may suggest that certain parasitized mollusks may be capable of accumulating and depositing more calcium in their shells and consequently possess heavier shells. To test this hypothesis, the following studies were carried out.

MATERIALS AND METHODS

Two models were employed in these studies: The *Nitocris – Acanthatrium* and the *Physa – Echinostoma* models. The first involved snails naturally parasitized by sporocysts; the second involved snails experimentally infected with rediae.

The specimens of *Nitocris dilatatus* used in this study were collected in Sinking Creek, where Virginia State Route 700 crosses it, in Giles County, Virginia. A survey carried out during the summer of 1963 had revealed that 21 % of these snails from this site were parasitized by a virgulate xiphidiocercaria identified as that of the lecithodendriid *Acanthatrium* (=*Prosthodendrium* in part) *anaplocami*. The life cycle of this trematode has been reported by Etges (1960). This cercaria develops in sporocysts embedded in the intertubular spaces within the digestive gland of the molluscan host.

Snails collected in Sinking Creek were brought into the laboratory of the Mountain Lake Biological Station, University of Virginia, where they were isolated individually in distilled water. Those snails which emitted *A. anaplocami* cercariae were separated from the nonemitting specimens and from those which emitted other species of cercariae. After a starvation period of 48 hours to empty their alimentary tracts, the shells of the specimens used during this study were removed and examination of the digestive glands of nonemitting snails confirmed that they were not parasitized. If parasites were found, it was acertained that these were *A. anaplocami* and the hosts were grouped with the infected ones. No snails infected with two or more species of larval trematodes were found. A total of 35 parasitized snails comprised the experimental group for the chemical studies; 43 uninfected snails served as the controls. Five additional infected and 7 uninfected snails were used in the histological and histochemical studies. All of the specimens of *N. dilatatus* were from the same population and were essentially of the same age and size.

The specimens of *Physa sayii* used for experimental infections were raised from eggs in the laboratory. All were 6 weeks old at the time they were each exposed to 20 miracidia of *Echinostoma revolutum* hatched from eggs produced by adults maintained in ducklings. They were 11 weeks old when processed. All of the specimens of *Physa sayii* were maintained at 22°C in spring water and fed on lettuce leaves. A total of 33 infected and 25 uninfected *P. sayii* were employed in the weight studies (Table 4) but only 18 of the 25 nonparasitized specimens were used in the chemical studies (Table 5). In addition, 5 parasitized and 6 nonparasitized snails were used in the histological and histochemical studies.

Both species of snails, infected and uninfected, were subjected to identical analyses. The procedures carried out were as follows.

Determination of Weights of Snails and Shells.—Each snail of both uninfected series used in the ionic calcium concentration studies was removed from the finger bowl, rinsed in triple-distilled water, dried on soft tissue paper, and weighed on a Mettler (Model H4) balance resting on a preweighed aluminum planchet. Subsequently, the shell of each snail was crushed in the planchet and all of the shell fragments were removed under a stereomicroscope with watchmaker's forceps. The hemolymph seeping from the soft tissues was collected in the planchet; that adhering to the pieces of shell was not collected. The weight of each snail without its shell was then determined. The shell fragments were rinsed in triple-distilled water, placed on a preweighed planchet, and dried under an infrared lamp. The weights of the dry shell samples were then ascertained.

In the case of the infected snails of both species used in the ionic calcium concentration studies, a slight modification of the above procedure was followed. Specifically, after each whole snail was weighed and its shell was removed, the soft tissues, resting on a preweighed planchet, were placed under a stereomicroscope and as many sporocysts in the case of *A. anaplocami*-infected *N. dilatatus*, and rediae in the case of *E. revolutum*-infected *P. sayii*, as possible were removed from the digestive gland by use of watchmaker's forceps and a micropipette. It was not possible to remove all of the sporocysts and rediae but a large number, at least 80 %, was removed. After this was completed, the weight of each snail without its shell was determined. The shell of each snail of the infected series was also rinsed, dried, and weighed.

The sporocysts and rediae removed from each infected snail were also weighed, in the following manner. Each group of sporocysts or rediae was washed three times in triple-distilled water in a preweighed centrifuge tube to remove any molluscan hemolymph adhering to the exterior. The water was removed each time with a Pasteur pipette after centrifugation at a slow speed to concentrate the germinal sacs. Finally, the tube plus the larvae were weighed. Thus the wet weight of each group of sporocysts or rediae was ascertained.

Preparation of Tissues for Ca^{++} Determination.—After the weights of the soft tissues of snails of both series and species and the sporocysts and rediae were determined, the individual snails and the parasites from each host were homogenized in 1 ml of triple-distilled water under ice to minimize any release of bound calcium. The homogenates were subsequently centrifuged at 1,500 rpm in a clinical centrifuge for 5 minutes, after which the supernatant solutions were saved for analysis. If analysis was not performed immediately, the extracts were stored in the frozen state. In no instance was a sample stored for more than 72 hours.

Method for Determining Ca^{++} Concentration.—The technique employed for Ca^{++} determination was a modification of that given by Ferro and Ham (1957). This procedure is based on the quantitative precipitation of calcium by chloranilic acid. Chloranilic acid is an orange-red crystal prepared from chloranil, which has been employed successfully by Tyner (1948) for the quantitative determination of calcium in plant materials.

After the calcium is precipitated as calcium chloranilate, the latter is dissolved in a solution of the tetrasodium salt of ethylenediamine tetraacetic acid (EDTA) which liberates the water-soluble sodium chloranilate to form a clear pink color. The calcium ions, which are substituted by sodium in the choranilate compound, are sequestered by EDTA and are thus prevented from precipitating the sodium chloranilate (Fales, 1953; Kenny and Toverud, 1954; Gehrke *et al.*, 1955; Young and Sweet, 1955; Zak and Hindman, 1955). The resulting color, which is indicative of the amount of Ca^{++} in the original calcium chloranilate compound, is stable, Beer's Law is applicable over a wide range (Ferro and Ham, 1957), and the color can be quantified with a spectrophotometer.

This procedure was modified so that only 0.5 ml of the aqueous extract was required for quantitative Ca^{++} determinations. To carry out this procedure, three solutions were required: (1) the chloranilic acid solution; (2) a 50% isopropyl alcohol solution; and (3) the EDTA solution. The first was prepared by adding 1,000 ml of distilled water to 4 g of choranilic acid (Eastman Organic Chemicals, Rochester, N.Y.) and shaking the mixture vigorously for 10 minutes after which it was allowed to stand with occasional shaking during the next hour. The mixture was then allowed to stand overnight during which time the excess acid crystals settled out. The solution was subsequently filtered through Whatman No. 42 paper just prior to use. The EDTA (tetrasodium salt; purchased as Versene from Dow Chemical Co., Framingham, Mass.) solution was prepared by dissolving 75 g of the salt in 200 ml of distilled water and the volume was then adjusted to 1,000 ml.

To 0.5 ml of each extract sample, placed in a 12-ml, heavy-walled centrifuge tube, was added 0.25 ml of the chloranilic acid solution. After thorough mixing and standing for 30 minutes, the content was centrifuged at 1,800 rpm in a free-swinging, clinical centrifuge for 10 minutes. The supernatant solution was subsequently drawn off by suction and the tube was set upside down on filter paper for 5 minutes to drain off the remaining supernatant solution. The lip of the tube was then dried and the precipitate at the bottom was washed in 5 ml of 50 % isopropyl alcohol. The content was then recentrifuged at 1,800 rpm for 10 minutes and the supernatant solution was again removed by suction and draining. After addition of 0.05 ml of distilled water the precipitate was broken up and evenly suspended in the water. Subsequently, 1.1 ml of the EDTA solution was added to each tube after which a coverglass was placed over the top of the tube and the contents were agitated until the precipitate was completely dissolved. The solution was then transferred into a cuvette and read in a spectrophotometer (Spectronic 20) at 520 mμ. All of the readings were made within 4 hours after final processing of the solution.

As controls, 0.5 ml samples of "Lab-trol" (Dade Reagents, Inc., Miami, Florida) were processed in identical fashion.

In addition to analyzing the test extracts and controls, standards were prepared and read in an identical manner. The standards consisted of a series of 5 dilutions ranging from 0.01 to 0.05 mg per 0.5 ml made from a stock solution of reagent grade calcium carbonate. The stock solution was prepared by adding 0.249 g of $CaCO_3$ to 9 ml of 1 N HCl in a 1 liter volumetric flask and allowing the $CaCO_3$ to dissolve, after which sufficient distilled water was added to make a total volume of 1,000 ml. The calcium concentration in the stock solution was thus 10 mg per 100 ml ($\times$ 0.249/0.2497) or 0.1 mg/ml. After the percentage transmittances of the standards were determined, these were plotted against the Ca^{++} concentrations on semilogarithmic paper. The Ca^{++} concentration per 0.5 ml for each of the unknowns was determined from the resulting linear graph.

Histological and Histochemical Procedures.—The 5 parasitized and 7 nonparasitized *Nitocris dilatatus* and the 5 parasitized and 6 nonparasitized *Physa sayii* used for histological and histochemical studies were fixed in 10 % neutral formalin for 24 hours after their shells had been removed. The tissues were embedded in Tissuemat and sectioned at 8 μ in thickness. Every third slide was stained with (1) Mallory's triple connective tissue stain, (2) Delafield's hematoxylin, and (3) McGee-Russell's (1955) calcium red method for calcium.

RESULTS

Weights and Calcium Data.—The weights of whole snails, soft tissues, and dry shells of the 45 nonparasitized and 35 parasitized *Nitocris dilatatus* and 33 parasitized and 25 nonparasitized *Physa sayii* are given in Tables 1 and 4, respectively.

TABLE 1. *Weights of whole specimens, soft tissues, and dried shells of nonparasitized* Nitocris dilatatus *and those parasitized by* Acanthatrium anaplocami *and sporocysts removed from the latter.*

	Parasitized snails	*Nonparasitized snails*	*Sporocysts*	*Statistical significance at 5% level*
No. of specimens	35	45		
Mean wt. of whole snails (g±SD)	0.396±0.144	0.329±0.137		+
Mean wt. of soft tissues (g±SD)	0.126±0.055	0.114±0.045		–
Mean dry wt. of shells (gm±SD)	0.221±0.084	0.162±0.071		+
Mean wt. in each parasitized snail (gm±SD)			0.024±0.011	

TABLE 2. *Ca^{++} concentrations in 0.5 ml of extract, soft tissues per snail, and per gram of soft tissues of nonparasitized* Nitocris dilatatus *and those parasitized by* Acanthatrium anoplocami.

	No. of snails	*Ca^{++}conc./0.5 ml aqueous extract (mg/0.5 ml extract) mean±SD*	*Ca^{++}conc. in soft tissues/ snail (mg) mean±SD*	*Ca^{++}conc./g of soft tissues (mg) mean±SD*
Parasitized	*35*	0.046±0.020	0.092±0.042	0.730±0.170
Nonparasitized	*45*	0.027±0.014	0.054±0.029	0.473±0.052
Statistical significance at 5% level		+	–	+

TABLE 3. *Ca^{++} concentrations in 0.5 ml of extract of whole sporocysts of* Acanthatrium anoplocami, *in all specimens of sporocysts removed from each snail, and per gram of sporocyst tissue.*

Ca^{++}conc./0.5 ml aqueous extract (mg/0.5 ml extract) mean±SD	*Ca^{++}conc. in all sporocysts removed from each snail (mg) mean±SD*	*Ca^{++}conc./g of sporocyst (mg) mean±SD*
0.007±0.004	0.013±0.004	0.003±0.005

The Ca^{++} concentration in the 0.5 ml of aqueous extract of each snail, the extrapolated Ca^{++} concentration in the soft tissues of each snail, and that in each gram of soft tissue of both the uninfected and infected series of both molluscan species, are tabulated in Tables 2 and 5, respectively. The same values for the sporocysts of *A. anaplocami* and the rediae of *E. revolutum* are given in Tables 3 and 6, respectively.

TABLE 4. *Weights of whole specimens, soft tissues, and dried shells of nonparasitized* Physa sayii *and those parasitized by* Echinostoma revolutum *and rediae removed from the latter.*

	Parasitized snails	*Nonparasitized snails*	*Rediae*	*Statistical significance at 5% level*
No. of specimens	*33*	*25*		
Mean wt. of whole snails (g±SD)	0.335±0.016	0.217±0.090		+
Mean wt. of soft tissues (g±SD)	0.085±0.030	0.086±0.024		−
Mean dry wt. of shells (g±SD)	0.156±0.002	0.085±0.040		+
Mean wt. in each parasitized snail (g±SD)			0.031±0.012	

TABLE 5. *Ca^{++} concentrations in 0.5 ml extract, soft tissues per snail, and per gram of soft tissues of non-parasitized* Physa sayii *and those parasitized by* Echinostoma revolutum.

	No. of snails	*Ca^{++}conc./0.5 ml aqueous extract (mg/0.5 ml extract) mean±SD*	*Ca^{++}conc. in soft tissues/ snail (mg) mean±SD*	*Ca^{++}conc./g of soft tissues (mg) mean±SD*
Parasitized	*33*	0.044±0.012	0.088±0.024	1.023±0.014
Nonparasitized	*18*	0.016±0.004	0.032±0.008	0.376±0.026
Statistical significance at 5% level		+	+	+

TABLE 6. *Ca^{++} concentrations in 0.5 ml of extract of whole rediae of* Echinostoma revolutum, *in all specimens of rediae removed from each snail, and per gram of redial tissue.*

Ca^{++}conc./0.5 ml aqueous extract (mg/0.5 ml extract) mean+SD	*Ca^{++}conc. in all rediae removed from each snail (mg) mean±SD*	*Ca^{++}conc./g of redia (mg) mean±SD*
0.010±0.020	0.020±0.032	0.606±0.023

Histological and Histochemical Observations.—The histological composition, especially the types of cells comprising the digestive gland acini of gastropods, has been studied extensively (Barfurth, 1881, 1883; Frenzel, 1886; Yung, 1888; Biedermann and Moritz, 1899; Krijgsman, 1928; Baecker, 1932; Fretter, 1952; van Weel, 1950; Guardabassi and Ferreri, 1953; Billett and McGee-Russell, 1955; Pan, 1958; Abolinš-Krogis, 1960). Among the cells comprising the digestive gland of uninfected *N. dilatatus*, a few calcium cells can be seen (Figs. 12, 13). These include relatively few (6–12) calcium spherites, each measuring 10μ (10–15μ) in diameter. In sections treated with either Mallory's or hematoxylin stains, these spherites appear greenish-yellow. That these are indeed calcium-containing spherites was verified by their characteristic red color when treated with the calcium red method.

In addition to calcium cells, ferment or secretion cells are present and comprise the digestive gland acini of uninfected snails. These cells are quite conspicuous as they enclose large, yellowish-brown, ovoid secretory globules within their cytoplasm (Figs. 11, 12, 13). Each of these globules measures $13 \times 8\mu$ ($10–15\mu \times 6–9\mu$). According to the various authors cited above, the primary function of these cells is to collect, deposit, and secrete waste products accumulated in the digestive gland. Occasionally, large, yellowish-brown, secretory globules can be observed excreted into the lumen of the acinus (Figs. 12, 13). These globules do not include calcium as they do not stain with calcium red.

Hepatic cells are the most numerous of the cell types comprising each acinus (Fig. 11). They are elongate and have conspicuous nuclei. Their cytoplasm is finely granular. Essentially none of these cells contains calcium as determined by the calcium red method.

In *Nitocris dilatatus* parasitized by *Acanthatrium anaplocami*, both sporocysts and motile, fully developed cercariae cause the displacement and rupturing of the host's digestive gland tubules (Figs. 7, 8, 9). The resulting histopathology is comparable to that in *Helisoma trivolvis* parasitized by the sporocysts of *Glypthelmins pennsylvaniensis* (see Cheng and Snyder, 1962). Most of the acinar tubules are dissociated and the individual cells are ruptured. Critical examination of the ruptured cells has revealed that none of the injured calcium cells includes calcium spherites (Fig. 10).

It is also rather conspicuous that the number of secretory globules in the gastropod's secretion cells, especially those situated in the proximity of the parasites, is greatly increased (Figs. 7, 8, 14, 15). Furthermore, among the remaining intact calcium cells, the number of calcium spherites is conspicuously increased, ranging from 16 to 23 in the cytoplasm of each cell (Figs. 16, 17). These spherites, averaging 13μ in diameter, are somewhat larger than those found in uninfected snails. They tend to collect at the luminal border of each calcium cell (Fig. 18) and are sporadically discharged into the lumen of the tubules. Once they reach the lumen they are apparently degraded chemically, since, although fine calcium deposits can be observed in the lumina of such tubules in sections treated by the calcium red method, only rarely are typical calcium spherites found in the proximal portion of the tubules where the main duct leading from the digestive gland joins the alimentary tract.

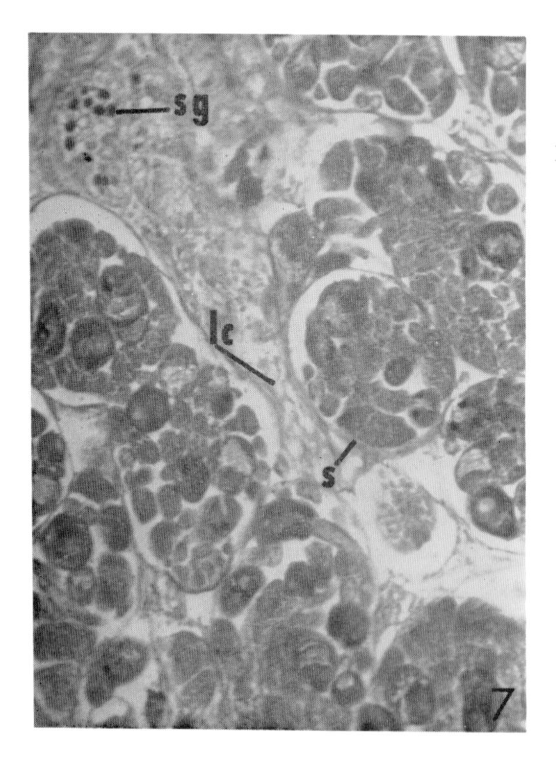

Fig. 7. Photomicrograph of section of digestive gland of *Nitocris dilatatus* infected with *Acanthatrium anaplocami* sporocysts. Notice displacement and breakdown of host cells (*lc*) and appearance of large number of secretory globules (*sg*) in ferment cells in vicinity of parasites. (Mallory's triple stain. ×10 obj.)

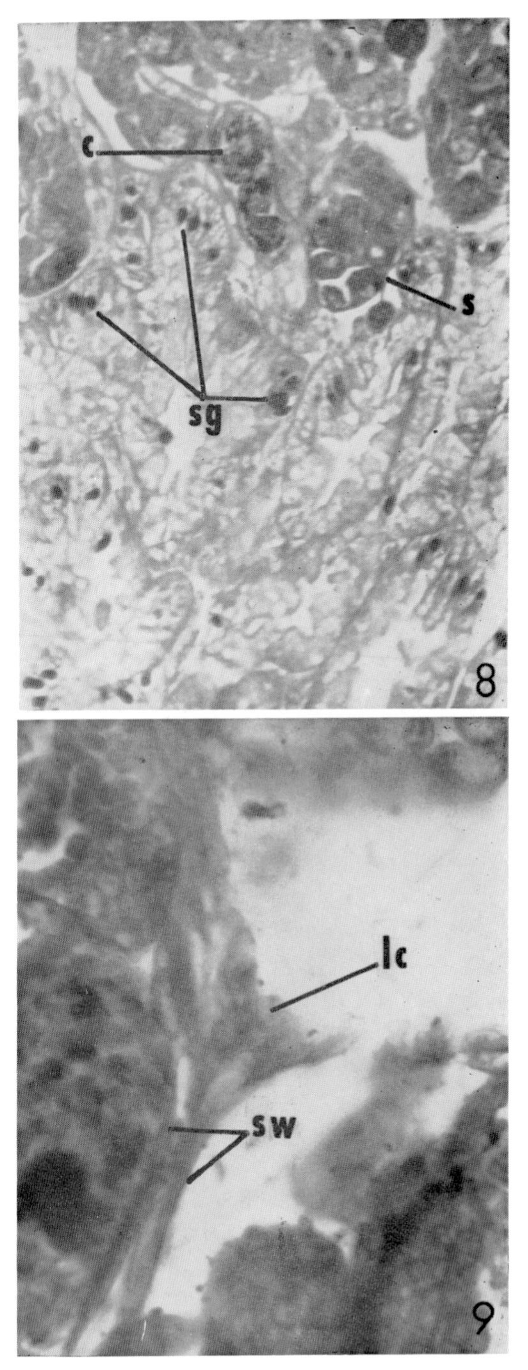

Fig. 8. Photomicrograph of section of digestive gland of *Nitocris dilatatus* infected with *Acanthatrium anaplocami* sporocysts. Notice presence of both sporocysts (*s*) and free cercaria (*c*) and the accumulation of secretory globules (*sg*) in ferment cells in proximity of parasites. (Delafield's hematoxylin. ×10 obj.)

Fig. 9. Photomicrograph showing lysed host digestive gland cells (*lc*) in infected *Nitocris dilatatus*. (Mallory's triple stain. ×40 obj.)

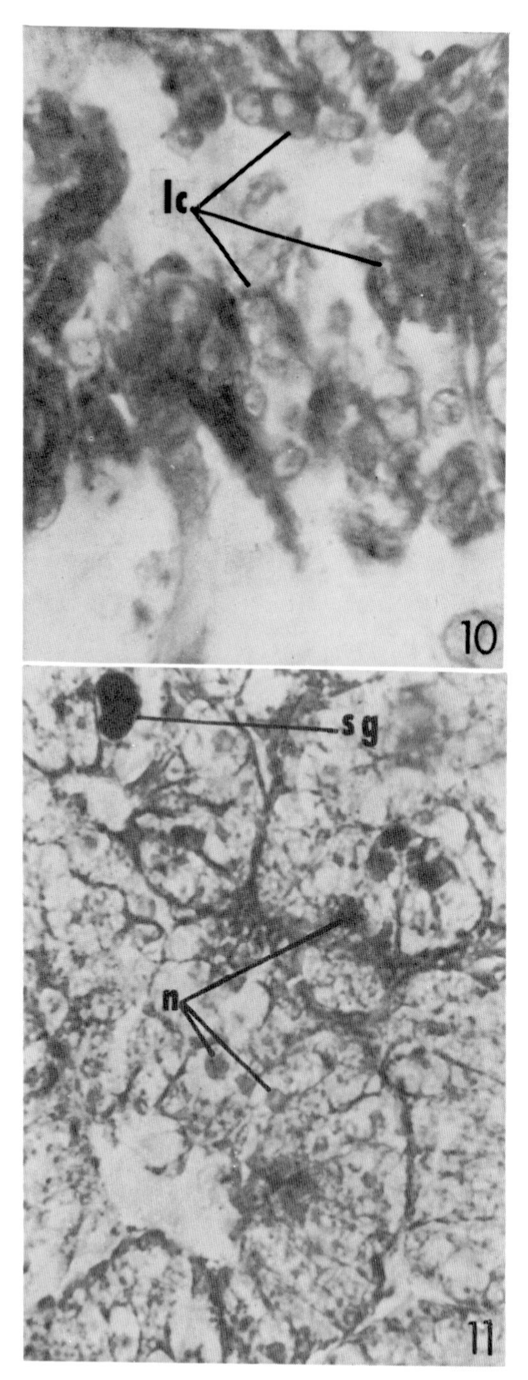

Fig. 10. Photomicrograph showing dissociated and ruptured digestive gland cells (*lc*) in *Nitocris dilatatus* infected with *Acanthatrium anaplocami* sporocysts. (Delafield's hematoxylin. ×40 obj.)

Fig. 11. Photomicrograph of cross-section through digestive gland of uninfected *Nitocris dilatatus*. Notice presence of secretory globules (*sg*) in ferment cells and large number of digestive cells with conspicuous nuclei (*n*). (Mallory's triple stain. ×40 obj.)

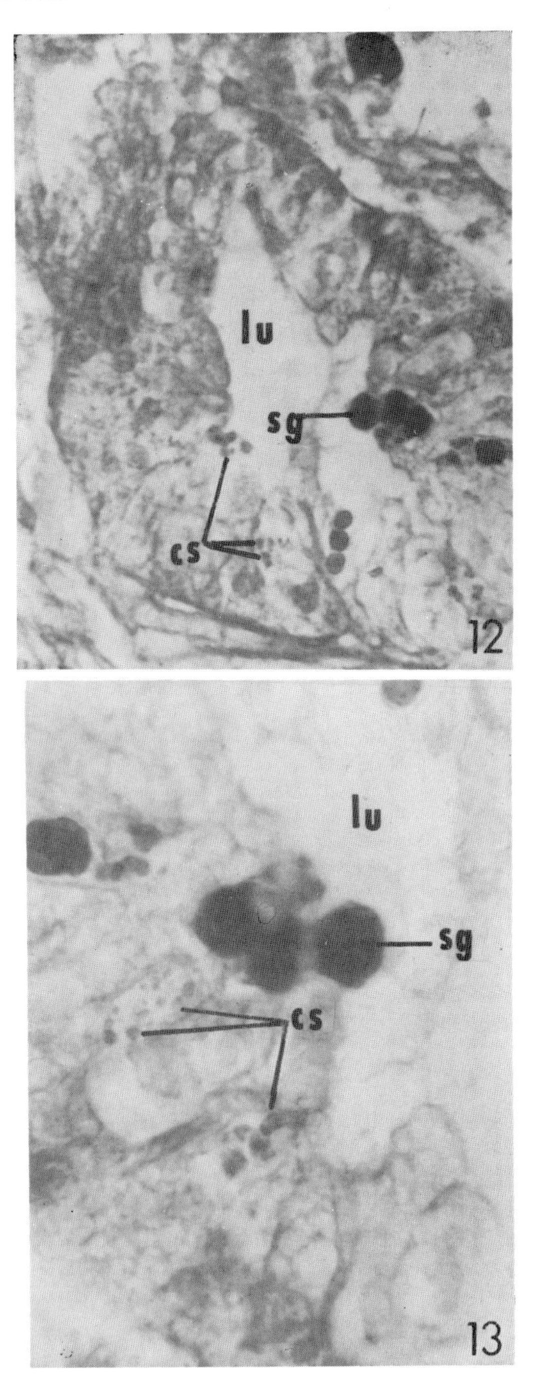

Fig. 12. Photomicrograph of cross-section through digestive gland tubules of uninfected *Nitocris dilatatus*. Notice presence of secretory globules (*sg*) and calcium spherites (*cs*). The lumen (*lu*) of the tubule is in the middle. (Mallory's triple stain. ×40 obj.)

Fig. 13. Photomicrograph of cross-section through portion of digestive gland tubules of uninfected *Nitocris dilatatus*. Notice presence of secretory globules (*sg*) entering tubular lumen (*lu*) and calcium spherites (*cs*) in calcium cells. (Delafield's hematoxylin. ×90 obj.)

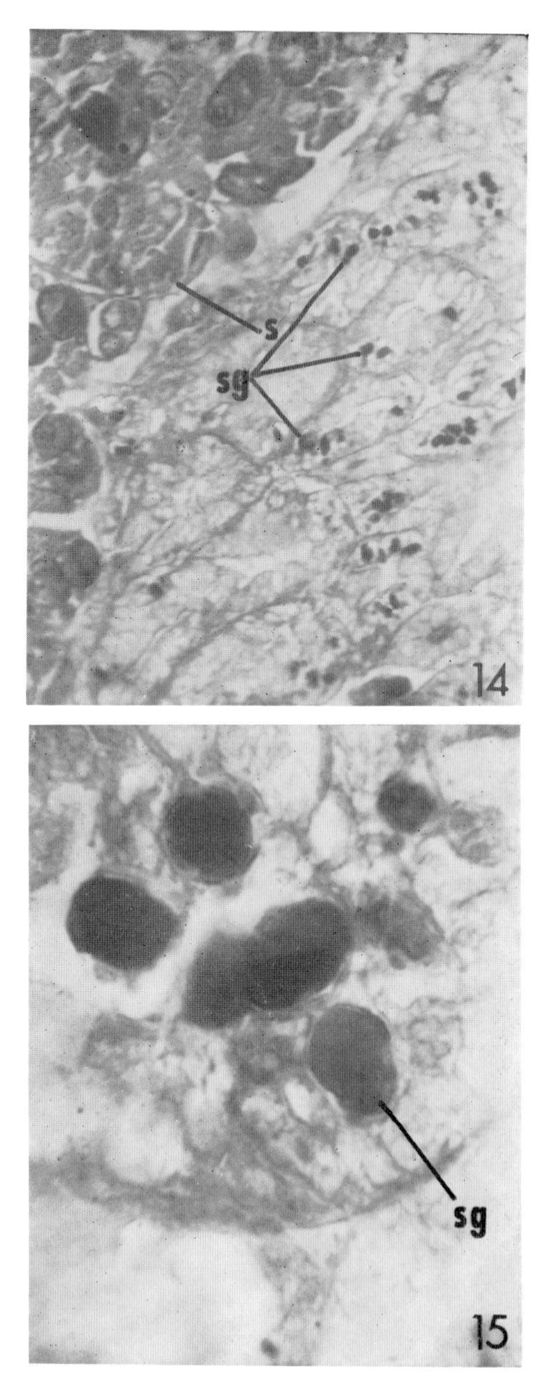

Fig. 14. Photomicrograph of section through digestive gland of *Nitocris dilatatus* infected with *Acanthatrium anaplocami*. Notice increased number of secretory globules (*sg*) in vicinity of sporocysts (*s*). (Mallory's triple stain. ×10 obj.)

Fig. 15. Photomicrograph showing large numbers of secretory globules in digestive gland of *Nitocris dilatatus* parasitized by *Acanthatrium anaplocami* sporocysts. (Mallory's triple stain. ×90 obj.)

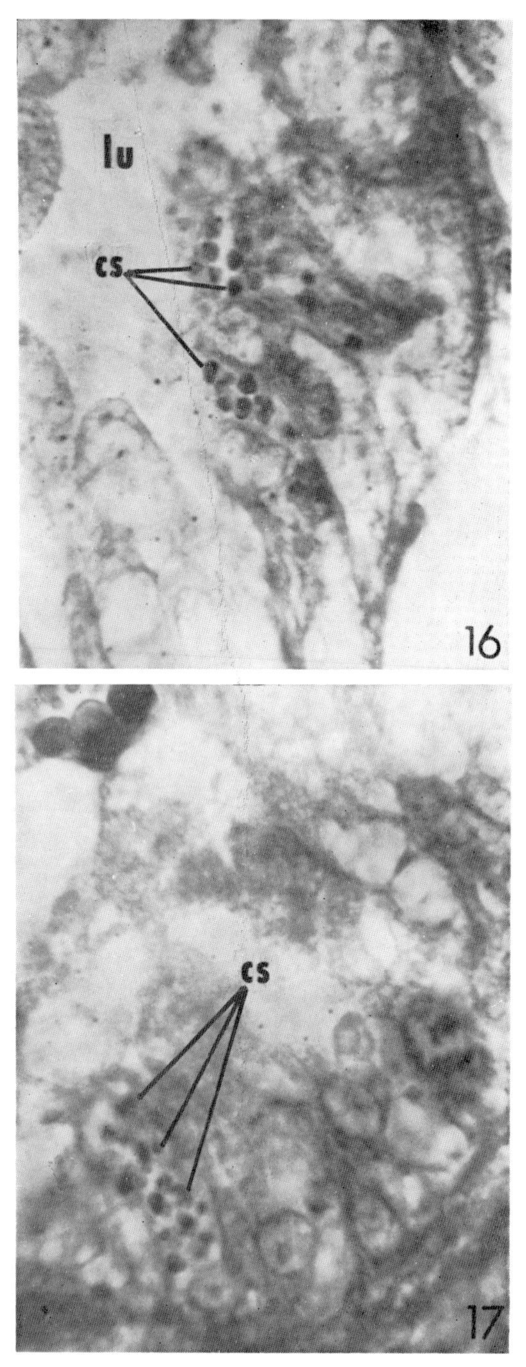

Fig. 16. Photomicrograph of section of digestive gland of *Nitocris dilatatus* infected with *Acanthatrium anaplocami* sporocysts. Notice larger and greater number of calcium spherites (*cs*) in calcium cells. (Oil red 0. ×90 obj.)

Fig. 17. Photomicrograph of section of digestive gland of *Nitocris dilatatus* infected with *Acanthatrium anaplocami* sporocysts. Notice larger and greater number of calcium spherites (*cs*) in calcium cell. (Mallory's triple stain. ×90 obj.)

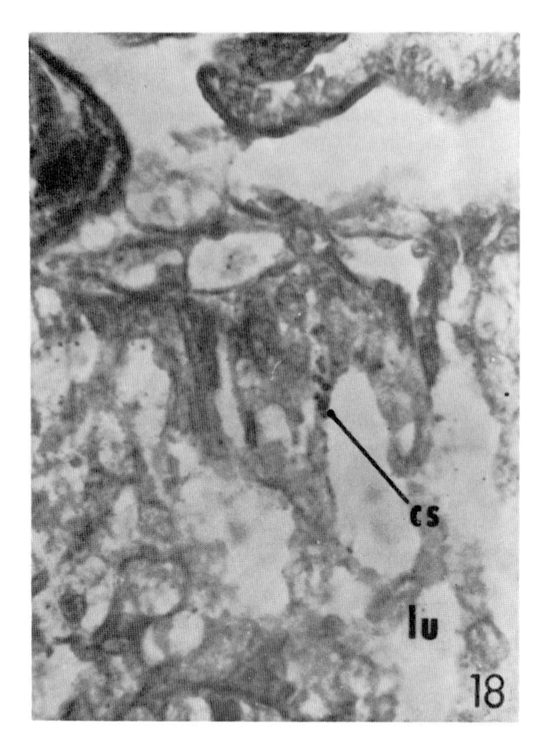

Fig. 18. Photomicrograph of section of digestive gland tubule of *Nitocris dilatatus* infected with *Acanthatrium anaplocami* sporocysts. Notice arrangement of calcium spherites (*cs*) along luminal border (*lu*) of calcium cell. (Oil red 0. ×40 obj.)

Except for size differences of the components, the normal histology of the digestive gland of *Physa sayii* is essentially the same as that of *N. dilatatus*. The histopathological changes in the digestive gland of this snail parasitized by *Echinostoma revolutum* rediae are more dramatic, in that rediae can actively devour host cells. The shifts in quantity and positions of the secretory globules and calcium spherites within specific acinar cells, however, are essentially the same as those encountered in *N. dilatatus* parasitized by *A. anaplocami*.

DISCUSSION AND CONCLUSIONS

It is evident from the data presented in Table 1 that the specimens of *Nitocris dilatatus* parasitized by *Acanthatrium anaplocami* sporocysts, averaging 0.378 ± 0.144g in weight, are significantly heavier than uninfected specimens, with the latter averaging 0.329 ± 0.137g. As indicated in Table 4, the specimens of experimentally infected *Physa sayii*, averaging 0.335 ± 0.016g, are

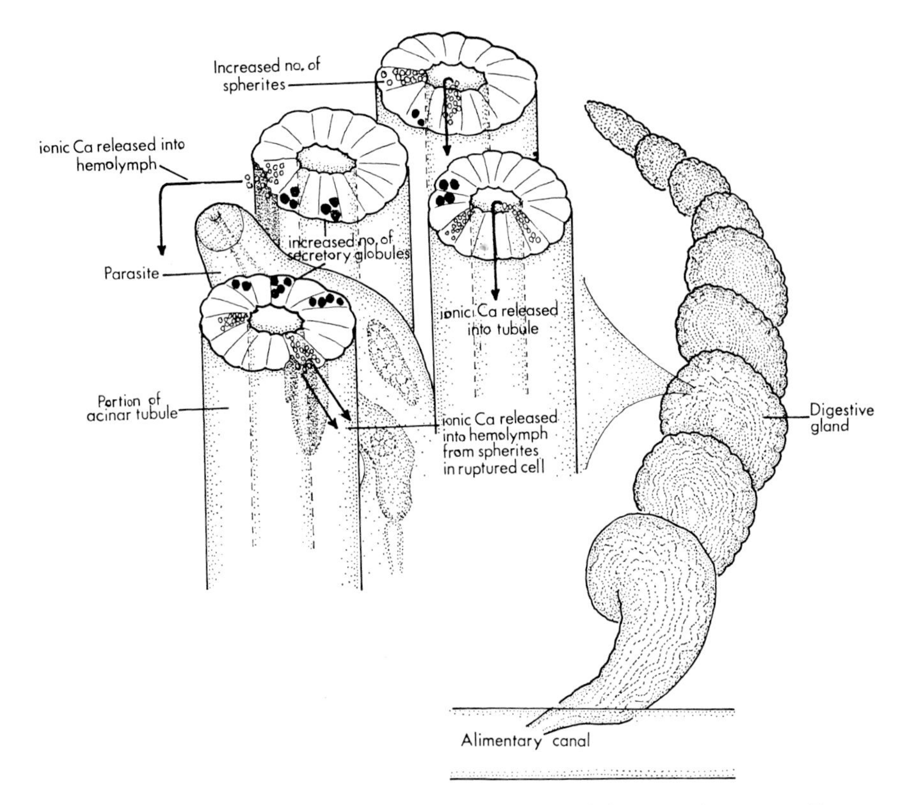

Fig. 19. Diagrammatic drawing showing parasite in intertubular space between molluscan digestive gland acini. Notice increased number of secretory globules and calcium spherites. Also notice the release of ionic calcium from calcium cells into tubular lumen and into hemolymph in intertubular space from calcium cells ruptured by parasite.

also heavier than the uninfected controls which averaged 0.217±0.090 g. On the other hand, the weights of the soft tissues of both species of parasitized snails (averaging 0.126±0.055 g in the case of *N. dilatatus*, and 0.085±0.030 g in the case of *P. sayii*) are not significantly heavier than those of their uninfected counterparts, with the latter averaging 0.114±0.045 g in the case of *N. dilatatus* and 0.086±0.024 g in the case of *P. sayii*. However, the shells of both species of parasitized snails, averaging 0.221±0.084 g in the case of *N. dilatatus* and 0.156±0.002 g in the case of *P. sayii*, are heavier than those of nonparasitized ones, with the shells of nonparasitized *N. dilatatus* averaging 0.162±0.071 g and those of nonparasitized *P. sayii* averaging 0.085±0.040 g. The discrepancies between the true weights of the shells and the weights obtained when the soft tissue weights are subtracted from the gross weights of each snail of both species, can be attributed to the small amounts of hemolymph lost when the shells were removed.

As stated earlier, Abdel-Malek (1952) had reported that specimens of *Helisoma corpulentum* parasitized by trematode larvae were heavier than nonparasitized ones. The results presented herein indicate that although the gross weights of both species of parasitized snails are greater than those of their nonparasitized counterparts, the mean weights of the soft tissues of both categories of snails of both species are not significantly different. This could only mean that the greater total weights are due to heavier shells. This, as indicated in Tables 1 and 4, is the case.* Thus, the quantitative data presented support the earlier observations by Etges (1961a,b) that parasitized snails have heavier shells.

Relative to the Ca^{++} concentrations presented in Tables 2 and 5, these cannot be considered as absolute concentrations but should be considered as relative ones, since most probably some calcium was lost during the extraction procedure. Nevertheless, the data presented do permit a comparison of the relative concentrations of Ca^{++} in the soft tissues of parasitized and nonparasitized snails of both species. As indicated, although there is no significant difference between the mean Ca^{++} concentrations in the soft tissues of snails of the uninfected and infected series of *N. dilatata*, the Ca^{++} concentration per gram of soft tissue of parasitized snails (0.730 ± 0.170 g) is significantly greater than that of nonparasitized specimens (0.473 ± 0.052 g) (Table 2), and the ionic calcium concentration per 0.5 ml of aqueous extract of parasitized and nonparasitized *N. dilatata* is significantly different, being 0.046 ± 0.020 g in parasitized and 0.027 ± 0.014 g in nonparasitized snails.

The situation in parasitized and nonparasitized *Physa sayii* is slightly different (Table 5). The mean Ca^{++} concentrations in 0.5 ml of aqueous extract, in the soft tissues per snail, and in each gram of soft tissue of both series of snails, are all significantly different, being in each case greater in parasitized specimens. A possible explanation for these differences in both species of snails has been revealed by the histological and histochemical studies and is presented below.

Barfurth (1883), Yung (1888), Biederman and Moritz (1899), Cuénot (1892, 1899), Krijgsman (1928), Baecker (1932), Fretter (1952), van Weel (1950), Guardabassi and Ferreri (1953), Billett and McGree-Russell (1955), Abolinš-Krogis (1960, 1963a) have all described calcium cells in the digestive gland of mollusks in which calcium, entering the body, is stored in the form

* Cheng *et al.* (1966), in an abstract based on preliminary work, reported that there was no statistically significant difference between the shell weights of parasitized and nonparasitized *Nitocris dilatatus*. This earlier report is being emended as the result of additional work.

of calcium phosphate globules or "calcium spherites." Recent studies by Grzycki and Krolikowska (1956) and Abolinš-Krogis (1960, 1963a,b), involving histochemistry and analytical chemistry, have confirmed the presence of calcium in these cells. Their findings have been confirmed herein. Furthermore, Abolinš-Krogis (1963a,b) has reported that in addition to calcium phosphate, ionic calcium, an acid mucopolysaccharide, RNA, xanthine, probably hypoxanthine and pteridines, lipids, and proteins are also present in calcium spherites. During certain periods of stress, such as during shell repair, Sioli (1935) has suggested that some yet undetermined enzyme liberates the stored calcium phosphate in ionic form or as calcium proteinate from the calcium cells, while the phosphate remains within the cell, perhaps in some organic combination. This appears to be substantiated by Mead (1950) who has found that the phosphate (P_2O_5) concentration in *Achatina fulica* is significantly greater (35.34 %) in the visceral mass, which includes the digestive gland, than in the foot and mantle (11.78 %).

Abolinš-Krogis (1960) has reported that the deposition of calcium and the formation of calcium spherites are increased in calcium cells during stress, a fact my histological and histochemical studies have confirmed. Also, she has noted that the calcium from the digestive gland is released into the hemolymph in ionic form and reaches the damaged site during shell repair as such. The conversion of calcium phosphate, which is readily detected as spherical calcium spherites, into ionic calcium is suggested by my histochemical studies. They indicate that the spherites in parasitized snails are released from the calcium cells into the acinar lumina or when the cells are ruptured and the conspicuous absence of spherites in the proximal portion of the digestive gland tubules and in the tissues.

The data presented above indicate a phenomenon of fundamental importance relative to this study. That is, calcium is stored in the digestive gland of snails and when placed under stress, including parasitic stress, calcium deposition in the storage (calcium) cells is increased. When released, the calcium is transported through the soft tissues in the ionic form. Thus the greater concentration of Ca^{++} per gram of soft tissues in parasitized snails of both species can be explained by the increased amount of stored calcium in calcium cells, but especially by the release of Ca^{++} into the hemolymph and tissues because of parasitic stress.

Relative to the mechanism responsible for the release of calcium in parasitized snails, it is known that trematode larvae invading the digestive gland of mollusks do cause lysis and mechanical destruction of the constituent cells (Faust, 1917, 1920; Faust and Hoffman, 1934; Hurst, 1927; F. G. Rees, 1934; W. J. Rees, 1936; Agersborg, 1924; Dubois, 1929; Pratt and Barton, 1941; Cort *et al.*, 1941; Lal and Premvati, 1955; Cheng and James, 1960; Cheng and Snyder, 1962; Cheng, 1962, 1963). When this occurs, it follows that the increased amount of calcium in calcium cells is released into the tissues via the hemolymph. The calcium found in the tissues includes the free calcium from the calcium cells as well as that resulting from the initially bound calcium which becomes dissociated from the phosphate moeity and occurs free in the tissues as bivalent ions (Sioli, 1935; Abolinš-Krogis, 1960). It is suggested that much of the Ca^{++} in the hemolymph and tissues becomes incorporated in the nacre-secreting mantle epithelium and is eventually deposited in the shell, thus resulting in the heavier shells of parasitized snails (Fig. 19).

In view of my demonstration that there is greater calcification in the shells of parasitized *A. dilatata* and *P. sayii*, a reinvestigation of the assumed greater growth of young mollusks parasitized by trematode larvae appears to be warranted to determine whether the enhanced growth, as determined by shell measurements, is actually more rapid growth of the whole organism or merely increased shell deposition.

Until experimental, quantitative data are available, no explanation can be offered at this time for two additional phenomena which have been reported. First, Wesenberg-Lund's (1934) observation that the shells of parasitized *Radix auricularis* are thinner than those of nonparasitized specimens and are often "ballooned." Second, Pesigan *et al.*'s (1958) and Moose's (1963) observations that *Oncomelania nosophora* parasitized by *Schistosoma japonicum* are retarded in their growth, thus suggesting the possibility of less calcium in their shells. Perhaps the availability of calcium in the environment could be responsible in part.

Finally, the increased number of secretory globules in the secretion cells of the digestive glands of parasitized *A. dilatata* and *P. sayii*, especially in those areas in the immediate proximity of larval trematodes, suggests that another aspect of the hosts' reaction to larval trematodes is the rapid removal of metabolic products secreted or excreted by the parasites. The nature of such waste products and the exact mechanism by which they are removed by the host remains to be elucidated.

SUMMARY

The literature pertaining to enhanced growth in parasitized invertebrates and vertebrates is reviewed critically. Although the data indicating parasite-induced enhanced growth in mammals are convincing, those reports suggesting so-called gigantism in mollusks parasitized by larval trematodes have been based primarily on field observations and hence are subject to alternative interpretations. The available experimental data are limited to *Biomphalaria glabrata* infected with *Schistosoma mansoni* and the enhanced growth, as determined by shell size, is reported to be limited to actively growing snails.

Quantitative weight and Ca^{++} determinations of the whole snail, shell, and soft tissues of *Nitocris dilatata* naturally infected with *Acanthatrium anaplocami* and laboratory-reared *Physa sayii* experimentally infected with *Echinostoma revolutum* have revealed that the increased total weights of parasitized snails can be accounted for by more heavily calcified shells. Furthermore, histological and histochemical studies have revealed that this hypercalcification is due to an initial increase of the number and size of calcium spherites in snails under parasitic stress, followed by the release of ionic calcium into the hemolymph and tissues when spherite-containing cells are destroyed by the parasites as a normal process. The increased ionic calcium is eventually deposited in the shells.

In view of this finding, reinvestigation of enhanced growth of parasitized mollusks as determined by shell measurements appears warranted to ascertain whether the greater shell sizes are due to growth of the whole organism or to greater calcium deposition in shells.

ACKNOWLEDGMENTS

I am deeply grateful to Dr. Randall W. Snyder, Jr., Medical School, Jefferson University, and Mr. Alan B. Blumenthal, California Institute of Technology, who assisted me during 1963 when some the studies reported herein were carried out. These studies were supported by a grant from the National Institutes of Health (AI-3443) which has since been terminated.

REFERENCES

Abdel-Malek, E. T. 1952. Morphology, bionomics and host-parasite relations of Planorbidae (Mollusca: Pulmonata). Ph.D. Thesis, Univ. Michigan. 226 p.

Abolinš-Krogis, A. 1960. The histochemistry of the hepatopancreas of *Helix pomatia* (L.) in relation to the regeneration of the shell. Ark. Zool. 13:159–201.

Abolinš-Krogis, A. 1963a. The morphological and chemical basis of the initiation of calcification in the regenerating shell of *Helix pomatia* (L.). Acta Univ. Upsaliensis 20:1–22.

Abolinš-Krogis, A. 1963b. Some features of the chemical composition of isolated cytoplasmic inclusions from the cells of the hepatopancreas of *Helix pomatia* (L.). Ark. Zool. 15:393–429.

Agersborg, H. P. K· 1924. Studies on the effects of parasitism upon the tissues I. With special reference to certain gastropod molluscs. Quart. J. Microscop. Sci. 68:361–401.

Andrade, R. M. De. 1962. Ecologia de "Australorbis glabratus" em belo horizonte, Brasil. III: Indices de infeccao natural por "Schistosoma mansoni" segundo os diametros dos caramujos. Rev. Brasileira Biol. 22:383–390.

Baecker, R. Z. 1932. Die Mikromorphologie nous *Helix pomatia* und einigen anderen Stylommatophoran. Ergebnisse Anat. EntwGesch. 29:449–585.

Barfurth, D 1881. Der Kalk in der Leber der Helicinen und seine biologische Bedeutung. Zool. Anzeig. 4:20–23.

Barfurth, D. 1883. Ueber den Dau und die Thätigkeit der Gasteropodenleber. Arch. Mikroscop. Anat. 22:473–524.

Beiderman, W. and Moritz, P. 1899. Beiträge zur vergleichenden Physiologie der Verdauung III. Ueber die Function der sogenannten "Leber" der Mollusken. Arch. Ges. Physiol. 75:1–86.

Billett, F. and McGee-Russell, S. M. 1955. The histochemical localization of β-glucoronidase in the digestive gland of the Roman snail (*Helix pomatia*). Quart. J. Microscop. Sci. 96:35–48.

Cheng, T. C. 1962. The effects of parasitism by the larvae of *Echinoparyphium* Dietz (Trematoda: Echinostomatidae) on the structure and glycogen deposition in the hepatopancreas of *Helisoma trivolvis* (Say). Amer. Zool. 2:513.

Cheng, T. C. 1963. The effects of *Echinoparyphium* larvae on the structure of and glycogen deposition in the hepatopancreas of *Helisoma trivolvis* and glycogenesis in the parasite. larvae. Malacologia 1:291–303.

Cheng, T. C. 1967. Marine molluscs as hosts for symbioses (Adv. Marine Biol , Vol. 5). Academic Press, London. 424 p.

Cheng, T. C. 1970. Symbiosis. Pegasus Press, New York. 242 p.

Cheng, T. C. and James, H. A. 1960. The histopathology of *Crepidostomum* sp. infection in the second intermediate host, *Sphaerium striatinum*. Proc. Helminthol. Soc Wash. 27:67–68.

Cheng, T. C. and Snyder, R. W., Jr. 1962. Studies on host-parasite relationships between larval trematodes and their hosts. I. A. review. II. The utilization of the host's glycogen by the intramolluscan larvae of *Glypthelmins pennsylvaniensis* Cheng, and associated phenomena. Trans. Amer. Microscop. Soc. 81:209–228.

Cheng, T. C., Snyder, R. W., Jr , Rourke, A. W. and Blumenthal, A. B. 1966. Ionic calcium concentrations in nonparasitized *Nitocris dilatatus* and those parasitized by the larvae of *Prosthodendrium* (*Acanthatrium*) *anaplocami* (Trematoda). Amer. Zool. 6:225.

Chernin, E. 1960. Infection of *Australorbis glabratus* with *Schistosoma mansoni* under bacteriologically sterile conditions. Proc. Soc. Exp. Biol. Med. 105:292–296.

Cort, W.W., Olivier, L. and McMullen, D. B. 1941. Larval trematode infection in juveniles and adults of *Physa parkeri* Currier. J. Parasitol. 27:123–147.

Croll, N. A. 1966. Ecology of parasites. Harvard Univ. Press, Cambridge, Mass. 136 p.

Cuénot, L. 1892. Études physiologiques sur les gastéropodes pulmonés. Arch. Biol. 12:683–740.

Cuénot, L. 1899. La fonction excretrice du foie des gastéropodes pulmonés. Critique d'un travail de Beidermann et Moritz. Arch. Zool. Exp. 7:25–28.

Dubois, G. 1929. Les cercaires de la region de Neuchâtel. Bull. Soc. Neuchâtel, Sci. Natur. 53 n.s., 2:3–177.

Etges, F. J. 1960. On the life history of *Prosthodendrium* (*Acanthatrium*) *anaplocami* n. sp. (Trematoda: Lecithodendriidae). J. Parasitol. 46:235–240.

Etges, F. J. 1961a. *Cercaria reynoldsi* n. sp. (Trematoda: Echinostomatidae) from *Helisoma anceps* (Menke) in Mountain Lake, Virginia. Trans. Amer. Microscop. Soc. 80:221–226.

Etges, F. J. 1961b. Contributions to the life history of the brain fluke of newts and fish, *Diplostomulum scheuringi* Hughes, 1929 (Trematoda: Diplostomatidae) J. Parasitol- 47:453–458.

Fales, F. W. 1953. A micromethod for the determination of serum calcium. J. Biol. Chem. 204:577–585.

Faust, E. C. 1917. Life-history studies on Montana trematodes. Ill. Biol. Monogr. 4:1–121.

Faust, E. C. 1920. Pathological changes in the gastropod liver produced by fluke infection. Bull. Johns Hopkins Hosp. 31:79–84.

Faust, E. C. and Hoffman, W. A. 1934. Studies on schistosomiasis mansoni in Puerto Rico. III. Biological studies. 1. The extra-mammalian phases of the life cycle. Puerto Rico J. Public Health Trop. Med. 10:1–47.

Ferro, P. V. and Ham, A. B. 1957. A simple spectrophotometric method for the determination of calcium. Amer. J. Clin. Pathol. 28:208–217.

Fisher, F. M., Jr. and Sanborn, R. C. 1962a. Production in insect juvenile formone by the microsporidian parasite *Nosema.* Nature 194:1193.
Fisher, F. M., Jr. and Sanborn, R. C. 1962b. Observations on the susceptibility of some insects to *Nosema* (Microsporidia: Sporozoa). J. Parasitol. 48:926–932.
Fisher, F. M., Jr. and Sanborn, R. C. 1964. *Nosema* as a source of juvenile hormone in parasitized insects. Biol. Bull. 126:235–252.
Frenzel, J. 1886. Mikrographie der Mitteldarmdrüse (Leber) der Mollusken. Ester Thiel. Allgeneine Morphologie und Physiologie des Drüsenepithels. Nova Acta Ksl Leop.-Carol. Deutsch Akad. Naturforsch. 48:83–196.
Fretter, V. 1952. Experiments with P^{32} and I^{131} on species of *Helix*, *Arion*, and *Agriolimax*. Quart. J. Microscop. Sci. 93:133–146.
Gehrke, C. W., Affsprung, H. E. and Lee, Y. C 1955. Direct ethylenediamine tetraacetate titration methods for magnesium and calcium. Anal Chem. 26:1944–1948.
Grzycki, S. and Krolikowska, I. 1956. Glycogen, desoxyribonucleaic acid and calcium salts in cells of the digestive gland of the snail (*Helix pomatia* L.). Ann. Univ. Mariae Curie-Sklodowska 11(Sect. C):153–160·
Guardabassi, A. and Ferreri, E. 1953. Contributo allo studio dell'assorpimento dei lipidi: esperimenti su tratti intestinali di *Helix pomatia*, isolati e sopraviventi in soluzione fisiologica. Arch. Soc. Biol. Bologna 37:287–305.
Harlow, D. R., Mertz, W. and Mueller, J. F. 1964. Effects of *Spirometra mansonoides* infection on carbohydrate metabolism. II.An insulin-like activity from the sparganum. J. Parasitol. 50(Sect. 2):55.
Hoshina, T. and Ogino, C. 1951. Studien ueber *Gymnophalloides tokiensis* Fujita, 1925. I. Ueber die einwirkung der Larvalen Trematoda auf die chemische Komponente und das Wachstum von *Ostrea gigas* Thunberg J. Tokyo Univ. Fisheries 38:335–350.
Hurst, C. T. 1927. Structural and functional changes produced in the gastropod mollusk, *Physa occidentalis*, in the case of parasitism by the larvae of *Echinostoma revolutum*. Univ. Calif. Pub. Zool. 29:321–404.
Kenny, A. D. and Toverud, S. U. 1954. Noninterference of phosphate in ethylenediamine tetraacetate method for serum calcium. Anal. Chem. 26:1059.
Krijgsman, B. J. Z. 1928. Arbeitsrhythmus der Verdauugadrüsen bei *Helix pomatia* II. Teil: Sekretion, Resorption und Phagozytose. Z. Vergl. Physiol. 8:187–280.
Lal, M. B. and Premvati. 1955. Studies in histopathology. Changes induced by larval monostome in the digestive gland of the snail, *Melanoides tuberculatus* (Müller). Proc. Indian Acad. Sci. Sect. B, No. 2:193–299.
Lincicome, D. R. 1963. Chemical basis of parasitism. Ann. N. Y. Acad. Sci. 113:360–380.
Lincicome, D. R. and Shepperson, J. R. 1961. Increased rate of growth of experimental hosts associated with foreign autonomous cells. The Physiologist 4:65.
Lincicome, D. R. and Shepperson, J. R. 1963. Increased rate of growth of mice infected with *Trypanosoma duttoni* J. Parasitol. 49:31–34.
Lincicome, D. R., Rossan, R. N. and Jones, W. C. 1960 Rate of body weight gain of rats infected with *Trypanosoma lewisi*. J. Parasitol. 46(Sect 2):42.
Lincicome, D. R., Rossan, R. N. and Jones, W. C. 1963. Growth of rats infected with *Trypanosoma lewisi*. Exp. Parasitol. 14:54–65.
Lysaght, A. M. 1941. The biology and trematode parasites of the gastropod *Littorina neritoides* (L.) on the Plymouth breakwater. J. Marine Biol. Ass. U.K. 25:41–67.
McGee-Russell, S.M. 1955. A new reagent for the histochemical and chemical detection of calcium. Nature 175:301–302.
Mead, A. R. 1950. The giant African snail problem (*Achatina fulica*) in Micronesia. Final Report Invert. Consult. Comm. Micronesia, Pacific Sci. Board, Nat. Res. Council. 55 p.
Menzel, R. W. and Hopkins, S. H. 1955a. The growth of oysters parasitized by the fungus *Dermocystidium marinum* and by the trematode *Bucephalus* cuculus. J. Parasitol. 41:333–342.

Menzel, R. W. and Hopkins, S. H. 1955b. Effects of two parasites on the growth of oysters. Proc. Nat. Shellfisheries Ass. 45:184–186.

Moose, J. W. 1963. Growth inhibition of young *Oncomelania nosophora* exposed to *Schistosoma japonicum*. J. Parasitol. 49:151–152.

Mueller, J. F. 1962. Parasite-induced weight gain in laboratory mice infected with *Sparganum mansonoides*. J. Parasitol. 48(Sect. 2):45.

Mueller, J. F. 1963a. Parasite-induced weight gain in mice. Ann. N.Y. Acad. Sci. 113:217–233.

Mueller, J. F. 1963b. Further studies on parasite-induced weight gain in mice. J. Parasitol. 48(Sect. 2):18.

Mueller, J. F. 1965a. Further studies on parasitic obesity in mice, deer mice, and hamsters. J. Parasitol. 51:523–531.

Mueller, J. F. 1965b. Food intake and weight gain in mice parasitized with *Spirometra mansonoides*. J. Parasitol. 51:537–540.

Mueller, J. F. 1968. Growth stimulating effect of experimental sparganosis in thyroidectomized and hypophysectomized rats, and comparative activity of different species of *Spirometra*. J. Parasitol. 54:795–801.

Mueller, J. F. and Reed, P. 1968. Growth stimulation induced by infection with *Spirometra mansonoides* spargana in propylthiouracil-treated rats. J. Parasitol. 54:51–54.

Meyer, F., Kimura, S. and Mueller, J. F. 1965. Stimulation of lipogenesis in hamsters by *Spirometra mansonoides*. J. Parasitol. 51(Sect. 2):57.

Mueller, J. F. 1966. Host-parasite relationships as illustrated by the cestode *Spirometra mansonoides*, p. 15–58. *In* J. E. McCauley (ed.) Host-parasite relationships. Oregon State Univ. Press, Corvallis.

Neumann, L. G. 1905. A treatise on the parasites and parasitic diseases of the domesticated animals, 2nd ed. (George Flemming, trans.; James MacQueen, ed.) Bailliere, Tindall and Cox, London.

Pan, C. T. 1958. The general histology and topographic microanatomy of *Australorbis glabratus*. Bull. Mus. Comp. Zool. (Harvard) 119:237–299.

Pan, C. T. 1962. The course and effect of infection with *Schistosoma mansoni* in *Australorbis glabratus*. J. Parasitol. 48(Sect. 2):20.

Pan, C. T. 1965. Studies on the host-parasite relationship between *Schistosoma mansoni* and the snail *Australorbis glabratus*. Amer. J. Trop. Med. Hyg. 14:931–976.

Pesigan, T. P., Hairston, N. G., Tauregui, J. J., Garcia, E. G., Santos, A. T., Santos, B. C. and Besa, A. A. 1958. Studies on *Schistosoma japonicum* infection in the Philippines. 2. The molluscan host. Bull. WHO 18:481–578.

Pratt, I. and Barton, G. D. 1941. The effects of four species of larval trematodes upon the liver and ovotestis of the snail, *Stagnicola emarginata angulata* (Sowerby). J. Parasitol. 27:283–288.

Rees, F. G. 1934. *Cercaria patellae* Lebour, 1911, and its effects on the digestive gland and gonads of *Patella vulgata*. Proc. Zool. Soc. London pp. 45–53.

Rees, W. G. 1936. The effect of parasitism by larval trematodes on the tissues of *Littorina littorea* (Linné). Zool. Soc. (London), Proc.:357–368.

Rothschild, M. 1936. Gigantism and variation in *Peringia ulvae* Pennant, 1777, caused by infection with larval trematodes. J. Marine Biol. Ass. U.K. 30:537–546.

Rothschild, M. 1938. Further observations on the effect of trematode parasites of *Peringia ulvae* (Pennant, 1777). Novitates Zool. 41:84–102.

Rothschild, M. 1941a. The effect of trematode parasites on the growth of *Littorina neritoides* (L.). J. Marine Biol. Ass. U.K. 25:84–102.

Rothschild, M. 1941b. Observations on the growth and trematode infections of *Peringia ulvae* (Pennant, 1777) in a pool in the Tamar Saltings, Plymouth. Parasitology 33:406–415.

Rothschild, M. and Clay, T. 1952. Fleas, flukes and cuckoos. Macmillan, London, 305 p.

Rothschild, A. and Rothschild, M. 1939. Some observations on the growth of *Peringia ulvae* (Pennant, 1777) in the laboratory. Novitates Zool. 41:240–247.

Sadun, E. H., Williams, J. S., Meroney, F. C. and Mueller, J. F. 1965. Biochemical changes in mice infected with spargana of the cestode, *Spirometra mansonoides*. J. Parasitol. 51:532–536.

Sioli, H. 1935. Über den Chemismus der Reparatur von Schalendefektion bei *Helix pomatia*. Zool. Jahrb. Abt. Allgem. Zool. Physiol. Tiere 54:507–534.

Smyth, J. D. 1962. Introduction to animal parasitology. Charles C Thomas, Springfield, Ill. 470 p.

Steelman, S. L., Morgan, E. R., Cuccaro, A. J. and Glitzer, M. S. 1970. Growth hormone-like activity in hypophysectomized rats implanted with *Spirometra mansonoides* sparganа. Proc. Soc. Exp. Biol. Med. 133:269–273.

Tyner, E. H. 1948. Determining small amounts of calcium in plant materials. Anal. Chem. 20:76–80.

Van Weel, P. B. 1950. Contribution to the physiology of the glandula media intestini of the African giant snail, *Achatina fulica* Per., during the first hours of digestion. Physiol. Comp. Oecol. 2:1–19.

Wesenberg-Lund, C. J. 1934. Contributions to the development of the Trematoda Digenea. Part II. The biology of the freshwater cercariae in Danish freshwaters. K. Danske Vidensk. Selsk. Skr. 9R, 5:1–223.

West, A. F. 1960. The biology of a species of *Nosema* (Sporozoa: Microsporidia) parasitic in the flour beetle *Tribolium confusum*. J. Parasitol. 46:745–754.

Whitlock, J. H. 1949. The relationship of nutrition to the development of the trichostrongyloidoses. Cornell Vet. 39:146–182.

Young, A. and Sweet, T. R. 1955. Complexes of eriochrome black T with calcium and magnesium. Anal. Chem. 27:418–420.

Yung, E. 1888. Contributions a l'histoire physiologique de l'éscargot (*Helix pomatia*). Mém. Couron. Acad. Roy. Bélg. 49:1–119.

Zak, B. and Hindman, W. H. 1955. Spectrophotometric titration of calcium and magnesium, p. 55c. *In* Amer. Chem. Soc. 128th Meeting. (Abstr.)

Zischke, J. A. and Zischke, D. P. 1965. The effects of *Echinostoma revolutum* larval infection on the growth and reproduction of the snail host *Stagnicola palustris*. Amer. Zool. 5:707–708.

The Goodness of Parasitism: A New Hypothesis

DAVID RICHARD LINCICOME

Department of Zoology
Howard University
Washington, D.C.

INTRODUCTION

A philosophic view of parasitism and how it is related to commensalism, phoresis, mutualism, and symbiosis has been expressed (Lincicome 1963). The structural basis of parasitism was accepted as chemical where the host-parasite interrelationship was the result of molecular exchanges.

A further view of the basic nature of parasitism was arrived at by additional experimental evidences. Parasitism, one of life's great phenomena, possesses a quality of goodness that has largely been overlooked because man has been so possessed with the disease aspects of this association. The goodness of the parasitism hypothesis was originally expressed by the author in the Twentieth Anniversary Address before the Midwestern Conference of Parasitologists in June, 1968.

The purpose of the present paper is to present the experimental evidence, both published and unpublished, that led to the hypothesis of the basic quality of goodness in parasitism.

METHODS AND MATERIALS

New Experimental Methods

In vitro approaches to experimental study of parasitic animals seemed undesirable for several reasons although these were highly productive for fungi, bacteria, and many free-living animal groups (e.g., Protozoa). It was clearly desirable to be able to examine the parasitic animal in its natural habitat (the host), for some forms changed their morphogenetic expression when removed from their natural environment into an *in vitro* culture. In some cases, changes in morphologic form also included metabolic alterations which resulted in a totally different organism. In order to avoid these inevitable changes due to *in vitro* culturing, it seemed necessary to devise means to manipulate the host experimentally. Two methods were developed:

the nutritional imbalance procedure (Lincicome, 1953) and the heterologous host technique (Lincicome 1955, 1958a,b).

The Nutritional Imbalance Procedure.—This technique depended upon the ready availability of a host with well known nutritional requirements; which could be nutritionally imbalanced by single factors; whose metabolic well-being could be restored by returning the factor originally removed; and whose metabolic state could be monitored by accurate checking techniques. The albino laboratory rat satisfied these criteria for an experimental host.

The Heterologous Host Technique.—A homologous host is defined as one in which a parasite would normally grow either experimentally or in the feral state. A heterologous host is one in which the parasite does not grow under usual or normal circumstances. A heterologous host animal might therefore be used as a kind of *in vivo* test tube to which would be added factors from the homologous host in order to induce parasite growth. Such factors could be fractions, extractions, and so on from organs or tissues of the homologous host. In theory, the successful application of the heterologous host technique (Lincicome, 1963) depends upon the degree of host-specificity shown by the parasite. An organism developing in a wide selection of vertebrate animals would not be expected to be a good experimental model. On the other hand, an organism having a limited number of vertebrate hosts in which it developed would be useful to study by this method.

Experimental Models

Trypanosoma lewisi and the albino laboratory rat served as the experimental model for application in both the nutritional imbalance and heterologous host technique. Either the Sprague-Dawley rat or the Holtzmann rat was used exclusively because these animals are gentle to handle, highly standardized, and readily available at weanling or older weights. Body weights of any specific group of rats were closely matched with ± 1.0g standard deviation of the group average. The trypanosome was the "L" form long maintained by continuous syringe passage in rats under laboratory conditions (Lincicome and Watkins, 1965). The trypanosome was originally received from the late Dr. Elery R. Becker of Iowa State College, Ames, Iowa, about 1941 or 1942. Until 1965 it was subpassaged by whole blood transfers, but since 1965 it has been transferred using washed trypanosomes suspended in physiologic saline solution.

Trypanosoma duttoni and the mouse constituted another experimental model in the nutritional imbalance technique. This trypanosome has been maintained for many years by syringe passage in albino mice (Lincicome and

Shepperson, 1961, 1963). Albino mice (general purpose) obtained from the animal laboratories of the National Institutes of Health in Bethesda, Maryland, were used in some studies while a beige mouse developed in my laboratory (Stirewalt *et al.,* 1965; Lincicome *et al.,* 1965) was used in other experiments.

Trichinella spiralis in the albino rat constituted still another experimental model in the nutritional imbalance technique. The trichina worm had been maintained through periodic transfer to albino rats over a period of more than 20 years (Lincicome and Fergusson, 1964).

Experimental Hosts

Female Sprague-Dawley albino rats were used in the investigations reported here. They varied in weight according to plan. They were supplied by the animal laboratories of the National Institutes of Health. More than 225 rats of 50–52 g in weight were used to study pyridoxine metabolism in *T. lewisi* infections; 292 weighing 50–60 g were used to study thiamine in *T. lewisi* infections; and 1,184 weighing 80 ± 1 g were used in pyridoxine and *Trichinella spiralis* studies.

More than 1,500 female Swiss albino mice were used in *T. duttoni* infections for pantothenate studies. These were also obtained through the generosity of the National Institutes of Health.

Countless other Sprague-Dawley rats and Swiss albino mice have also been employed for routine transfer of stock trypanosomes and trichina worms, and for collection of serum.

Experimental Diets

Three semisynthetic experimental diets were used: those deficient in (1) thiamine, (2) pyridoxine, or (3) pantothenate. Appropriate diets, fully restored with respect to the components eliminated for experimental purposes, were used as controls. These were, in part, prepared in the laboratory. In some instances whole commercial animal foods were also used as additional control diets (Purina Laboratory Chow, Ralston-Purina Company, St. Louis, Missouri).

Thiamine-deficient Diets.—The diets used to examine the metabolic role of thiamine in *T. lewisi* infections were prepared according to Lincicome (1953). The essential components of the full-complement (complete) control diet are listed in Table 1. Thiamine-deficient diets were identical except for the elimination of thiamine. The composition of the "Salts Mixture IV" listed in Table 1 is given in Table 2.

Pyridoxine-deficient Diets.—The diets used to examine the metabolic role of pyridoxine in *T. lewisi* and *Trichinella spiralis* infections are listed in

TABLE 1. *Composition of semisynthetic control diets for thiamine studies (After Lincicome, 1953)*

Component	*Amount/10 kg Ration*
Basal Portion (in grams)	
Commercial sucrose	7,300
Hot alcohol-extracted, vitamin-free casein	1,800
Salt Mixture IV	400
Corn Oil	500
Vitamin Supplements (in milligrams)	
Riboflavin	30.0
Thiamine	30.0
Pyridoxine	20.0
Nicotinic acid	200.0
Calcium pantothenate	200.0
Folic acid	2.5
Biotin	1.0
Inositol	1,000.0
Choline chloride	10,000.0
p-Aminobenzoic acid	2,500.0
Menadione	4.0

TABLE 2. *Composition of Salt Mixture IV*

Component	*Grams*
$CaCO_2$	1200.0
$CaHPO_4$	248.0
K_2HPO_4	1290.0
$MgSO_4 \cdot 7H_2O$	408.0
NaCl	670.0
$FeC_6H_5O_7 \cdot 5$	110.0
KI	3.2
$MnSO_4 \cdot 4H_2O$	20.0
$CuSO_4 \cdot 5H_2O$	1.2
$ZnCl_2$	1.0

Tables 3 and 4. This diet differed from the control diet only by the omission of pyridoxine. Both diets were prepared and purchased commercially (Nutritional Biochemicals, Cleveland, Ohio).

Pantothenate-deficient Diets.—The basic components of the diets used to examine the metabolic role of pantothenate in *T. duttoni* infections are the same as those for pyridoxine (Tables 3 and 4). Both diets were prepared and purchased commercially. The pantothenate component was omitted from the control diet.

Enzyme Assays

Transketolase.—Transketolase activities were measured as a function of thiamine metabolism in rat liver and kidney homogenates (Lincicome and

TABLE 3. *Composition of diets for pyridoxine and pantothenate studies*

Main Components	*per cent*
Vitamin B complex casein	18
Sucrose	68
Vegetable oil	10
USP salt mixture No. 2	4
Vitamin Fortification Mixture Added	*grams per 100 lb of diet*
Vitamin A concentrate	4.50
Vitamin D concentrate	0.25
Alpha tocopherol	5.00
Ascorbic acid	45.0
Inositol	5.00
Choline chloride	75.00
Menadione	2.25
p-Aminobenzoic acid	5.00
Niacin	4.50
Riboflavin	1.00
Pyridoxine hydrochloride	1.00
Thiamine hydrochloride	1.00
Calcium pantothenate	3.00
	milligrams per 100 lb of diet
Biotin	20.00
Folic acid	90.00
Vitamin B_{12}	1.35

TABLE 4. *Composition of USP Salt Mixture No. 2*

Component	*per cent*
Calcium biphosphate	13.58
Calcium lactate	32.70
Ferric citrate	2.97
Magnesium sulfate	13.70
Potassium phosphate, dibasic	23.98
Sodium biphosphate	8.72
Sodium chloride	4.35

Shepperson, 1965). The methods used were those described by Brin *et al.* (1960) and Brin (1962). Rat liver and kidney homogenate assays for enzyme activity were performed at intervals of 16, 23, 30, and 37 days on diets when *T. lewisi* infections were 1, 2, 3, and 4 weeks old, respectively. Transketolase activity was expressed as milligrams of hexose formed per gram of tissue per hour.

Aminotransferases.—The spectrophotometric method for measuring l-alanine:2-oxoglutarate aminotransferase E. C. 2.6.1.2 (glutamic pyruvic transaminase or GPT) and l-aspartate:2-oxoglutarate aminotransferase

E. C. 2.6.1.1. (glutamic oxalacetic transaminase or GOT) was that recommended in a commercially purchased enzyme kit (TC-R/S No. 15964; C. F. Boeringer und Soehne Gmbh., Mannheim, Germany). Activities were expressed in milliunits where 1 unit of enzyme activity was that quantity catalyzing the transformation of 1 micromole of the substrate per minute at 30°C.

Aminotransferase assays were performed on the sera of rats fed control and pyridoxine-deficient diets and inoculated with *T. lewisi*, homogenates of *T. lewisi*, and metabolic cell products of this trypanosome (Warsi, 1968). Assays of GPT and GOT were also performed on sera from rats fed control and pyridoxine-deficient diets, and inoculated with living larvae of *Trichinella spiralis*, homogenates of larvae, and larval metabolic products (Sen, 1969).

Microbiological Assay of Pantothenate and Pyridoxine

Pantothenate.—Microbiological assays of pantothenate in tissues of mice which were fed control and pantothenate-deficient diets and inoculated with *T. duttoni*, were performed by the method of Baker *et al.* (1960), using *Lactobacillus arabinosus* cell populations as a direct function of vitamin content. Such turbidometric determinations were also performed on homogenates of the trypanosomes. Data were expressed as nanograms per milliliter of plasma, as micrograms per gram of fresh liver, or as micrograms per 10^8 trypanosomes.

Pyridoxine.—Microbiological assays of pyridoxine in liver tissue of rats fed a control diet, pyridoxine-deficient diet, or pyridoxine-supplemented diet were performed according to the method of Freed (1966). *Saccharomyces carlsbergensis* 4228 was the biologic agent, and was originally obtained from the American Type Culture Collection, Rockville, Maryland. Pyridoxine assays were also performed on homogenates of *T. lewisi*. Data were expressed as micrograms of pyridoxine per gram of fresh tissue or per 10^8 trypanosomes.

Oxygen Uptake by Tissue Slices

As an additional metabolic parameter of these studies, oxygen uptake was charted for tissue slices from rats and mice fed vitamin-deficient diets.

Preparation of Tissue Slices.—Tissue slices were prepared by the method of Majno and Bunker (1957). Small blocks of liver approximately 8 × 8 mm in size were cut from various lobes of the liver and sliced into sections approximately 0.2 mm thick. Tissue slices were then rinsed in several changes of cold Ringer's phosphate buffer (pH 7.2–7.4). Excess fluid was removed from tissue slices by blotting them on filter paper. One hundred milligrams of liver slices were weighed on a torsion balance scale with a sensitivity of 0.01 g. Four or 5 samples of liver slices were weighed for each animal (Lee and

Lincicome, 1970a). One sample was used for nitrogen determination (Lang, 1958).

Measurement of Oxygen Uptake.—Oxygen consumption was measured by the conventional Warburg technique (Umbreit *et al.*, 1957). Two milliliters of Ringer's phosphate glucose, pH 7.4, were placed in the main compartment of 15 ml reaction flasks. For adsorption of carbon dioxide (CO_2), 0.2 ml of freshly prepared potassium hydroxide (KOH) was placed in the center well. Tissue slices having a combined weight of 100 mg were placed in the flask containing the reaction medium. Oxygen consumption was measured over an interval of 1 hour at 37°C. The gas phase was air. Data in the studies on thiamine metabolism were expressed as microliters of oxygen consumed per hour/per milligram of tissue dry weight (Lincicome and Shepperson, 1965). For the pantothenate studies, the data were expressed as microliters of oxygen per milligram of nitrogen per hour (Lee, 1969; Lee and Lincicome, 1970a).

Thiamine.—Oxygen uptake of rat liver slices was studied in two experiments and rat kidney slices in one experiment. At weekly intervals over a four week period, beginning a week after inoculation of *T. lewisi*, oxygen consumption of rat liver slices was measured. Twelve different rats served as donors each week, 48 rats in all were sacrificed. In another experiment, oxygen uptake by liver and kidney slices was measured at three weekly intervals using 12 rats the first week, 12 the second, and 18 during the third. Only 6 control diet rats were used the fourth week. Rats fed control and thiamine-deficient diets were used.

Pantothenate.—Oxygen uptake of liver slices from mice fed control or pantothenate-deficient diets was studied in two experiments on days 15, 23, 30, 39, 47, and 55.

Measurement of Pyruvate Utilization

Pyruvate utilization by slices of liver from pantothenate-deficient mice was determined by the enzymatic spectrophotometric method of Segal *et al.* (1956). Liver slices were permitted to respire in 15 ml Warburg reaction vessels. The suspending medium was buffered Ringer's phosphate with 5 μmoles of sodium pyruvate added per liter. The reaction was carried out for 1 hour at 37°C under a gas phase of air; at the end of the hour it was stopped by addition of 0.2 ml of 100 % trichloroacetic acid (TCA) per flask. Pyruvate was then assayed in the TCA filtrate.

Preparation of Homogenates and Metabolic Products

Trypanosomes.—Blood which contained dividing trypanosomes was obtained by aseptic heart puncture before any demonstrable ablastin had formed

(Thillet and Chandler, 1957). The trypanosomes were separated by centrifugation (Lincicome and Watkins, 1963) and washed 3 times with physiologic saline solution. They were then suspended in equal parts of serum (either mouse or rat) and physiologic saline solution so that there were 10^8 cells/ml of medium. This cell suspension was then incubated for 24 hours in a waterbath at 27.5°C. With addition of 0.0025 g of glucose per milliliter after 12 hours, the trypanosomes remained actively motile for the whole 24 hours.

After incubation, the majority of cells was separated by centrifugation and the supernatant solution containing the metabolic products of the trypanosomes was filtered through a fritted glass filter of fine porosity and stored in a deep-freeze storage box until ready for use.

The separated trypanosomes were then washed in physiologic saline solution and triturated by repeated freezing and thawing. In the experiments on *T. duttoni*, mice on complete, pantothenate-deficient and pairfed control diets, were given 1 ml of metabolic products intraperitoneally at 3-day intervals. Other mice on the same dietary regimen were given similar injections of 1 ml of triturated trypanosome cells (homogenate) having 10^8 cells per ml. In other experiments, the volumes were reduced to 0.25 ml but the quantities of metabolic products or homogenate remained constant. Inoculations began on day 15 after the initiation of each experiment.

Trichinella Larvae.—Homogenates of *Trichinella* larvae were prepared by the method of Agosin and Aravena (1959). Larvae isolated from donor rats were suspended in 0.15 M potassium chloride (KCl) so that there were 4×10^4 per ml. Two milliliters of this suspension were placed in plastic cups and stored at 0°C. Trituration of larvae in frozen cubes was performed at 2–4°C with a prechilled mortar and pestle followed by homogenization in a Waring blender for 1 minute at 8–10°C. Antibiotics were then added (500 units of penicillin and 500 micrograms of streptomycin per milliliter) before passage through a glass filter of fine porosity and storage at 0°C.

Rats were inoculated with 1 ml of *Trichinella* homogenate intraperitoneally at 5-day intervals. The homogenate was checked for living larvae by feeding homogenate to young rats and testing for living larvae 30–35 days later (Sen, 1969). Metabolic products of *Trichinella* were collected by the method of Mills and Kent (1965). Each milliliter of the final solution contained the metabolic products of 4×10^4 larvae. Experimental rats were each inoculated intraperitoneally with 1 ml of the metabolic products solution at 5-day intervals (Sen, 1969).

Preparation of Inoculum of Trypanosomes or *Trichinella* Larvae

Trypanosomes.—Trypanosomes were harvested and prepared for experiments by the method of Lincicome and Watkins (1963). Inocula varied from 10^2 to 10^6 cells according to specific protocols.

Trichinella.—A donor rat (about 30 days after inoculation) served as the source of worms. The carcass was cut into pieces and ground mechanically before homogenization in a Waring blender (Sen, 1969). Subsequently, larvae were recovered using a modified Baermann apparatus (Kagan, 1960; Sadun, 1955) after peptic digestion, and then washed several times in sterile physiologic saline solution.

Larvae were then counted, after suspension in 2 % agar, by a standard dilution technique and a Stoll pipette. Five larvae per gram of body weight were administered via an esophageal tube to female rats weighing 80 grams.

Measurement of Populations and Reproductive Development of Trypanosomes

Cell populations in peripheral circulation blood of mice and rats and reproductive development of trypanosomes were studied in these hosts fed control (complete) and experimental (vitamin deficient) diets.

Beginning the day after inoculation of trypanosomes, wet blood films of tail blood were prepared daily to determine the time of subsequent appearance of trypanosomes in peripheral circulation. Subsequently, the parasitemias were measured by estimation of the density of trypanosomes by duplicate hemacytometer counts (Lincicome and Hill, 1965).

Numbers of trypanosomes, their lengths, and coefficient of variability in length were used as parameters for judging reproductive development by the technique of Taliaferro and Taliaferro (1922) (Lee and Lincicome, 1970b; Lincicome and Shepperson, 1965; Warsi, 1968).

Measurement of Host Growth

Weight gains and food consumption were determined for all rats and mice under the varied conditions of diet control in these studies. Each animal was individually weighed on a laboratory balance scale with a sensitivity of 0.5 g. Rats were usually weighed at intervals of five days. Gains were expressed as percentages relative to initial weights.

Feeding and Care of Hosts

All experimental hosts were housed individually in suspended wire-bottom cages and fed the appropriate diets from metal feeding cups designed to minimize spillage. Most animals were permitted to feed at will, but pairfed animals were offered control diets daily in amounts equal to that consumed by their experiment (vitamin deficient) mates. The daily intake per animal was determined by subtracting the amount of food remaining in the tared feeding cup from the amount given the previous day. All animals were provided water freely. Water bottles and feeding cups were cleaned daily and cages were scrubbed frequently to minimize algal growth and to prevent possible coprophagy.

RESULTS

Evidence for the hypothesis of the goodness of parasitism accumulated from two general experimental sources. One dealt with contributions of the parasite; the other concerned the host. The former concerned molecular factors supplied by the dependent cell or organism to the host; the latter concerned molecular substances the host gave the parasite.

EVIDENCE FOR CONTRIBUTIONS BY THE PARASITE

The evidence concerning the parasite's contribution to its organic environment (the host) has accumulated from experimental studies of *Trypanosoma duttoni* in the mouse, of *T. lewisi* in the rat, and of *Trichinella spiralis* in the rat. These studies have utilized the metabolic imbalance idea (Lincicome, 1953) in a variety of ways. Categories of data obtained included:

1. Body weight gains of well-fed hosts.
2. Body weight gains of metabolically-imbalanced hosts.
3. Measurements of transketolase and aminotransferases.
4. Pantothenate levels in host tissues.
5. Pantothenic acid content of *Trypanosoma duttoni* cells.
6. Pyridoxine levels in host tissues.
7. Pyridoxine content of *Trypanosoma lewisi* cells.
8. Pyruvate levels and utilization by host tissues.
9. Homogenates.
10. Metabolic products.
11. Food intake of hosts.
12. Longevity of hosts.
13. Respiration of host tissues.

Weight Gains of Well-fed Hosts

Body Weight Gains of Well-fed Mice Infected with Trypanosoma duttoni.—In Table 5 are given the details of one representative experiment showing the

TABLE 5. *Average per cent body weight gains (±SE) of mice inoculated with* Trypanosoma duttoni, *Experiment I. (After Lincicome and Shepperson, 1963)*

Treatment	*Day*						
	8	17	25	34	44	52	58
Control	15±2	31±5	45±5	66±5	69±4	83±6	91±6
Infected (small inoculum)	28±3	62±6	71±7	74±7	80±8	95±9	103±10
Infected (large inoculum)	31±2	67±4	75±5	84±5	96±5	99±6	110±6

percentage gains in body weight of mice inoculated with *T. duttoni* (Lincicome and Shepperson, 1963). Table 6 summarizes three such experiments but shows the increments of percentage gain of trypanosome-infected mice over non-infected ones. The initial weights of these female albino mice are shown in Table 7. For this kind of experiment it is important to select animals with initial body weights having a narrow range of variation. Infected mice were inoculated with two levels of trypanosomes (Table 8). There appeared to be no consistent relationship between the degree of growth stimulation and the number of trypanosomes initially introduced (Table 6). All infected mice developed strong infections (Table 9) but there was no marked disparity in character resulting from smaller and larger inocula between the two populations. Tables 9, 11, and 12 present confirming data that mice inoculated with *T. duttoni* are indeed significantly heavier even though a semisynthetic diet

TABLE 6. *Increments of body weight gains (%) of* Trypanosoma duttoni-*infected mice over uninfected control mice. Student's* t-*test probabilities in parentheses (After Lincicome and Shepperson, 1963)*

Experiment	*Day*	*8*	*17*	*25*	*34*	*44*	*52*	*58*
I	SI[a]	13 (<0.01)	31 (<0.01)	26 (<0.01)	8 (0.1–0.5)	11 (0.1–0.5)	12 (0.1–0.5)	12 (0.1–0.5)
	LI[b]	16 (<0.1)	36 (<0.1)	30 (<0.01)	18 (0.02–0.05)	27 (<0.01)	16 (0.05–0.10)	19 (0.02–0.05)
	Day	*7*	*13*	*19*	*25*	*31*		
II	SI	0	1(NS)[c]	7(0.1)	5(<0.1)	8(<0.1)		
	LI	0	0	0	0	3(NS)		
	Day	*7*	*13*	*19*	*25*	*29*		
III	SI	0	8(0.1–0.5)	16(0.1–0.5)	9(NS)	5(NS)		
	LI	0	2(NS)	2(0.1–0.5)	0	0		

[a] Smaller inoculum
[b] Larger inoculum
[c] Not significant

TABLE 7. *Average initial body weights. SE and numbers of mice employed in parentheses (After Lincicome and Shepperson, 1963)*

Experiment	*Controls*	*Smaller Inoculum*	*Larger Inoculum*
I	11.2±0.2 (13)	11.6±0.1 (10)	11.4±0.2 (13)
II	10.6±0.1 (20)	10.6±0.2 (14)	10.6±0.2 (16)
III	9.6±0.3 (17)	9.6±0.2 (16)	10.2±0.2 (13)

replaced the commercially available whole laboratory chow of earlier work (Lincicome and Lee, 1970a; Lee, 1969). Infected mice had as much as 35% greater weight gain (Table 10) than did normal control animals.

TABLE 8. *Numbers of trypanosome cells* (Trypanosoma duttoni) *inoculated into female albino mice (Lincicome and Shepperson, 1963)*

Protocol	*Cells in smaller inoculum* $\times 10^4$	*Cells in larger inoculum* $\times 10^5$
I	4.0	4.0
II	5.3	5.3
III	3.8	3.8

TABLE 9. *Average hemacytometer estimations of* Trypanosoma duttoni *populations in peripheral tail blood of mice. A factor of 2,000 will convert all data to numbers of cells/mm*^{3} *of blood (Lincicome and Shepperson, 1963)*

Day	*Exp. I*		*Exp. II*		*Exp. III*	
	S[a]	*L*[b]	*S*	*L*	*S*	*L*
4	3	3	<1	<1		
5			<1	<1	<1	2
6			1	2	2	3
7	4	5	2	3	4	6
8	4	3	5	6	8	14
9			5	6	5	9
10			5	7	3	4
11	3	3	10	8	5	4
12	2	2	8	5	10	6
13			10	4	8	7
14			4	4	8	7
15			3	3	6	4
16			<1	2	4	3
17	<1	1	<1	<1	2	2
18	<1	<1	<1	<1	1	1
19			<1	<1	<1	<1
20	+[c]	+	<1	<1	<1	<1
21	—[d]	—	—	<1	<1	—
22				<1	—	

[a] Smaller inoculum
[b] Larger inoculum
[c] Refers to presence of trypanosomes only
[d] No trypanosomes demonstrable

TABLE 10. *Average per cent body weight gains (±S.D.) of mice given a semisynthetic full complement diet and infected with* Trypanosoma duttoni, *Experiment I (Lincicome and Lee, 1970)*

	Day 5	*10*	*15*	*20*	*25*	*30*	*35*	*40*	*45*	*50*
A	25±3	37±3	41±4	67±3	89±5	114±3	143±4	171±7	186±5	212±6
B	30±2	36±1	40±2	63±3	80±6	93±5	121±6	136±3	162±2	189±7
C	0	1	1	4	9	21	22	35	24	23
D	N.S.	N.S.	N.S.	S*	S	S	S	S	S	S

A=Six infected mice
B=Six uninfected mice
C=Percentage increase over controls
D=Statistical significance (S) at 1% Levels
N.S.=not significant
*=Significant at 5% level

TABLE 11. *Average per cent body weight gains (±S.D.) of mice given a semisynthetic full complement diet and infected with* Trypanosoma duttoni, *Experiment II (After Lee, 1969)*

	Day 5	*10*	*15*	*20*	*25*	*30*	*35*	*40*	*45*	*50*
A	30±3	48±5	57±2	84±2	122±4	145±4	166±2	179±5	208±3	226±5
B	28±3	49±4	60±2	87±5	120±6	133±2	148±2	165±3	195±2	206±4
C	2	0	0	0	2	12	18	14	13	20
D	N.S.	N.S.	N.S.	N.S.	N.S.	S	S	S	S	S

A=Six infected mice
B=Six uninfected mice
C=Percentage increase over controls
D=Statistical significance (S) at 1% level
N.S.=not significant

TABLE 12. *Average per cent body weight gains (±S.D.) of mice given a semisynthetic full complement diet and infected with* Trypanosoma duttoni, *Experiment III (After Lee, 1969)*

	Day 5	*10*	*15*	*20*	*25*	*30*	*35*	*40*	*45*	*50*
A	36±2	45±4	74±3	89±2	119±7	145±5	165±6	184±2	219±7	232±3
B	39±5	40±6	72±2	87±3	116±4	130±4	152±2	163±3	190±2	219±5
C	0	5	2	2	3	15	13	21	29	13
D	N.S.	N.S.	N.S.	N.S.	N.S.	S	S	S	S	S

A=Six infected mice
B=Six uninfected mice
C=Percentage increase over controls
D=Statistical significance (S) at 1% level
N.S.=not significant

Body Weight Gains of Well-fed Rats Inoculated with Trypanosoma lewisi.—The original observations of increased growth of a host inoculated with a parasitic organism were made by Lincicome *et al.* (1960, 1963) and Lincicome and Shepperson (1961). The numbers and initial body weights of rats used in these experiments are shown in Table 13, representing weanling (56g), mature (184g), and intermediate (77–111g) sizes. Six experiments were performed with beginning dates in February, April, May, June, October, and November, respectively, thus representing each season. The number of trypanosomes ranged from 0.12×10^6 to 5.2×10^6 cells (Table 14). Weanling rats grow exponentially (Table 15) but grew even faster when inoculated with *T. lewisi*. Older, mature rats have a reduced rate of growth, and the stimulation provided by the trypanosome was therefore small. Table 16 summarizes the results of these six experiments.

In experiments dealing with thiamine metabolism, study of the control rats (Table 17) which were fed a full complement, semisynthetic diet provided further evidence of stimulation of body growth (Table 18) in those animals carrying *T. lewisi* (Lincicome and Shepperson, 1965).

TABLE 13. *Numbers and initial average weights (±SE) of female albino rats used in experiments on body weight gains of rats inoculated with* Trypanosoma lewisi *(After Lincicome* et al., *1960, 1963)*

	Average body weights in grams		
Experiment number	*Control*	*Smaller Inoculum*	*Larger Inoculum*
1	84±0.9[a] (9)[b]	84±0.8 (9)	84±0.7 (9)
2	56±0.7 (9)	56±0.6 (10)	56±0.3 (10)
3	180±1.7 (10)	180±0.2 (7)	180±0.8 (10)
4	184±0.9 (10)	184±1.1 (9)	185±0.9 (9)
5	111±1.5 (10)	110±1.5 (11)	110±1.4 (12)
6	77±0.9 (12)	78±0.9 (12)	78±0.9 (13)

[a] Standard error
[b] Number of rats employed indicated in parentheses

TABLE 14. *Number of trypanosome cells inoculated into female albino rats for study of body weight gains (After Lincicome* et al., *1960, 1963)*

Experiment Number	*Number of trypanosomes inoculated* ($\times 10^6$) *Smaller inoculum*	*Larger inoculum*
1	0.2	4.0
2	0.2	4.0
3	0.2	4.0
4	0.2	4.0
5	0.52	5.2
6	0.12	1.2

Similarly, in pyridoxine metabolism studies, rats given a semisynthetic, full complement diet showed a range of 11–20 % stimulation of growth in 5 different experiments (Table 20). This was statistically significant compared to the control growth. The results of this series of experiments were more consistent than those of prior studies. All experimental animals were inoculated with 10^2 cells.

TABLE 15. *Cumulative percentage body weight gains of weanling female albino rats inoculated with* Trypanosoma lewisi *(After Lincicome* et al., *1960, 1963)*

	Controls					*Smaller inoculum*					*Larger inoculum*			
	Day					*Day*					*Day*			
Animal	*7*	*15*	*25*	*34*	*Animal*	*7*	*15*	*25*	*34*	*Animal*	*7*	*15*	*25*	*34*
1	25	78	136	181	10	26	83	135	172	20	22	57	131	177
2	20	72	140	186	11	34	85	134	180	21	33	110	183	221
3	32	76	111	151	12	40	93	127	155	22	32	97	172	232
4	26	82	139	180	13	36	93	151	195	23	18	80	146	196
5	30	90	135	190	14	35	100	155	213	24	27	91	154	199
6	19	59	109	161	15	37	100	165	203	25	23	78	148	207
7	21	61	112	160	16	31	79	134	185	26	22	76	142	192
8	35	98	165	211	17	42	114	168	220	27	25	80	156	210
9	18	58	90	124	18	37	85	121	167	28	30	86	148	210
					19	34	97	156	199	29	21	75	137	184
$\overline{X}$	25	75	126	172		35	93	145	189		25	83	152	203
SE	2	5	7	8		1	3	5	6		2	4	5	5

TABLE 16. *Increment of growth increases over controls (%) of female albino rats inoculated with* Trypanosoma lewisi *(After Lincicome* et al., *1963) See* TABLE *13 for weights of hosts*

Experiment 1

Inoculum	*Day 8*	*14*	*20*	*28*
0.2×10^6	0	3	2	4
$t=$		0.689	0.349	0.541
4.0×10^6	0	1	2	2
$t=$		0.194	0.310	0.240

Experiment 4

Inoculum	*Day 7*	*13*	*19*	*25*
0.2×10^6	2	3	1	1
t	not calculated			
4.0×10^6	0	0	0	0
$t=$	not calculated			

Experiment 2

Inoculum	*Day 7*	*7*	*15*	*25*
0.2×10^6	10	18	19	17
$t=$	4.072[a]	3.201[a]	2.170[a]	1.566
4.0×10^6	0	8	26	31
$t=$		1.40	2.94[a]	3.089[a]

Experiment 5

Inoculum	*Day 7*	*13*	*22*	*27*
5.2×10^5	2	6	10	12
$t=$	1.0	1.22	1.59	1.843[a]
5.2×10^6	3	5	8	9
$t=$	1.666	1.292	1.626	1.625

Experiment 3

Inoculum	*Day 7*	*15*	*23*	*31*
0.2×10^6	0	2	2	1
$t=$	not calculated			
4.0×10^6	2	2	2	1
$t=$	not calculated			

Experiment 6

Inoculum	*Day 7*	*13*	*19*	*28*	*36*
12×10^4	1	6	10	13	14
$t=$	0.534	2.10[a]	2.646[a]	2.677[a]	2.698[a]
12×10^6	1	3	5	9	11
$t=$	0.567	1.041	1.149	1.619	1.598

[a] Significant at 5% or below.

Body Weight Gains of Well-fed Rats Inoculated with Trichinella spiralis.—Eight experiments were performed using female, albino rats infected with *Trichinella spiralis.* In the first three experiments the average weights of the animals was 80 ± 4g, and the results of analysis of growth of these rats were not statistically significant. In experiments 4,5,6,7, and 8 only those rats whose initial body weights fell within a very narrow range of 80 ± 1g were studied. Table 21 gives the results of these latter experiments. These animals were maintained on either a commercial pellet diet or a semi-synthetic but full complement (complete) ration and were given 5 trichina larvae per gram of body weight. Their body weights were examined over a period of 55 days. Those rats inoculated with *Trichinella* were significantly heavier than normal controls (Table 21). This weight advantage ranged up to 28 %.

Weight Gains of Metabolically Imbalanced Hosts

Body Weight Gains of Mice Fed a Pantothenate-deficient Diet and Inoculated with Trypanosoma duttoni.—Table 22 shows the results of one experiment representative of three (Lincicome and Lee, 1970a) in which infected pantothenate-deficient mice had significantly better weight gains

TABLE 17. *Distribution, initial body weights (±SE), and number of female albino rats employed in studies on thiamine metabolism in* Trypanosoma lewisi *infections (After Lincicome and Shepperson, 1965)*

Experiment No.	*1*	*2*	*3*	*4*	*5*	*6*	*Total no. of Rats*
Control diet	50±0.6	50±1.8	59±0.1	52±0.3	54±0.6	48±0.4	45
Noninfected rats	(4)	(5)	(6)	(10)	(10)	(10)	
Infected rats	50±1.5	51±0.4	59±0.09	52±1.0	54±0.3	47±0.5	
	(3)	(5)	(5)	(10)	(10)	(10)	43
Thiamine deficient	50±0.4	51±0.5	59±0.1	53±0.4	55±0.9	48±0.2	
Noninfected rats	(4)	(4)	(6)	(10)	(10)	(10)	44
Infected rats	50±1	50±0.1	59±0.1	52±0.7	56±0.1	47±0.3	
	(4)	(5)	(6)	(10)	(10)	(10)	45
Pair-fed	49±1.0	51±0.5	59±0.1	52±0.5	54±1.0	48±0.6	
Noninfected rats	(4)	(4)	(6)	(10)	(10)	(10)	44
Infected rats	49±0.6	50±0.2	59±0.1	53±0.9	55±0.8	47±1	
	(4)	(5)	(6)	(10)	(10)	(10)	45
23% Fat control diet	—	—	59±0.1	—	—	—	
Noninfected rats			(6)				
Infected rats	—	—	59±0.1	—	—	—	
			(6)				6
23% Fat thiamine-deficient diet	—	—	59±0.1	—	—	—	
Noninfected rats			(6)				6
Infected rats	—	—	59±0.1	—	—	—	
			(6)				6
Total No. of rats	*23*	*28*	*59*	*60*	*60*	*60*	*290*

than normal uninoculated control mice. The advantage in growth was initially demonstrated by the fourth day after the beginning of the experimental period and increased to a maximum advantage of nearly 20 % on the fortieth day. The infected mice on the fiftieth day, long after the trypanosomes had disappeared from the peripheral, circulating blood, still showed a 10 % increment of growth beyond the controls.

Other controls were used in this series of experiments. Tables 23 and 24 show additional evidence supporting the observations in Table 22. Mice given a supplement of pantothenic acid to their pantothenate-deficient diet, showed a significant stimulation of growth as a result of infection with *T. duttoni*. Similarly mice pairfed as caloric controls showed the same phenomenon as the wholly-deficient animals: marked stimulation of growth associated with infection with the trypanosome which was of the order observed for the experimental animals (Table 22).

TABLE 18. *Average per cent body weight gains (±SE) of female albino rats fed a full complement, semi-synthetic ration in studies on thiamine metabolism in* Trypanosoma lewisi *infections (After Lincicome and Shepperson, 1965). See* TABLE *17 for numbers and weights of animals*

		Day 15	21	27	33
Expt. 1					
Controls		26±6	61±4	88±3	108±5
Experimentals		30±6	75±2	93±6	118±7
% Increase over controls		4	14	5	10
	$t=$	0.3635	1.6958	0.6558	0.9833
	$p=$	0.8	0.2	0.6	0.4
Expt. 2					
Controls		14±2	37±4	63±3	85±3
Experimentals		18±4	40±4	61±3	82±3
% Increase over controls		4	3	0	0
	$t=$	0.5914	0.3707	0	0
	$p=$	0.6	0.8	0	0
Expt. 3					
Controls		11±2	34±3	54±2	74±2
Experimentals		10±2	33±3	51±2	70±3
% Increase over controls		0	0	0	0
	$t=$	Not calculated			
Expt. 4					
Controls		15±3	36±3	60±2	77±1
Experimentals		17±3	40±3	63±2	82±1
% Increase over controls		2	4	3	5
	$t=$	0.2785	0.6786	0.5326	1.6051
	$p=$	0.8	0.6	0.6	0.2
Expt. 5					
Controls			11±2	30±2	51±2
Experimentals			13±2	31±2	52±2
% Increase over controls			2	1	1
	$t=$		0.5557	0.2088	0.1683
	$p=$		0.7	0.9	0.9
Expt. 6					
Controls		18±4	42±3	61±4	77±3
Experimentals		18±4	46±4	71±5	96±3
% Increase over controls		0	4	10	19
	$t=$	—	0.4356	1.6620	4.6360
	$p=$		0.7	0.2	0.01

Body Weight Gains of Rats Fed a Thiamine-deficient Diet and Inoculated with Trypanosoma lewisi.—Table 19 shows that thiamine-deficient rats inoculated with *T. lewisi* (Lincicome and Shepperson, 1965) have up to an 11 % body weight advantage over uninoculated animals. Thiamine-deficient animals lost weight throughout the experimental period, but this loss is reduced in animals inoculated with trypanosomes; this indirectly indicates that the dependent cell is supplying a factor which under these controlled

TABLE 19. *Average per cent body weight gains (±SE) of female albino rats fed a thiamine-deficient ration and infected with* Trypanosoma lewisi. *(After Lincicome and Shepperson, 1965). See* TABLE *17 for numbers and weights of animals*

	Day 15	*21*	*27*	*33*
Expt. 1				
Controls	5±0.6	−0.15±3.0	−12±2.0	−22±1.0
Experimentals	6±1.0	0.12±1.0	−9±2.0	−20±2.0
% Advantage over controls	1	0	3	2
$t=$	0.6710	—	0.8776	0.7239
$p=$	0.6	—	0.5	0.6
Expt. 2				
Controls	5±0.8	−0.6±1	−15±2	−26±0.3
Experimentals	6±0.5	3±1	−7±0.9	−15±0.9
% Advantage over controls	1	4	8	11
$t=$	0.7176	2.2448	2.4095	5.5000
$p=$	0.6	0.1	0.1	0.01
Expt. 3				
Controls	−1.17±0	−7±1	−13±2	−28±1
Experimentals	2±1	−6±2	−13±2	−24±2
% Advantage over controls	3	1	0	4
$t=$	1.5741	0.4977	—	1.5372
$p=$	0.2	0.7	—	0.2
Expt. 4				
Controls	5±0.7	1±2	−14±2	−27±1
Experimentals	8±1	6±2	−10±1	−21±1
% Advantage over controls	3	5	4	6
$t=$	1.3176	1.0833	1.0404	2.0389
$p=$	0.3	0.4	0.4	0.2
Expt. 5				
Controls		−10±1	−25±2	−35±1
Experimentals		−6±1	−19±1	−30±0.5
% Advantage over controls		4	6	5
$t=$		1.2019	1.3047	4.9593
$p=$		0.3	0.3	0.01
Expt. 6				
Controls	2±0.1	−3±1	−20±3	−30±1
Experimentals	6±1	−2±2	−19±2	−19±0.7
% Advantage over controls	4	1	1	3
$t=$	2.6596	0.3456	0.1834	2.0677
$p=$	0.1	0.8	0.9	0.2

conditions appears to be the thiamine molecule. Table 17 gives details of the structuring of these experiments.

Body Weight Gains of Rats Fed a Pyridoxine-deficient Diet and Inoculated with Trypanosoma lewisi.—In Table 25 it is shown that the body weight gain of pyridoxine-deficient rats is stimulated by as much as 16% if the hosts are

TABLE 20. *Average per cent body weight gains of female albino rats (±SD) given a full complement semisynthetic diet and inoculated with trypanosomes in pyridoxine metabolism studies of* Trypanosoma lewisi *infections. (After Warsi, 1969). Initial body weight=50–52g*

Expt. 1	*Day 7*	*14*	*21*	*28*	*35*	*42*	*49*	*No. of Rats Used*
Controls (uninfected)	—	—	112±5	167±5	180±4	226±9	257±8	6
Experimentals (infected)	—	—	113±3	173±4	195±6	243±7	277±10	6
Increase over controls			1	6	15	17	20	
Statistical significance at 1% level			NS	NS	S	S	S	
Expt. 2								
Controls (uninfected)	—	—	111±3	154±9	186±7	218±6	245±5	6
Experimentals (infected)	—	—	110±3	168±3	199±4	229±3	260±5	6
Increase over controls	—	—	−1	14	13	11	15	
Statistical significance at 1% level			NS	S	S	S	S	
Expt. 3								
Controls (uninfected)	28±3	75±4	129±4	162±5	190±6	228±4	265±5	6
Experimentals (infected)	27±4	73±3	130±3	169±2	200±4	236±3	276±2	6
Increase over controls	−1	−2	1	7	10	8	11	
Statistical significance at 1% level	NS	NS	NS	NS	S	S	S	
Expt. 4								
Controls (uninfected)	25±3	73±2	118±2	165±5	207±5	249±7	270±5	6
Experimentals (infected)	27±2	74±3	117±5	168±9	219±2	264±4	288±6	6
Increase over controls	2	1	−1	3	12	15	18	
Statistical significance at 1% level	NS	NS	NS	NS	S	S	S	
Expt. 5								
Controls (uninfected)	27±2	72±3	123±4	162±3	192±3	226±4	267±3	6
Experimentals (infected)	29±4	71±4	122±2	166±2	205±4	237±3	285±7	6
Increase over controls	2	−1	−1	4	13	11	18	
Statistical significance at 1% level	NS	NS	NS	NS	S	S	S	*Total=60*

NS=Not statistically significant S=Significant

TABLE 21. *Average per cent body weight gains of rats (±SD) fed commercial pellet or semi-synthetic, full-complement diets and inoculated with* Trichinella spiralis. *(After Sen, 1969). Initial body weights of rats=80±1g*

Expt. 4	*Day 10*	*20*	*30*	*40*	*55*	*No. of Rats Used*
Controls (uninfected)	53±2	93±4	118±3	139±2		10
Experimentals (infected)	53±3	98±3	123±1	144±2		10
Increase over control	0	5	5	5		
Statistical significance at 1% level	NS	S	S	S		
Expt. 5						
Controls (uninfected)	51±6	96±1	119±3	139±2		10
Experimentals (infected)	51±6	99±2	128±6	148±8		10
Increase over controls	0	3	9	9		
Statistical significance at 1% level	NS	S	S	S		
Expt. 6						
Controls (uninfected)	75±4	112±5	138±5	155±5	174±2	6
Experimentals (infected)	77±2	114±4	139±2	175±3	201±3	6
crease Inover controls	2	2	1	20	27	
Statistical significance at 1% level	NS	NS	NS	S	S	
Expt. 7						
Controls (uninfected)	67±6	109±3	138±5	152±10	167±6	6
Experimentals (infected)	61±8	110±13	136±4	164±9	193±10	6
Increase over controls	0	1	0	12	26	
Statistical significance at 1% level	NS	NS	NS	S	S	
Expt. 8						
Controls (uninfected)	70±8	114±5	137±5	156±3	172±8	6
Experimentals (infected)	61±7	110±5	139±2	166±7	200±9	6
Increase over controls	0	0	2	10	28	
Statistical significance at 1% level	NS	NS	NS	S	S	*Total=76*

NS=Not statistically significant S=Statistically significant at 5% level or less

TABLE 22. *Average per cent body weight gains (±SD) of mice given a pantothenate-deficient diet and infected with* Trypanosoma duttoni. *(After Lincicome and Lee, 1970)*

	Day 5	*10*	*15*	*20*	*25*	*30*	*35*	*40*	*45*	*50*
A	12±2	19±3	25±3	34±4	42±2	46±2	48±3	44±3	37±2	33±4
B	15±4	20±2	27±4	30±5	35±3	38±4	33±6	26±4	29±2	24±3
C	0	0	0	4	7	8	15	18	8	9
D	NS	NS	NS	NS	S	S	S	S	S	S

A=Infected
B=Uninfected
C=Percentage increase over controls
D=Statistical significant (S) at 1% level
N.S.=Not singificant

TABLE 23. *Average per cent body weight gains (±SD) of mice given a pantothenate-deficient diet supplemented with pantothenic acid. (After Lincicome and Lee, 1970)*

	Day 5	*10*	*15*	*20*	*25*	*30*	*35*	*40*	*45*	*50*
A	17±3	20±5	24±2	31±4	34±3	41±2	43±4	35±3	38±2	34±2
B	15±4	20±2	27±4	30±5	35±3	38±4	33±6	26±4	29±2	24±3
C	2	0	0	1	0	3	10	9	9	10
D	NS	NS	NS	NS	NS	S	S	S	S	S

A=Experimental
B=Control
C=Percentage increase over control
D=Statistical significance (S) at 1% level
NS=Not significant

TABLE 24. *Average per cent body weight gains (±SD) of pair-fed control mice infected with* Trypanosoma duttoni. *(After Lincicome and Lee, 1970)*

	Day 5	*10*	*15*	*20*	*25*	*30*	*35*	*40*	*45*	*50*
A	21±4	29±2	37±2	49±1	68±6	84±3	98±2	106±6	118±3	129±2
B	23±2	27±4	38±7	51±5	66±2	72±5	89±3	94±2	100±5	114±5
C	0	2	1	0	2	12	9	12	18	15
D	NS	NS	NS	NS	NS	S	S	S	S	S

A=Infected
B=Uninfected
C=Percentage increase over controls
D=Statistical significance (S) at 1% level
NS=Not significant

TABLE 25. *Average per cent body weight gains of pyridoxine-deficient rats (±SD) inoculated with* Trypanosoma lewisi. *(After Warsi, 1968)*

	Day 7	*14*	*21*	28	35	42	49
Expt. 1							
Controls (uninfected)	—	—	64±4	72±2	89±4	96±2	99±6
Experimentals (infected)	—	—	66±4	79±5	99±4	104±3	115±6
Increase over controls	—	—	2	7	10	8	16
Statistical significance at 1% level	—	—	NS	NS	S	S	S
Expt. 2							
Controls (uninfected)	—	—	64±4	75±4	83±3	94±4.6	97±4
Experimentals (infected)	—	—	65±4	80±5	94±3	109±4	111±5
Increase over controls	—	—	1	5	11	15	14
Statistical significance at 1% level	—	—	NS	NS	S	S	S
Expt. 3							
Controls (uninfected)	27±3	39±3	53±3	66±3	75±3	84±3	89±3
Experimentals (infected)	26±2	42±3	54±3	73±2	83±3	95±3	101±5
Increase over controls	1	3	1	7	8	11	12
Statistical significance at 5% level	NS	NS	NS	S	S	S	S
Expt. 4							
Controls (uninfected)	24±4	37±8	53±4	64±7	65±7	68±5	69±6
Experimentals (infected)	22±2	39±3	55±4	68±4	72±1	76±2	84±3
Increase over controls	−2	2	2	4	7	8	15
Statistical significance at 5% level	NS	NS	NS	S	S	S	S

NS=Not statistically significant S=Statistically significant

inoculated with *T. lewisi.* Observations in four experiments were consistently in this range 49 days after inoculation when the parasitemias were no longer detectable. These results indicate that the dependent cell supplies the pyridoxine molecule or other substance having pyridoxine-like activity which the host is able to utilize (Warsi, 1968). Pairfed control rats, as well as rats fed a pyridoxine-deficient diet to which an adequate amount of pyridoxine had been added, all showed significant weight gain stimulation of up to 13 % and 11 % respectively, provided they were inoculated with *T. lewisi*. These data are not presented here but may be found in A. A. Warsi's (1968) doctoral thesis.

Body Weight Gains of Rats Fed a Pyridoxine-deficient Diet and Inoculated with Trichinella spiralis.—In Table 26 are shown the principal results of

TABLE 26. *Average per cent body weight gains of rats (±SD) fed a pyridoxine-deficient diet and inoculated with* Trichinella spiralis. *(After Sen, 1969). Average initial weight of rats = 80±1g*

Expt. 6	*Day 15*	*25*	*35*	*45*	*55*	*No. of Rats Used*
Controls (uninfected)	16±4	23±3	26±4	31±1	36±1	6
Experimentals (infected)	16±3	23±3	35±2	56±2	74±1	6
Increase over controls	0	0	9	25	38	
Statistical significance at 1% level	NS	NS	S	S	S	
Expt. 7						
Controls (uninfected)	13±1	20±1	26±2	32±2	35±1	6
Experimentals (infected)	15±2	24±1	40±3	58±2	71±4	6
Increase over controls	2	4	14	26	36	
Statistical significance at 1% level	NS	NS	S	S	S	
Expt. 8						
Controls (uninfected)	15±2	20±1	26±3	30±1	35±1	6
Experimentals (infected)	15±2	21±2	31±3	49±4	65±3	6
Increase over controls	0	1	5	19	30	
Statistical significance at 1% level	NS	NS	NS	S	S	*Total=36*

NS=Not statistically significant
S=Statistically significant

three experiments on body weight gains of pyridoxine-deficient rats inoculated with *Trichinella*. The infected animals received 5 larvae per gram of body weight (Sen, 1969) and at the 55th day after inoculation showed up to 38 % better weight gains than uninoculated rats. Significantly different weight gains occurred at the 35th day after exposure, but not at the 25th. Pairfed controls not listed in Table 26 also showed significant stimulation of the growth, up to 22 % at the 55th day (Sen, 1969). Similarly, control animals on a pyridoxine-supplemented diet (the deficient diet supplemented with the same amount of the vitamin as the full complement diet) showed significant growth stimulation which shows that the re-supply of the vitamin did in fact produce the same kind of growth stimulation (Sen, 1969).

Transketolase and Aminotransferases

Transketolase Levels in Homogenates of Rat Liver and Kidney Tissue Infected with Trypanosoma lewisi.—The levels of transketolase activity in rat liver and kidney tissue are shown in Tables 27 and 28. This activity was

TABLE 27. *Transketolase activity of rat liver homogenates; data expressed as mg of hexose formed per gram of tissue per hour ±SE. (After Lincicome and Shepperson, 1965)*

Control diet	*Week 1*	*2*	*3*	*4*	*Homogenate sample*
		52±2.0	40±0.2	43±0.3	No addition
		52±0	42±0.8	43±0.9	With added thiamine
Uninfected	43±1.0	47±0.8	44±1.0		No addition
	42±0.9	45±0.3	44±0.2		With added thiamine
Infected	44±0.8	50±0.4	49±2.0	44±0.5	No addition
	44±0.2	51±0.3	48±0.7	42±0.6	With added thiamine
		41±0.3	40±0.5		No addition
		40±2.0	40±0.6		With added thiamine
Thiamine-deficient diet		18±0.7	12±1.0	9±0.7	No addition
		35±2.0	—	28±0.7	With added thiamine
Uninfected	24±1.0	9±1.0	6±0.4		No addition
	41±0.2	25±0.3	28±0.6		With added thiamine
Infected	24±2.0	23±0.9	18±0.8	12±0.3	No addition
	45±0.4	31±0.3	36±0.7	22±0.6	With added thiamine
		12±0.5	10±0.6		No addition
		28±0.8	31±0.9		With added thiamine
Pair-fed controls		27±0.4	20±0.7	18±0.5	No addition
		35±0.9	29±0.4	28±0.1	With added thiamine
Uninfected	33±0.7	20±0.7	18±0.6		No addition
	37±0.2	31±0.4	31±0.9		With added thiamine
Infected	33±0.8	25±0.6	22±0.5	14±0.6	No addition
	35±0.7	30±0.4	29±0.4	20±0.9	With added thiamine
		26±0.8	23±0.6		No addition
		35±0.3	31±0.2		With added thiamine

TABLE 28. *Transketolase activity of rat kidney homogenates; data expressed as mg of hexose formed per gram of tissue per hour ±SE. (After Lincicome and Shepperson, 1965)*

Control diet	*Week 1*	*2*	*3*	*4*	*Homogenate Sample*
		30±0.7	26±2.0	24±0.3	No addition
		29±0.8	24±0.8	23±1.0	With added thiamine
Uninfected	26±0.6	25±0.5	26±0.2		No addition
	29±0.9	27±0.8	26±0.6		With added thiamine
Infected	33±0.9	31±0.3	28±0.5	23±0.8	No addition
	32±1.0	31±0.4	28±0.2	21±0.3	With added thiamine
		29±0.4	28±0.9		No addition
		28±0.3	29±0.7		With added thiamine
Thiamine-deficient diet		6±1.0	3±0.1	2±0.5	No addition
		13±0.5	17±0.6	15±0.8	With added thiamine
Uninfected	8±0.9	5±0.8	4±0.6		No addition
	19±0.2	17±0.5	15±0.4		With added thiamine
Infected	12±0.7	9±0.2	5±0.3	4±0.3	No addition
	23±0.8	21±0.9	18±0.5	19±0.7	With added thiamine
		7±0.9	8±1.0		No addition
		13±0.7	15±0.2		With added thiamine
Pair-fed controls		13±0.6	9±0.1	8±0.4	No addition
		20±0.6	17±0.5	18±0.5	With added thiamine
Uninfected	19±0.9	11±1.0	7±0.4		No addition
	23±0.3	21±0.9	16±0.3		With added thiamine
Infected	14±0.9	10±1.0	11±0.9	12±0.6	No addition
	23±0.2	19±0.7	15±0.2	15±0.8	With added thiamine
		12±0.6	9±1.0		No addition
		19±0.5	20±0.8		With added thiamine

based upon the amount of hexose formed. Thiamine-deficient infected liver homogenates showed greater activity than did uninfected deficient tissue. During the second, third, and fourth weeks of the *T. lewisi* infection, 3–5 more milligrams of hexose were formed by infected, deficient tissue than by uninfected deficient tissue. Similar results (Table 28) were obtained for rat kidney homogenates. Hexose formed by homogenates of kidneys from rats fed the complete diet was about the same as for liver. Thiamine-deficient kidney homogenates formed more hexose (2–4mg) than did those of uninfected controls. In pairfed (inanition) controls, the transketolase (E.C. 2.2.1.1) activity decreased with time, but infected rats' tissues showed greater enzyme activity during the third and fourth weeks (Lincicome and Shepperson, 1965).

Aminotransferases in Serum of Rats Fed Pyridoxine-deficient Diets and Infected with Trypanosoma lewisi.—In Tables 29 and 30 are shown the ac-

TABLE 29. *Serum glutamic-pyruvic transaminase (GPT) levels in rats fed complete and pyridoxine-deficient diets and inoculated with* Trypanosoma lewisi *(±SD). (After Warsi, 1968)*

Day	Complete Diet		Pyridoxine-Deficient Diet	
	Uninfected M.U.[b]	Infected M.U.	Uninfected M.U.	Infected M.U.
1	11±0.5	—	12±1.0	—
7	12±1.1	—	9±1.5	—
14	11±1.0	—	6±1.5	—
21[a]	13±0.3	12±1.0	4±0.3	4±0.3
28	12±0.7	14±1.0	3.5±0.3	5±0.5
35	11±0.7	19±1.5	3.0±0.3	8.5±1.0
42	11±1.5	15±1.0	2.3±0.5	7±0.5
49	11±0.7		2.4±0.4	3±0.5

[a] Day 21 is the day of inoculation of trypanosomes
[b] M.U.=Milliunit of enzyme activity

tivities of two aminotransferases in the serum of rats fed complete and pyridoxine-deficient diets. Glutamic-pyruvic transaminase (GPT) (E.C. 2.6.1.2) levels in rats on a complete diet were steady throughout the study. In rats fed the deficient diet, the enzyme levels decreased as expected. But in both well-fed and deficient rats the enzyme levels in trypanosome-infected rats increased, showing that either the enzyme or its pyridoxine cofactor were probably contributed by the trypanosome.

These data are representative of a larger series of experiments (Warsi, 1968).

TABLE 30. *Serum glutamic-oxalacetic transaminase (GOT) levels in rats fed complete and pyridoxine-deficient diets and inoculated with* Trypanosoma lewisi *(±SD). (After Warsi, 1968)*

Day	Complete Diet		Pyridoxine Deficient Diet	
	Uninfected M.U.[b]	Infected M.U.	Uninfected M.U.	Infected M.U.
1	32±1.5	—	33±1.0	—
7	30±2.0	—	29±1.5	—
14	30±1.0	—	23±1.5	—
21[a]	33±2.3	32±2.2	19±1.5	19±1.3
28	32±1.4	35±1.2	19±1.5	26±1.5
35	32±1.0	44±1.8	16±1.2	28±1.4
42	30±1.4	34±1.3	15±1.5	22±1.5
49	31±1.1	32±1.2	13±1.0	15±1.2

[a] Day 21 is the day of inoculation of trypanosomes
[b] M.U.=Milliunit of enzyme activity

Aminotransferases in Serum of Rats Fed Pyridoxine-deficient Diets and Infected with Trichinella spiralis.—In Table 31 are shown the levels of GPT in the sera of rats maintained on complete and pyridoxine-deficient diets. During the period from the 28th to the 42nd day after exposure of animals to *Trichinella*, there were significant increases of this aminotransferase over that of controls in those rats fed a control, full complement diet. This was also true for animals on a pyridoxine-deficient diet except that on the 49th day infected animals also showed a higher level of enzyme activity. This is interpreted to mean that either *Trichinella* is supplying these enzymes intact or is supplying sufficient pyridoxine to activate host enzymes inactive because of lack of the cofactor.

TABLE 31. *Glutamic pyruvic transaminase (GPT) levels in serum of rats inoculated with* Trichinella spiralis *and fed semisynthetic complete and pyridoxine-deficient diets (data in milliunits ±SD). (After Sen, 1969)*

Full Complement Diet	*Day 14*	*21*	*28*	*35*	*42*	*49*
Controls (uninfected)	12±1	12±1	11±1	12±1	12±1	12±1
Experimentals (infected)	11±1	12±1	14±1	20±2	14±1	11±1
Increase over controls	0	0	3	8	2	0
Statistical significance at 1% level	NS	NS	S	S	S	NS
Pyridoxine-Deficient Diet						
Controls (uninfected)	9±1	7±1	5±1	4±1	4±1	3±1
Experimentals (infected)	9±1	8±1	10±1	11±1	12±1	10±1
Increase over controls	0	1	5	7	8	7
Statistical significance at 1% level	NS	NS	S	S	S	S

NS = Not statistically significant
S = Significant

Aminotransferases in Homogenates and Metabolic Products of Trichinella spiralis.—In Table 32 are given the results of assays of *Trichinella* larvae and their metabolic products. Both GPT and GOT were present in both homogenates and metabolic products of larvae. Though the levels of enzyme activity were low, there was greater activity of both enzymes in larval homogenates than in their metabolic products.

TABLE 32. *Glutamic oxalacetic transaminase (GOT) levels in serum of rats infected with* Trichinella spiralis *and fed complete and pyridoxine-deficient, semisynthetic diets (data in milliunits ±SD. (After Sen, 1969)*

Complete Diet	*Day 14*	*21*	*28*	*35*	*42*	*49*
Controls (uninfected)	32±3	32±4	31±4	32±3	31±2	30±3
Experimentals (infected)	32±2	29±1	49±3	58±5	39±1	32±3
Increase over controls	0	0	18	26	8	2
Statistical significance at 1% level	NS	NS	S	S	S	NS
Pyridoxine-deficient diet						
Controls (uninfected)	25±1	23±1	20±1	18±1	17±1	17±1
Experimentals (infected)	25±1	24±1	25±1	28±2	24±1	21±1
Increase over controls	0	1	5	10	7	4
Statistical significance at 1% level	NS	NS	S	S	S	S

NS=Not statistically significant
S=Significant

Pantothenate Levels in Host Tissues

Pantothenate Levels in Liver Tissue of Normal and Pantothenic Acid-deficient Mice.—The average initial body weights, distribution, and numbers of mice used in studies on pyruvate and pantothenate levels in plasma and liver tissue are shown in Table 33 (Lee and Lincicome, 1970b). Pantothenate levels in livers of mice infected with *T. duttoni* and fed complete, pantothenic acid-deficient and calorically restricted diets (pairfed controls) are shown in Table 34. In mice on diets from which pantothenate was withdrawn there was a steady decline in liver tissue as a function of time. In all animals infected with this trypanosome, there were significantly higher levels of pantothenate in liver tissue usually beginning approximately one month after inoculation.

TABLE 33. *Distribution, average initial body weights ±SD, and numbers of animals employed (in parentheses) in studies on pantothenate levels in plasma and liver tissue of normal and pantothenic acid-deficient mice. (After Lee and Lincicome, 1970b)*

Dietary Group	*Expt. No. 1*	*2*	*3*	*4*	*5*	*6*	*7*	*Total No. of Mice*
Control diet								
uninfected mice	9.1±0.1(30)	9.1±0.4(30)	9.0±0.2(30)	9.3±0.1(21)	9.2±0.5(21)	9.1±0.1(32)	9.0±0.5(32)	196
infected mice	9.2±0.5(30)	9.0±0.1(30)	9.5±0.3(21)	9.0±0.4(21)	9.0±0.4(21)	9.3±0.2(32)	9.6±0.3(32)	196
Pantothenic acid deficient diet								
uninfected mice	9.1±0.1(30)	9.0±0.3(30)	9.3±0.1(30)	9.4±0.2(21)	9.2±0.1(21)	9.0±0.2(32)	9.5±0.3(32)	196
infected mice	9.4±0.2(30)	9.2±0.2(30)	9.2±0.1(21)	9.6±0.3(21)	9.6±0.3(21)	9.0±0.2(32)	9.2±0.2(32)	196
Pair-fed								
uninfected mice	9.2±0.1(30)	9.0±0.2(30)	9.1±0.2(30)	9.1±0.3(21)	9.4±0.6(21)	9.8±0.2(32)	9.2±0.4(32)	196
infected mice	9.0±0.3(30)	9.1±0.2(30)	9.0±0.1(30)	9.6±0.2(21)	9.5±0.3(21)	9.5±0.3(32)	9.0±0.1(32)	196
Total No. Mice	180	180	180	126	126	192	192	*1176*

This correlates well with the fact that at this time the peak parasitemias have been reached and the trypanosomes are being destroyed by the hosts' immunologic mechanisms. This probably is the mechanism whereby significant quantities of pantothenate are released to the host.

TABLE 34. *Pantothenate levels in liver of mice inoculated with 10^6 cells of* Trypanosoma duttoni *and fed complete, pantothenic-acid-deficient, and calorically-restricted (pairfed) diets (data in μg/g fresh tissue ±SD). (After Lee and Lincicome, 1970b)*

	Day 15	*23*	*31*	*39*	*47*	*55*
			Complete Diet			
A	58±6	50±5	58±2	64±2	69±2	66±3
B	59±4	56±2	53±3	57±3	60±4	58±2
C	0	0	5	7	9	7
D	NS	NS	S	S	S	S
			Pyridoxine-Deficient Diet			
A	41±2	39±3	34±3	29±1	26±4	19±2
B	37±3	36±5	32±3	21±4	16±2	10±3
C	4	3	2	8	10	9
D	NS	NS	NS	S	S	S
			Pair-Fed Control Diet			
A	55±3	51±3	56±5	54±2	53±3	48±4
B	50±4	54±2	52±3	51±4	46±1	41±3
C	5	0	4	3	7	7
D	NS	NS	NS	NS	S	S

A=Infected
B=Uninfected
C=Increase over controls
D=Statistical significance at 1% level
NS=Not statistically significant
S=Significant

Pantothenate Levels in Plasma of Normal and Pantothenic Acid-deficient Mice.—Pantothenate levels in plasma (Table 35) show essentially the same situation as was obtained with liver tissue: reduction accompanying pantothenate deficiency diets, increased levels reflecting the presence of the trypanosome (Lee and Lincicome, 1970b).

TABLE 35. *Plasma pantothenate levels of mice infected with* Trypanosoma duttoni *and fed complete, pantothenic-acid-deficient, and calorically-restricted (pairfed) diets (data in μg/ml ±SD). (After Lee and Lincicome, 1970b)*

	Complete Diet					
	Day 15	*23*	*31*	*39*	*47*	*55*
A	386±9	377±6	451±5	408±3	470±7	419±12
B	393±6	384±14	428±10	381±7	433±5	387±8
C	0	0	23	27	37	32
D	NS	NS	S	S	S	S
	Pantothenic-Acid-Deficient Diet					
A	289±6	277±8	238±10	201±9	172±3	148±3
B	283±2	268±12	210±8	183±4	158±9	129±6
C	6	9	28	18	14	19
D	NS	NS	S	S	S	S
	Pairfed Control Diet					
A	380±8	365±3	317±11	378±4	355±9	327±5
B	389±7	373±5	301±8	332±15	314±12	291±6
C	0	0	16	46	41	36
D	NS	NS	NS	S	S	S

A = Infected
B = Uninfected
C = Increase over controls
D = Statistical significance at 1% level
NS = Not significant
S = Significant

Pyrodozine Levels in Host Tissues

In Livers of Rats Fed Complete, Pyridoxine-deficient, and Pyridoxine-supplemented Diets and Infected with Trypanosoma lewisi.—In Table 36 are shown the levels of pyridoxine found in livers of rats infected with *T. lewisi* and fed complete, deficient, and supplemented diets. These results are representative of two such experiments performed (Warsi, 1968). As expected, pyridoxine levels were lower in livers from rats on a deficient diet. The steady rise in level from an initial low value was characteristic and expected in animals placed originally on a deficient diet and then supplemented with normal quantities of the vitamin. In all infected animals regardless of diet, there was significantly more pyridoxine than in respective controls, beginning two weeks after inoculation. The trypanosome is, therefore, associated with higher levels of the vitamin in infected hosts. These levels appear to be correlated with maximal populations of trypanosomes in the peripheral circulating blood and their subsequent immunologic destruction.

In Livers of Rats Inoculated with Trichinella spiralis *and Fed Complete and Pyridoxine-deficient Diets.*—In Table 37 are shown the amounts of pyridoxine

TABLE 36. *Pyridoxine levels in livers of rats fed complete, pyridoxine-deficient, and pyridoxine-supplemented diets and infected with* Trypanosoma lewisi *(data in μg/g fresh tissue ±SD). (After Warsi, 1968)*

Group	*Day 1*[a]	*7*	*14*	*21*	*28*
		Complete Diet			
Controls (uninfected)	13.8±0.8	13.6±1.8	13.2±1.8	13.1±1.6	12.7±1.7
Experimentals (infected)	13.8±0.8	14.3±2.1	17.8±0.7	17.3±0.6	15.5±0.8
Increase over controls	0.0	0.7	4.6	4.2	2.8
Statistical significance at 1% level	NS	NS	S	S	S
		Pyridoxine-deficient Diet			
Controls (uninfected)	9.1±0.5	7.3±0.5	5.6±0.6	4.4±0.5	3.7±0.6
Experimentals (infected)	9.1±0.5	7.8±0.6	8.2±0.5	6.3±0.6	5.5±0.5
Increase over controls	0.0	0.5	2.6	2.0	1.8
Statistical significance at 1% level	NS	NS	S	S	S
		Pyridoxine-supplemented Diet			
Controls (uninfected)	8.0±0.8	8.5±0.5	9.2±0.6	11.4±0.5	12.5±0.5
Experimentals (infected)	8.0±0.8	8.8±0.6	11.2±0.5	12.4±0.4	12.5±0.5
Increase over controls	0.0	0.3	2.0	1.0	0.0
Statistical significance at 1% level	NS	NS	S	S	S

[a] Day 1 is the day of inoculation of 100 *T. lewisi* cells in experimental rats and corresponds to day 21 of deficiency

assayed in liver tissues from rats inoculated with *T. spiralis* and fed full-complement, pyridoxine-deficient, and pyridoxine-supplemented diets. Animals on the pyridoxine-deficient diet typically showed reduced pyridoxine levels in liver tissue, progressively so with the duration of the deficiency. Well-fed control rats showed a steady level throughout the course of the study. Infected animals, whatever their dietary regimen, showed increased levels of the vitamin, so it must be concluded that these levels are somehow associated with the *Trichinella*.

TABLE 37. *Pyridoxine levels in livers of rats inoculated with* Trichinella spiralis *and fed semisynthetic full-complement, pyridoxine-deficient, and pyridoxine-supplemented diets (data in μg/g liver). (After Sen, 1969)*

	Day 1	*7*	*14*	*21*	*28*
			Full-Complement Diet		
Controls (uninfected)	11.6±0.5	12.0±0.5	11.3±0.5	12.0±0.5	11.5±0.5
Experimentals (infected)	11.6±0.8	13.3±4.3	13.5±1.0	14.5±0.8	14.5±0.5
Increase over controls	0	1.3	2.2	2.5	3.0
Statistical significance at 1% level	NS	NS	S	S	S
			Pyridoxine-Deficient Diet		
Controls (uninfected)	8.5±0.5	7.3±0.3	6.3±0.3	5.3±0.3	5.0±0.7
Experimentals (infected)	8.5±0.1	7.5±0.3	7.1±0.8	6.9±0.3	6.7±0.3
Increase over controls	0	0.2	0.8	1.6	1.7
Statistical significance at 1% level	NS	NS	NS	S	S
			Pyridoxine-Supplemented Diet		
Controls (deficient)	8.5±0.8	7.3±0.5	6.3±0.3	5.5±0.3	5.0±1.0
Experimentals (supplemented)	9.0±0.8	14.5±0.8	16.0±0.5	16.6±0.5	17.0±0.5
Increase over controls	0.5	7.2	9.7	11.1	12.0
Statistical significance at 1% level	NS	S	S	S	S

NS = Not statistically significant
S = Significant

Pyruvate Levels and Utilization by Host Tissues

Blood Pyruvate Levels of Mice Infected With Trypanosoma duttoni *and Fed Complete, Pantothenate-deficient, and Calorically-restricted Diets.*—In Table 38 are shown the data on blood pyruvate levels in these mice. These are representative data from one of three such experiments performed (Lee and Lincicome, 1970c). Infection with *T. duttoni* in these mice is associated with significant increases in blood pyruvate levels at several assay periods. This may be the result of pyruvate production by the trypanosomes since it is

TABLE 38. *Blood pyruvate levels in mice infected with* Trypanosoma duttoni *and fed complete, pantothenate-deficient, and calorically-restricted (pairfed control) diets (data in mg % ±SD). (After Lee and Lincicome, 1970c)*

	Day 15	*23*	*31*	*39*	*47*
			Complete Diet		
A	0.31±0.02	0.45±0.05	0.42±0.05	0.53±0.03	0.34±0.01
B	0.33±0.03	0.27±0.01	0.36±0.03	0.30±0.03	0.25±0.03
C	0	0.18	0.06	0.23	0.09
D	NS	S	NS	S	S
			Pantothenate-Deficient Diet		
A	0.58±0.06	0.67±0.08	0.73±0.05	0.82±0.04	0.54±0.05
B	0.51±0.05	0.55±0.04	0.62±0.02	0.65±0.04	0.46±0.02
C	0.07	0.12	0.11	0.17	0.08
D	NS	NS	NS	S	NS
			Pairfed Control Diet		
A	0.42±0.02	0.51±0.03	0.59±0.02	0.55±0.02	0.36±0.03
B	0.37±0.03	0.42±0.02	0.38±0.06	0.47±0.01	0.29±0.02
C	0.06	0.09	0.21	0.08	0.07
D	NS	S	S	S	NS

A = Infected (experimentals)
B = Uninfected (controls)
C = Increase over controls
D = Statistical significance at 5% level
NS = Not statistically significant
S = Significant

known that these cells produce pyruvate. On the other hand, the reduced levels seen at several other assay periods may be a function of more pantothenate since significantly more pantothenate seems to be available to the host as a result of parasite destruction in the blood stream.

Pyruvate Utilization by Liver Slices From Trypanosoma duttoni *Infected Mice Fed Complete, Pantothenate-deficient, and Pairfed Control Diets.*—Two experiments were performed on pyruvate utilization involving a large number of mice as shown in Table 39. In Table 40 are presented the results of one of these experiments (Lincicome and Lee, 1970b). In neither of the two groups of control animals (those on complete and pairfed diets) was there a significant difference in pyruvate utilization by liver slices. While there were small differences in favor of the experimental group (mice on pantothenate-deficient diet) none of these was statistically significant. The experimental group, however, showed significantly more pyruvate utilized by tissue slices coming from *T. duttoni* infected mice. These results, therefore, indicate that pyruvate utilization by mouse liver slices is depressed in a pantothenate-deficiency, and infection with this trypanosome tends to restore pyruvate utilization to normal.

TABLE 39. *Distribution, average initial body weights ±SD, and numbers of animals employed (in parentheses) in studies on pyruvate utilization by liver slices from mice infected with* Trypanosoma duttoni. *(After Lincicome and Lee, 1970b)*

Dietary Group	*Expt. 1*	*Expt. 2*	*Total No. of Mice*
Control diet noninfected mice	9.1±0.1(32)	9.0±0.5(32)	64
Infected mice	9.3±0.2(32)	9.6±0.3(32)	64
Pantothenic-acid-deficient noninfected mice	9.3±0.4(32)	9.5±0.3(32)	64
Infected mice	9.0±0.2(32)	9.2±0.2(32)	64
Pairfed noninfected mice	9.8±0.2(32)	9.2±0.4(32)	64
Infected mice	9.5±0.3(32)	9.0±0.1(32)	64
Total No. of Mice	192	192	*384*

TABLE 40. *Pyruvate utilization by liver slices from* Trypanosoma duttoni-*infected mice fed complete, pantothenate-deficient, and pairfed control diets (data in μg of pyruvate utilized per 100 mg fresh tissue per hour ±SD). (After Lincicome and Lee, 1970b)*

	Day 15	*23*	*31*	*39*	*47*	*55*
			Complete Diet			
A	5.88±0.1	5.65±0.5	5.74±0.9	7.79±0.2	7.13±0.6	6.67±0.2
B	5.09±0.3	6.33±0.2	5.49±0.4	7.30±0.5	6.42±0.4	6.90±0.2
C	0	0	0.25	0.49	0.71	0
D	NS	NS	NS	NS	NS	NS
			Pantothenate-Deficient Diet			
A	2.82±0.2	2.77±0.4	3.28±0.1	3.86±0.3	2.88±0.2	2.68±0.2
B	2.50±0.2	2.10±0.8	2.73±0.1	2.69±0.4	2.03±0.2	1.74±0.3
C	0.32	0.67	0.55	1.17	0.85	0.94
D	NS	NS	S	S	S	S
			Pairfed Control Diet			
A	6.4±0.8	6.11±0.7	7.01±0.2	5.95±0.6	5.12±0.8	6.27±0.2
B	6.5±0.4	6.5±0.1	6.93±0.3	6.34±0.3	4.98±0.4	5.83±0.3
C	0	0	0.08	0	0.14	0.44
D	NS	NS	NS	NS	NS	NS

A = Infected
B = Uninfected
C = Increase over controls
D = Statistical significance at 1% level
NS = Not statistically significant
S = Significant

Pantothenic Acid Content of Trypanosomes

A series of five determinations was made to assay the pantothenic acid content of cells of *Trypanosoma duttoni.* Table 41 shows the average content of 10^8 cells (Lincicome and Lee, 1970b).

TABLE 41. *Pantothenic acid (PA) content of* Trypanosoma duttoni *cells (data in μg per 10^8 cells). (After Lincicome and Lee, 1970b)*

Sample	*μg PA*
1	0.22
2	0.27
3	0.19
4	0.21
5	0.26
$\bar{X} \pm SD$	*0.23±0.03*

Pyridoxine Content of Trypanosomes and Trichinella

In Table 42 are shown the results of 15 assays of the pyridoxine content of *Trypanosoma lewisi.* The average level of this vitamin per 2×10^8 cells was 0.55μg.

In Table 89 are given the results of two experiments assaying the pyridoxine content of homogenated larvae of *Trichinella.* Eight samples have an average of about 0.63μg per 4×10^4 larvae.

TABLE 42. *Pyridoxine content of* Trypanosoma lewisi *cells (data in μg/2×10^8 cells). (After Warsi, 1968)*

Sample	*Expt. 1*	*Expt. 2*	*Expt. 3*
1	0.50	0.55	0.50
2	1.00	0.50	0.50
3	0.50	0.55	0.55
4	0.50	0.50	0.55
5	0.50	0.50	0.55
$\bar{X}$	*0.60*	*0.52*	*0.53*

Homogenates

Trypanosoma lewisi.—Whole, washed cells of *Trypanosoma lewisi* were homogenized and then inoculated into rats fed complete and pyridoxine-deficient diets for observations on: body weight gains of rats, their food intake, and enzyme activities of rat tissues.

Table 43 gives the observations on body weight gains of rats fed complete and pyridoxine-deficient diets and inoculated with a whole body homogenate of *Trypanosoma lewisi.* The results were similar, if not identical, to those obtained when living cells of *T. lewisi* were inoculated: a significant stimulation of growth apparent about the 35th day for the complete diet and later

TABLE 43. *Average per cent body weight gains (±SD) of rats given complete and pyridoxine-deficient diets and inoculated with* Trypanosoma lewisi *homogenate. (After Warsi, 1968)*

Group	*Day 7*	*14*	*21*	*28*	*35*	*42*	*49*
				Complete Diet			
Controls (uninoculated)	27±2	72±3	123±4	162±3	192±3	226±4	267±3
Experimentals (inoculated)	30±2	71±2	124±4	164±4	200±3	233±2	280±5
Increase over controls	3	−1	1	2	8	7	13
Statistical significance at 1% level	NS	NS	NS	NS	S	S	S
				Pyridoxine-Deficient Diet			
Controls (uninoculated)	24±4	37±8	53±4	64±7	65±7	68±5	69±6
Experimentals (inoculated)	26±2	39±3	55±2	66±3	68±1	73±2	76±2
Increase over controls	2	2	2	2	3	5	7
Statistical significance at 5% level	NS	NS	NS	NS	NS	S	S

NS = Not statistically significant S = Significant

for the pyridoxine-deficient diet. The factor responsible for this phenomenon appears to be associated with the whole cell. Pyridoxine or some similarly acting substance appears to be one possible factor.

Table 44 shows the food intake of rats inoculated with a whole body homogenate of *T. lewisi* cells fed complete and pyridoxine-deficient diets. All animals receiving whole body homogenates experienced stimulation of the reflex and, therefore, ate more food. Though the anorexia associated with pyridoxine deprivation is relieved significantly by administration of whole-body trypanosome homogenate, the food intake of vitamin deficient animals still did not approximate that of the rats on a complete diet.

In Table 45 are shown the levels of glutamic-pyruvic transaminase (GPT) and in Table 46 are given the levels of glutamic-oxalacetic transaminase (GOT) in sera of rats fed complete, pyridoxine-deficient, and pyridoxine-supplemented diets, and inoculated with *T. lewisi* cell homogenates. Enzyme activities of both GPT and GOT were steady in animals fed the complete diet. In animals on pyridoxine-deficient diets, enzyme levels declined progressively to a low level due to lack of the cofactor pyridoxine. In rats supplemented with pyridoxine after an initial period of deprivation, enzyme activities declined and then advanced. The results of inoculation of whole-cell homogenates of *T. lewisi* showed that enzyme activities were comparable to those resulting from administration of living cells.

TABLE 44. *Food intake of rats inoculated with a whole-body homogenate of* Trypanosoma lewisi *cells and fed complete and pyridoxine-deficient diets (data in grams per day per rat ±SD). (After Warsi, 1968)*

Group	*Day 5*	*10*	*15*	*20*	*25*	*30*	*35*	*40*	*45*	*50*
					Complete Diet					
Controls (uninoculated)	7.4±.3	8.7±.3	10.2±.5	11.3±.4	12.4±.4	13.7±.1	14.3±.3	15.5±.2	16.1±.4	16.6±.3
Experimentals (inoculated)	7.2±.3	8.4±.4	10.4±.5	11.0±.5	12.3±.3	14.0±.2	15.2±.3	16.2±.3	17.3±.2	17.8±.2
Increase over controls	−0.2	−0.3	0.2	−0.3	−0.1	0.3	0.9	0.7	1.2	1.2
Statistical significance at 1% level	NS	NS	NS	NS	NS	NS	S	S	S	S
					Pyridoxine-Deficient Diet					
Controls (uninoculated)	7.0±.5	8.1±.2	7.3±.3	6.4±.3	5.5±.3	4.0±.2	3.1±.2	2.5±.2	2.2±.4	2.1±.2
Experimentals (infected)	7.3±.2	8.5±.2	7.1±.3	6.2±.3	5.2±.3	4.9±.5	4.2±.3	3.2±.3	3.4±.5	3.7±.4
Increase over controls	.3	.4	−.2	−.2	−.3	0.9	1.1	0.7	1.2	1.6
Statistical significance at 1% level	NS	NS	NS	NS	NS	S	S	S	S	S

NS = Not statistically significant
S = Significant

TABLE 45. *Glutamic-pyruvate transaminase (GPT) levels in serum of rats fed complete, pyridoxine-deficient, and pyridoxine-supplemented diets and inoculated with living cells of* Trypanosoma lewisi, T. lewisi *cell homogenates, and* T. lewisi *metabolic cell products (data in milliunits of enzyme activity ±SD). (After Warsi, 1968)*

Days	*Uninfected*	*Infected*	*Inoculated with Metabolic Products*	*Inoculated with Trypanosome Homogenates*
		Complete Diet		
1	14±0.7	—	—	—
7	13±0.7	—	—	—
14	14±1.0	—	—	—
21[a]	12±1.4	14±0.7	13±0.70	13±1.0
28	14±0.7	15±1.0	14±0.86	15±0.7
35	13±0.7	24±0.9	13±0.90	18±0.86
42	14±0.7	18±1.4	15±1.0	22±1.4
49	13±1.0	14±0.7	13±2.2	23±0.7
		Pyridoxine-Deficient Diet		
1	14±0.9	—	—	—
7	10±1.5	—	—	—
14	9±0.7	—	—	—
21[a]	6±1.4	7±0.9	7±0.9	6±0.9
28	5±0.5	7±0.9	7±1.0	7±0.9
35	4±0.7	11±1.2	4±1.4	9±1.4
42	3±1.2	6±1.0	4±0.9	9±0.7
49	3±0.7	5±0.9	5±1.5	8±0.9
		Pyridoxine-Supplemented Diet		
1	14±1.7	—	—	—
7	10±1.0	—	—	—
14	9±0.9	—	—	—
21[a]	6±0.7	6±0.5	7±1.0	7±0.9
28	7±0.9	9±0.9	7±0.9	8±0.7
35	10±0.7	17±2.0	10±0.9	13±1.7
42	10±0.9	16±0.9	11±0.9	14±0.9
49	12±0.9	13±1.0	14±1.4	15±0.9

[a] Day 21 marks the initiation of inoculations and pyridoxine supplement

Tables 47 and 48 show the results of an experiment in which denatured whole-cell homogenates of *T. lewisi* were administered to rats to determine the effect upon levels of GPT and GOT. The experiments were further controlled by interrupting the administration of the unheated homogenate. Denaturation by heating destroyed the stimulating factor in whole-cell homogenates as evidenced by the levels of both GPT and GOT. Administration of unheated homogenate was followed by a rise in enzyme activities and thereafter a decline when the administration was withdrawn.

TABLE 46. *Glutamic-oxalacetic transaminase (GOT) levels in serum of rats fed complete, pyridoxine-deficient, and pyridoxine-supplemented diets and inoculated with living* Trypanosoma lewisi *cells,* T. lewisi *cell homogenates, and* T. lewisi *metabolic cell products (data in milliunits of enzyme activity ±SD). (After Warsi, 1968)*

Day	*Uninfected*	*Infected*	*Inoculated with Metabolic Cell Products*	*Inoculated with Cell Homogenates*
		Complete Diet		
1	34±1.0	—	—	—
7	34±1.4	—	—	—
14	33±0.90	—	—	—
21[a]	34±1.2	34±1.20	33±0.79	34±0.50
28	34±0.87	36±0.56	32±1.0	34±0.71
35	34±0.75	49±1.4	33±1.5	39±0.92
42	32±1.56	39±0.56	35±0.74	41±0.70
49	34±0.80	35±0.70	37±2.0	42±1.70
		Pyridoxine-Deficient Diet		
1	35±1.5	—	—	—
7	29±1.4	—	—	—
14	25±1.9	—	—	—
21[a]	21±0.8	21±0.8	20±0.6	21±0.7
28	17±0.7	19±1.7	17±0.9	19±1.4
35	14±0.8	30±0.8	15±0.7	22±0.8
42	10±0.8	20±0.6	13±1.9	21±1.8
49	9±1.6	10±1.9	13±3.0	21±1.0
		Pyridoxine-Supplemented Diet		
1	34±0.7	—	—	—
7	29±1.7	—	—	—
14	24±0.9	—	—	—
21[a]	19±1.0	20±1.4	21±1.7	21±1.7
28	21±1.4	24±0.7	20±1.0	22±0.9
35	26±0.9	37±1.0	27±1.0	30±1.2
42	29±1.0	35±0.9	30±1.0	33±1.0
49	30±0.9	32±0.9	30±1.5	35±1.0

[a] Day 21 marks the initiation of inoculations and pyridoxine supplement

Trypanosoma duttoni.—Whole, washed cells of *T. duttoni* were homogenized and then inoculated into mice fed complete, pantothenate-deficient, and pairfed control diets for observations on body weight gains and food intake.

Table 49 details the body weight gains of mice given whole-cell homogenates of *T. duttoni.* Body weight gain was stimulated in these mice significantly beginning the 30th day after inoculation (Lincicome and Lee, 1970a) for those on a complete or the pairfed control diet. For those fed the pantothenate-deficient diet, growth stimulation was significantly apparent by the 25th day.

TABLE 47. *Glutamic-pyruvic transaminase (GPT) in serum of rats fed complete and pyridoxine-deficient diets and inoculated with denatured* Trypanosoma lewisi *cell homogenates (data in milli-units of enzyme activity ±SD). (After Warsi, 1968)*

Day	*Uninoculated Control*	*Inoculated with Heated Cell Homogenates*	*Inoculated with Cell Homogenates (Discontinued at Day 35)*
		Complete Diet	
1	14±0.7	—	—
7	13±0.7	—	—
14	14±1.0	—	—
21[a]	12±1.4	13±1.7	13±1.0
28	14±0.7	12±1.7	15±0.7
35	13±0.7	15±1.0	18±0.9[b]
42	14±0.7	15±1.4	16±1.0
49	13±1.0	14±2.0	14±1.2
		Pyridoxine-Deficient Diet	
1	14±0.9	—	—
7	10±1.5	—	—
14	9±0.7	—	—
21[a]	6±1.4	6±0.9	6±0.9
28	5±0.5	5±1.0	7±0.86
35	4±0.7	5±0.86	9±1.4[b]
42	3±1.2	4±0.9	6±1.0
49	3±0.7	4±0.9	4±0.7

[a] Initiation of inoculations
[b] Inoculation of cell homogenate discontinued here

The stimulation here is therefore comparable to that observed for living cells of this trypanosome.

Table 50 gives the data on intake of food of mice fed complete and pantothenate-deficient diets and inoculated with cell homogenates. Significant relief from the anorexia of the vitamin deficiency occurred in those animals on the pantothenate-deficiency diet, but even at best the daily consumption of food was not equivalent to the well-fed control animals. The action of homogenates of *T. duttoni* is similar, if not identical, to that of living cells of this trypanosome in relieving the anorexia of pantothenate deficiency and stimulating the appetite of the well-fed animal.

Trichinella spiralis.—Study of whole-body homogenates of *T. spiralis* centered around host body weight gains, host carcass and liver weights, host food consumption, and measurement of two aminotransferase enzyme systems of the host during administration of these homogenates.

TABLE 48. *Glutamic-oxalacetic transaminase (GOT) in serum of rats fed complete and pyridoxine-deficient diets and inoculated with denatured* Trypanosoma lewisi *cell homogenates (data in milliunits of enzyme activity ±SD). (After Warsi, 1968)*

Day	*Uninoculated Controls*	*Inoculated with Heated Cell Homogenates*	*Inoculated with Cell Homogenates (Discontinued at Day 35)*
		Complete Diet	
1	34±1.0	—	—
7	35±0.9	—	—
14	35±1.7	—	—
21[a]	34±0.9	33±1.0	34±1.0
28	33±1.2	35±1.4	35±1.0
35	34±0.9	36±1.4	41±1.2[b]
42	32±2.0	33±0.7	36±0.9
		Pyridoxine-Deficient Diet	
1	34±0.84	—	—
7	30±0.86	—	—
14	24±1.4	—	—
21[a]	20±1.0	20±1.0	21±1.0
28	17±0.7	18±1.0	19±0.7
35	13±0.86	15±1.4	22±2.0[b]
42	8±1.0	9±0.7	15±1.0
49	8±0.7	6±0.9	10±0.7

[a] Initiation of inoculations
[b] Inoculation of cell homogenate discontinued here

Table 51 presents the data on body weight gains of rats fed complete and pyridoxine-deficient diets and inoculated with homogenates of *Trichinella*. Though there was a progressive increase in growth stimulation associated with the homogenate, this was not statistically significant until the 45th day. At that time, the rats receiving homogenate were 25 % larger than uninoculated animals. Rats fed a pyridoxine-deficient diet were significantly larger at the 35th day, and this advantage continued until the 55th day of the experiment when they were 33 % larger than similar animals receiving no homogenate. Homogenates of the whole bodies of larval *Trichinella* elicited, therefore, the same response as did the living larvae (Sen, 1969).

Weight gains of rat carcasses and livers from animals fed complete and pyridoxine-deficient diets and inoculated with live, as well as homogenated, *Trichinella* are shown in Table 52. At the 55th day, when all animals were autopsied, carcasses and livers of all animals given *Trichinella* were significantly larger. Both live worms and homogenates of worms appeared to be

TABLE 49. *Average per cent body weight gains ±SD of mice given complete, pantothenate-deficient, and pairfed control diets and inoculated with* Trypanosoma duttoni *whole cell homogenates. (After Lincicome and Lee, 1970a)*

	Day 5	*10*	*15*	*20*	*25*	*30*	*35*	*40*	*45*	*50*
					Complete Diet					
A	27±3	34±3	42±5	65±4	85±2	110±4	137±2	165±5	185±2	204±3
B	30±2	26±1	40±2	63±3	80±6	93±5	121±6	136±3	162±2	189±7
C	0	0	2	2	5	17	16	29	23	15
D	NS	NS	NS	NS	NS	S	S	S	S	S
					Pantothenate-Deficient Diet					
A	15±2	16±4	28±3	32±3	43±2	50±4	51±2	46±5	40±3	36±2
B	15±4	20±2	27±4	30±5	35±3	38±4	33±6	26±4	29±2	24±3
C	0	0	1	0	8	12	18	20	11	12
D	NS	NS	NS	NS	S	S	S	S	S	S
					Pairfed Control Diet					
A	18±5	25±5	34±3	47±8	65±4	85±2	96±1	109±2	113±6	124±3
B	23±2	27±4	38±7	51±5	66±2	72±5	89±3	94±2	100±5	114±5
C	0	0	0	0	0	13	7	15	13	10
D	NS	NS	NS	NS	NS	S	S	S	S	S

A = Inoculated
B = Uninoculated
C = Percentage increase over controls
D = Statistical significance at 1% level
NS = Not statistically significant
S = Significant

about equal with respect to response of the host on either diet regime (Sen, 1969).

In Tables 53 are given the data on food intake of rats fed complete and pyridoxine-deficient diets and inoculated with *Trichinella spiralis* whole-body homogenates (Sen, 1969). No significant differences were noted for rats fed the complete diet. On a pyridoxine-deficient diet, rats given homogenates ate significantly more food beginning at the 35th day.

In Table 54 and 55 are given the data on GPT and GOT aminotransferases in sera of rats fed complete and pyridoxine-deficient diets and inoculated with homogenates of *Trichinella*. In hosts on either complete or incomplete diets, the serum levels of both enzymes were significantly raised in hosts receiving homogenates. The interpretation is that both homogenate and living worms supply the same factor to the host.

In regard to *parasite aminotransferase systems* see Table 88 for these enzymes in homogenates of larval worms.

TABLE 50. *Food intake of mice fed complete and pyridoxine-deficient diets and inoculated with* Trypanosoma duttoni *whole cell homogenates (data in grams per mouse per day ±SD). (After Lincicome and Lee, 1970a)*

	Day 5	*10*	*15*	*20*	*25*	*30*	*35*	*40*	*45*	*50*
					Complete Diet					
A	1.0±0.3	1.4±0.4	1.9±0.3	2.6±0.2	3.5±0.2	3.7±0.3	4.2±0.2	4.6±0.2	5.0±0.2	5.0±0.4
B	1.1±0.4	1.5±0.2	1.7±0.1	2.5±0.2	2.8±0.2	2.8±0.3	3.1±0.3	3.4±0.2	3.8±0.1	4.0±0.5
C	0	0	0.2	0.1	0.7	0.9	1.1	1.2	1.2	1.0
D	NS	NS	NS	NS	S	S	S	S	S	S
					Pyridoxine-Deficient Diet					
A	1.3±0.2	1.4±0.2	1.6±0.2	1.5±0.4	1.8±0.3	2.2±0.5	2.3±0.2	2.0±0.3	1.9±0.1	1.7±0.3
B	1.2±0.2	1.1±0.2	1.4±0.4	1.2±0.4	1.5±0.3	1.7±0.2	2.1±0.3	1.6±0.2	1.2±0.2	1.0±0.2
C	0.1	1.3	0.2	0.3	0.3	0.5	0.2	0.4	0.7	0.7
D	NS	NS	NS	NS	NS	NS	NS	S	S	S

A = Inoculated
B = Uninoculated
C = Increase over controls
D = Statistical significance at 1% level
NS = Not statistically significant
S = Significant at 5% level

TABLE 51. *Average per cent body weight gains ±SD of rats fed semisynthetic complete and pyridoxine-deficient diets and inoculated with* Trichinella spiralis *whole-body homogenates. (After Sen, 1969)*

Group	*Day 15*	*25*	*35*	*45*	*55*
			Complete Diet		
Controls (uninoculated)	75±4	112±6	138±6	155±5	174±3
Experimentals (inoculated)	77±2	114±4	144±1	182±4	198±3
Increase over controls (%)	2	2	6	27	24
Statistical significance at 1% level	NS	NS	NS	S	S
		Pyridoxine-Deficient Diet			
Controls (uninoculated)	16±4	23±3	26±4	31±1	36±1
Experimentals (inoculated)	16±1	25±2	40±2	52±4	72±3
Increase over controls (%)	0	2	14	21	36
Statistical significance at 1% level	NS	NS	S	S	S

NS = Not statistically significant
S = Significant

Metabolic Products

The metabolic products of *Trypanosoma lewisi, Trypanosoma duttoni*, and larval *Trichinella spiralis* were collected and administered to mice and rats for a series of measurements of host body weight gains, host food consumption, and aminotransferase systems.

Trypanosoma lewisi.—Body weight gains and food consumption were measured in rats fed complete and pyridoxine-deficient diets with and without inocula of metabolic products. No significant differences could be demonstrated between rats receiving trypanosome metabolic products and those receiving none (Table 56) on either dietary regime, nor were there any differences in food intake (Table 57). These results argue for the view that metabolic products of *T. lewisi* do not contain the same stimulatory factor present in trypanosome homogenates.

Trypanosoma duttoni.—Tables 58 and 59 give the data for body weight gains and food consumption of mice fed complete and pantothenate-deficient

TABLE 52. *Average per cent weight gains ±SD of carcasses and livers from rats fed semisynthetic complete and pyridoxine-deficient diets and inoculated with live* Trichinella spiralis *and* T. spiralis *whole-body homogenates. Autopsies on 55th day of experiment. (After Sen, 1969)*

	Carcass		*Liver*	
Group	*Live* T. spiralis	*Worm homogenates*	*Live* T. spiralis	*Worm homogenates*
		Complete Diet		
Controls (uninoculated)	223±15	223±15	72±2	73±9
Experimentals (inoculated)	240±13	243±12	125±10	124±12
Increase over controls (%)	27	20	53	51
Statistical significance at 1% level	S	S	S	S
		Pyridoxine-Deficient Diet		
Controls (uninoculated)	54±12	24±12	28±10	28±12
Experimentals (inoculated)	80±12	84±8	62±5	50±4
Increase over controls (%)	26	60	34	22
Statistical significance at 1% level	S	S	S	S

S = Significant

diets and inoculated with metabolic products of *T. duttoni.* No significant differences could be demonstrated between mice with or without metabolic products regardless of diet. This is interpreted to mean that the metabolic products of *T. duttoni* as prepared and administered do not contain the stimulatory factor found in whole-cell homogenates of this trypanosome.

Trichinella spiralis.—Body weight gains, food consumption, and two serum aminotransferase levels were measured for rats fed complete and pyridoxine-deficient diets.

Table 60 presents the data for body weight gains. No significant differences could be demonstrated for rats on a complete diet. On a pyridoxine diet, rats receiving larval metabolic products grew significantly faster than uninoculated control animals. This is interpreted to mean that a growth stimulating factor similar to that in larval homogenates is present in larval metabolic products.

Table 61 presents the data on host food consumption, indicating that rats given a pyridoxine-deficient diet and larval metabolic products eat more food

TABLE 53. *Food intake of rats fed semisynthetic complete and pyridoxine-deficient diets and inoculated with* Trichinella spiralis *whole-body homogenates (data in grams per rat per day ±SD). (After Sen, 1969)*

	Day 15	*25*	*35*	*45*	*55*
			Complete Diet		
Controls (uninoculated)	11.5±0.7	13.0±1.0	14.4±0.5	15.6±1.0	17.3±1.0
Experimentals (inoculated)	11.9±1.2	13.6±0.6	15.4±1.4	17.2±1.2	18.3±0.6
Increase over controls	0.4	0.6	1.0	1.6	1.0
Statistical significance at 5% level	NS	NS	NS	NS	NS
			Pyridoxine-Deficient Diet		
Controls (uninoculated)	5.9±1.4	5.1±0.7	4.4±0.6	3.9±0.7	3.6±0.7
Experimentals (inoculated)	5.9±0.2	5.6±0.8	5.4±0.1	7.7±1.0	9.8±0.5
Increase over controls	0	0.5	1.0	3.8	6.2
Statistical significance at 5% level	NS	NS	S	S	S

NS=Not statistically significant
S=Significant

than do control animals. No differences were apparent among rats on a complete diet. This substantiates the observation on rat body weight gains: that in metabolically imbalanced (pyridoxine) hosts, the stimulatory factor may be demonstrated.

Tables 62 and 63 show that the levels of both GPT and GOT aminotransferases in serum of rats metabolically imbalanced (pyroxidine) reflect the presence, in larval metabolic products, of a factor increasing the activities of these enzymes.

See Table 88 for aminotransferase levels in metabolic products of larval *Trichinella.*

Host Food Intake

Food consumption by hosts was measured for two purposes: to ascertain whether greater food intake was associated with greater weight gains, and to determine whether concomitantly with administration of homogenates and metabolic products of trypanosomes or *Trichinella* there was greater consumption of food. The latter aspects of food intake were presented in a

TABLE 54. *Glutamic-pyruvic transaminase (GPT) in serum of rats fed complete and pyridoxine-deficient diets and inoculated with whole-body homogenates of* Trichinella spiralis *(data in milli-units of enzyme activity ±SD) (After Sen, 1969).*

	Day 14	*21*	*28*	*35*	*42*	*49*
			Complete Diet			
Controls (uninoculated)	12±1	12±1	11±1	12±1	12±1	12±1
Experimentals (inoculated)	14±1	16±1	17±1	16±1	14±1	13±1
Increase over controls	2	4	6	4	2	1
Statistical significance at 1% level	NS	S	S	S	S	NS
			Pyridoxine-Deficient Diet			
Controls (uninoculated)	9±1	7±1	5±1	4±1	4±1	3±1
Experimentals (inoculated)	10±1	13±1	12±1	12±1	16±1	9±1
Increase over controls	1	6	7	8	12	6
Statistical significance at 1% level	NS	S	S	S	S	S

NS=Not statistically significant

previous section. The former aspects are now presented for infections with *Trypanosoma lewisi, T. duttoni*, and *Trichinella spiralis.*

Trypanosoma lewisi.—Table 64 tabulates the food intake of rats fed complete, pyridoxine-deficient, and supplemented pyridoxine-deficient diets, and inoculated with *T. lewisi.* Infected animals consistently ate more food than did uninfected rats, even though the latter were deficient in pyridoxine or supplemented with this vitamin (Warsi, 1968).

Trypanosoma duttoni.—Table 65 tabulates the food intake of mice infected with this trypanosome and fed complete, pantothenate-deficient, and supplemented pantothenate-deficient diets. Infected mice showed significantly greater food intake (Lincicome and Lee, 1970a).

Trichinella spiralis.—Table 66 tabulates the food intake of rats fed complete, pyridoxine-deficient, and supplemented pyridoxine-deficient diets and

TABLE 55. *Glutamic-oxalacetic transaminase (GOT) in serum of rats fed complete and pyridoxine-deficient diets and inoculated with whole-body homogenates of* Trichinella spiralis *(data in milliunits of enzyme activity ±SD). (After Sen, 1969)*

	Day 14	*21*	*28*	*35*	*42*	*49*
			Complete Diet			
Controls (uninoculated)	32±3	32±4	31±4	32±3	31±2	30±3
Experimentals (inoculated)	38±2	39±1	46±1	42±3	39±0	38±2
Increase over controls	6	7	15	10	8	8
Statistical significance at 1% level	S	S	S	S	S	S
			Pyridoxine-Deficient Diet			
Controls (uninoculated)	25±1	23±1	20±1	18±1	17±1	17±1
Experimentals (inoculated)	28±2	27±3	28±1	25±1	20±1	20±1
Increase over controls	3	4	8	7	3	3
Statistical significance at 1% level	S	S	S	S	S	S

NS=Not statistically significant
S=Statistically significant

infected with *Trichinella.* Rats fed a complete diet showed no significant differences in intake. Rats fed the pyridoxine-deficient diet showed greater food intake if infected with *Trichinella.* The rats supplemented with the vitamin showed that the response to the added pyridoxine was an improvement in their foood intake.

Host Longevity

Host longevity as a parameter for judging a molecular contribution of a parasite to the metabolic economy of its host can only be employed where the control condition results in death at a predictable time after institution of the imbalance. Thiamine and pantothenate deficiencies lend themselves to this purpose.

Thiamine.—Table 67 lists the day of death for rats infected with *Trypanosoma lewisi.* Metabolically imbalanced hosts carrying the trypanosome lived longer.

TABLE 56. *Average per cent body weight gains ±SD of rats fed complete and pyridoxine-deficient diets and inoculated with* Trypanosoma lewisi *metabolic cell products. (After Warsi, 1968)*

	Day 7	*14*	*21*	*28*	*35*	*42*	*49*
				Complete Diet			
Controls (uninoculated)	25±3	73±2	118±2	165±5	207±5	249±7	270±5
Experimentals (inoculated)	26±4	70±7	119±3	164±6	209±3	251±7	275±6
Increase over controls (%)	1	−3	1	−1	2	2	5
Statistical significance at 1% level	NS	NS	NS	NS	NS	NS	NS
				Pyridoxine-Deficient Diet			
Controls (uninfected)	24±4	37±8	53±4	64±7	65±7	68±5	69±6
Experimentals (infected)	26±1	40±3	55±6	61±2	64±4	66±4	73±2
Increase over controls (%)	2	3	2	−3	−1	−2	4
Statistical significance at 1% level	NS	NS	NS	NS	NS	NS	NS

NS = Not statistically significant

Pantothenic Acid—Table 68 lists the average day of death for pantothenate-deficient mice inoculated with living *Trypanosoma duttoni* cells, homogenates of *T. duttoni* cells, and metabolic products of *T. duttoni* cells. Infected mice and mice receiving homogenates lived significantly longer than did control uninoculated mice. Those animals receiving metabolic products of *T. duttoni* cells lived essentially the same length of time as control mice. The interpretation of this representative experiment (Lincicome and Lee, 1970a) is that the trypanosome supplied the factor, presumably pantothenate or some molecule with comparable metabolic action, that prolonged the life of these mice. This factor was apparently absent in metabolic products of the cell.

Respiration of Host Tissues

Endogenous respiratory activity was charted for rat and mouse tissues from animals infected with trypanosomes. Metabolic imbalances using thiamine and pantothenate as specific factors were employed for *Trypanosoma lewisi* and *T. duttoni*.

Thiamine Metabolic Imbalance.—Table 69 shows the endogenous oxygen uptake rates for rat liver and kidney tissue slices in a thiamine deficiency

TABLE 57. *Food intake of rats fed complete and pyridoxine-deficient diets and inoculated with* Trypanosoma lewisi *metabolic cell products (data in grams per rat per day ±SD). (After Warsi, 1968)*

	Day 5	*10*	*15*	*20*	*25*	*30*	*35*	*40*	*45*	*50*
					Complete Diet					
Controls (uninfected)	7.3±.4	8.5±.4	10.4±.3	11.6±.2	13.0±.3	13.6±.2	14.6±.7	15.5±.4	16.3±.4	17.0±.5
Experimentals (infected)	7.4±.2	8.4±.2	9.8±.2	10.9±.3	12.3±.6	13.2±.4	14.4±.4	15.7±.2	17.0±.3	17.5±.5
Increase over controls	0.1	−0.1	−0.6	−0.7	−0.7	−0.4	−0.2	0.2	0.7	0.5
Statistical significance at 1% level	NS	NS	NS	NS	NS	NS	NS	NS	NS	NS
					Pyridoxine-Deficient Diet					
Controls (uninoculated)	6.5±.3	7.9±.4	7.2±.2	6.3±.2	5.2±.2	4.0±.2	3.1±.3	2.3±.5	2.1±.4	2.0±.3
Experimentals (inoculated)	6.8±.5	7.7±.2	7.0±.2	6.1±.15	4.9±.2	3.9±.15	3.0±.2	3.1±.4	2.9±.4	2.8±.4
Increase over controls	0.3	−0.2	−0.2	−0.2	−0.3	−0.1	−0.1	0.8	0.8	0.8
Statistical significance at 1% level	NS	NS	NS	NS	NS	NS	NS	NS	NS	NS

NS = Not statistically significant

TABLE 58. *Average per cent body weight gains ±SD of mice fed complete, pantothenate-deficient, and pairfed control diets and inoculated with* Trypanosoma duttoni *metabolic products. (After Lincicome and Lee, 1970a)*

	Day 5	*10*	*15*	*20*	*25*	*30*	*35*	*40*	*45*	*50*
					Complete Diet					
A	28±4	31±4	45±6	60±5	84±5	95±3	119±4	130±4	166±5	193±4
B	30±2	36±1	40±2	63±3	80±6	93±5	121±6	136±3	162±2	189±7
C	0	0	5	0	4	2	0	0	4	4
D	NS	NS	NS	NS	NS	NS	NS	NS	NS	NS
					Pantothenate-Deficient Diet					
A	18±5	21±3	26±2	28±4	36±4	40±3	40±8	32±7	25±3	21±2
B	15±4	20±2	27±4	30±5	35±3	38±4	33±6	26±4	29±2	24±3
C	3	1	0	0	1	2	7	6	0	0
D	NS	NS	NS	NS	NS	NS	NS	NS	NS	NS
					Pairfed Control Diet					
A	20±2	25±4	37±3	49±3	62±3	74±3	86±4	96±2	104±3	107±5
B	23±2	27±4	38±7	51±5	66±2	72±5	89±3	94±2	100±5	114±5
C	0	0	0	0	0	2	0	2	4	3
D	NS	NS	NS	NS	NS	NS	NS	NS	NS	NS

A = Inoculated
B = Uninoculated
C = Percentage increase over controls
D = Statistical significance at 1% level
NS = Not statistically significant

state. Tissues from infected rats generally respired at greater rates than did tissues from noninfected hosts.

Pantothenate Metabolic Imbalance.—Table 70 shows the endogenous oxygen uptake of mouse liver slices from hosts infected with *T. duttoni* and fed pantothenate-deficient diets. Tissues from infected hosts had higher endogenous rates regardless of the diets employed.

EVIDENCE OF THE CONTRIBUTIONS OF THE HOST

Growth of *Trypanosoma lewisi* in the Heterologous Mouse Host

Table 71 shows the prolongation of the survival of *T. lewisi* in the albino mouse by administration of rat serum (Lincicome, 1957; 1958a,b,c).

Table 72 gives the results of experiments determining the efficacy of various levels of food intake in promoting establishment and development of *T. lewisi* in mice. Mice fed 1 g of laboratory chow appeared to be better hosts for *T. lewisi* than were mice fed either 2 or 3 g of food per day.

TABLE 59. *Food intake of mice fed complete and pantothenate-deficient diets and inoculated with* Trypanosoma duttoni *metabolic products (data in grams per mouse per day ± SD). (After Lincicome and Lee, 1970a)*

	Day 5	*10*	*15*	*20*	*25*	*30*	*35*	*40*	*45*	*50*
					Complete Diet					
A	1.0±0.2	1.8±0.6	2.0±0.3	2.5±0.5	2.9±0.2	3.2±0.2	3.4±0.4	3.9±0.1	4.1±0.4	4.3±0.3
B	1.1±0.4	1.5±0.3	1.7±0.1	2.5±0.2	2.8±0.2	2.9±0.3	3.1±0.3	3.4±0.3	3.8±0.1	4.0±0.5
C	0	0.3	0.3	0	0.1	0.3	0.3	0.5	0.3	0.3
D	NS	NS	NS	NS	NS	NS	NS	NS	NS	NS
					Pantothenate-Deficient Diet					
A	1.2±0.2	1.2±0.4	1.4±0.2	1.6±0.5	1.7±0.2	1.5±0.3	1.8±0.3	1.4±0.2	1.2±0.3	1.1±0.2
B	1.2±0.2	1.1±0.2	1.4±0.4	1.2±0.6	1.5±0.3	1.7±0.2	2.1±0.3	1.6±0.2	1.2±0.2	1.0±0.2
C	0	0.1	0	0.4	0.2	0	0	0	0	0.1
D	NS	NS	NS	NS	NS	NS	NS	NS	NS	NS

A = Inoculated
B = Uninoculated
C = Increase over controls
D = Significance at 1% level
NS = Not statistically significant

TABLE 60. *Average per cent body weight ±SD gains of rats fed complete and pyridoxine-deficient diets and inoculated with* Trichinella spiralis *metabolic products. (After Sen, 1969)*

	Day 15	*25*	*35*	*45*	*55*
			Complete Diet		
Controls (uninoculated)	75±4	112±5	135±5	155±5	174±3
Experimentals (inoculated)	77±4	112±4	136±6	159±2	174±4
Increase over controls (%)	2	0	1	4	0
Statistical significance at 5% level	NS	NS	NS	NS	NS
			Pyridoxine-Deficient Diet		
Controls (uninoculated)	16±4	23±3	26±4	31±1	36±1
Experimentals (inoculated)	16±2	26±4	39±1	50±3	64±3
Increase over controls (%)	0	3	13	19	28
Statistical significance at 1% level	NS	NS	S	S	S

NS = Not statistically significant S = Significant

TABLE 61. *Food intake of rats fed semisynthetic complete and pyridoxine-deficient diets and inoculated with* Trichinella spiralis *metabolic products (data in grams per rat per day ±SD). (After Sen, 1969)*

	Day 15	*25*	*35*	*45*	*55*
			Complete Diet		
Controls (uninoculated)	11.5±0.7	13.0±1.0	14.4±0.5	15.6±1.0	17.3±1.0
Experimentals (inoculated)	11.8±2.0	13.7±0.9	15.1±0.5	16.6±0.3	17.5±1.0
Increase over controls	0.3	0.7	0.7	1.0	0.2
Statistical significance at 5% level	NS	NS	NS	NS	NS
			Pyridoxine-Deficient Diet		
Controls (uninoculated)	5.9±1.4	5.1±0.7	4.4±0.5	3.9±0.7	3.6±0.7
Experimentals (inoculated)	5.6±1.3	5.7±0.6	5.5±1.1	7.6±1.1	9.3±0.9
Increase over controls	0	0.6	1.1	3.7	5.7
Statistical significance at 1% level	NS	NS	S	S	S

NS = Not statistically significant S = Significant

TABLE 62. *Glutamic-pyruvic transaminase (GPT) in serum of rats fed complete and pyridoxine-deficient diets and inoculated with* Trichinella spiralis *metabolic products (data in milliunits of enzyme activity ±SD). (After Sen, 1969)*

	Day 14	*21*	*28*	*35*	*42*	*49*
			Complete Diet			
Controls (uninoculated)	12±1	12±1	11±1	12±1	12±1	12±1
Experimentals (inoculated)	12±1	12±1	12±1	13±1	12±1	11±1
Increase over controls	0	0	1	1	0	0
Statistical significance at 5% level	NS	NS	NS	NS	NS	NS
			Pyridoxine-Deficient Diet			
Controls (uninoculated)	9±1	7±1	5±1	4±1	4±1	3±1
Experimentals (inoculated)	10±1	7±1	7±1	5±1	4±1	4±1
Increase over controls	1	0	2	1	0	1
Statistical significance at 1% level	NS	NS	S	NS	NS	NS

NS = Not statistically significant
S = Significant

The trypanosome appeared to grow only in those mice to which normal rat serum (1 ml) had been given on a daily basis. Parasitemias developed and persisted for as long as two weeks (Tables 71 and 72).

Effect of Continued Development in the Heterologous Host

It seemed necessary to determine next whether there might be sufficient metabolic changes in *T. lewisi* during its development in the heterologous host to permit it to grow in this host ultimately, without whatever it was in rat serum initiating the development originally. To do this, it was necessary to subpassage the trypanosome from one generation mouse passage to the next, testing at each passage to determine if the trypanosome would indeed grow in the heterologous mouse without rat serum supplementation, and whether it then would develop in its normal homologous host.

Three cell lines (A, B, and C) of *T. lewisi* were established and studied experimentally (Lincicome, 1959a,b; 1960). Two isolates (A and B) of *T. lewisi* from the same rat stock source were serially transferred in calorically-restricted mice (1 g/day) supplemented daily with 1 ml of rat serum per day.

TABLE 63. *Glutamic-oxalacetic transaminase (GOT) in serum of rats fed complete and pyridoxine-deficient diets and inoculated with* Trichinella spiralis *metabolic products (data in milliunits of enzyme activity ±SD). (After Sen, 1969)*

	Day 14	*21*	*28*	*35*	*42*	*49*
			Complete Diet			
Controls (uninoculated)	32±3	32±4	31±4	32±3	31±2	30±1
Experimentals (inoculated)	31±1	32±2	34±2	31±2	30±3	31±2
Increase over controls	0	0	3	0	0	1
Statistical significance at 5% level	NS	NS	NS	NS	NS	NS
			Pyridoxine-Deficient Diet			
Controls (uninoculated)	25±1	23±1	20±1	18±1	17±1	17±1
Experimentals (inoculated)	26±1	28±2	27±2	22±2	20±1	17±1
Increase over controls	1	5	7	4	3	0
Statistical significance at 1% level	NS	S	S	S	S	NS

NS=Not statistically significant
S=Significant

The A strain was transferred consecutively through 300 mice over a period of more than 3 years and was voluntarily discontinued. The B strain died out spontaneously after 43 consecutive passages in mice. The developmental histories of these 2 isolates were analyzed and compared with respect to duration of the parasitemic period, interval to the next subsequent passage in mice, day of death of each host animal, proportion of host animals that died, intensity of parasitemia in mouse tail blood, interval required for development of the observed maximal parasitemia, and duration of maximal parasitemia.

Strain A appeared to have become progressively adapted to the mouse as judged by (1) a decrease in parasitemic period with successive transfer associated with a progressive increase in trypanosome population, and (2) by declines in the interval required for development of the observed maximal parasitemia and in the duration of this maximal response. There did not appear to be any correlation of the percentage of animals that died with any other factor. The B strain did not appear to have adapted itself to mice as judged by the foregoing criteria.

TABLE 64. *Food intake of rats fed complete, pyridoxine-deficient, and pyridoxine-supplemented diets and infected with* Trypanosoma lewisi *(data in grams per rat per day ±SD). (After Warsi, 1968)*

	Day 5	*10*	*15*	*20*	*25*	*30*	*35*	*40*	*45*	*50*
Complete Diet										
Controls (uninfected)	7.3±0.2	8.7±0.2	10.4±0.3	11.1±0.3	12.0±0.3	13.0±0.2	13.8±0.2	14.7±0.2		
Expeirmentals (infected)	7.5±0.4	8.7±0.5	10.6±0.5	11.7±0.5	12.7±0.1	13.9±0.4	15.0±0.2	15.2±0.3		
Increase over controls	0.2	0.0	0.2	0.6	0.7	0.9	1.2	0.5		
Statistical significance at 1% level	NS	NS	NS	NS	S	S	S	NS		
Pyridoxine-Deficient Diet										
Controls (uninfected)	6.4±0.2	8.3±0.2	7.2±0.3	6.1±0.3	5.7±0.3	4.7±0.2	4.4±0.2	4.1±0.2		
Experimentals (infected)	6.2±0.3	8.1±0.5	7.3±0.2	6.5±0.3	6.9±0.6	5.8±0.5	5.5±0.5	5.0±0.3		
Increase over controls	−0.2	−0.2	0.1	0.4	1.2	1.1	1.1	0.9		
Statistical significance at 1% level	NS	NS	NS	NS	S	S	S	S		
Pyridoxine-Supplemented Diet										
Controls (infected)	6.4±0.2	7.8±1.2	7.5±0.7	5.5±0.8	4.7±0.5	5.8±0.1	7.0±0.3	8.7±0.2	10.0±0.2	12.5±0.7
Experimentals (infected)	6.2±0.5	8.2±0.3	7.4±0.3	5.6±0.3	5.7±0.5	7.0±0.3	8.2±0.3	9.8±0.2	11.0±0.3	14.4±0.4
Increase over controls	−0.2	0.4	−0.1	0.1	1.0	1.2	1.2	1.1	1.0	1.9
Statistical significance at 1% level	NS	NS	NS	NS	NS	S	S	S	S	S

NS = Not statistically significant S = Significant

TABLE 65. *Food intake of mice fed complete, pantothenate-deficient, and pantothenate-deficient but pantothenate supplemented diets and infected with* Trypanosoma duttoni *(data in grams per mouse per day ±SD). (After Lincicome and Lee, 1970a)*

	Day 5	*10*	*15*	*20*	*25*	*30*	*35*	*40*	*45*	*50*
					Control Diet					
A	1.1±0.3	1.3±0.6	1.9±0.2	2.8±0.4	3.3±0.1	3.8±0.2	4.1±0.2	4.6±0.1	4.8±0.2	5.1±0.3
B	1.1±0.4	1.5±0.3	1.7±0.1	2.5±0.2	2.8±0.2	2.9±0.3	3.1±0.3	3.4±0.2	3.8±0.5	4.0±0.5
C	0	0	0.2	0.3	0.5	0.9	1.0	1.2	1.0	1.1
D	NS	NS	NS	NS	NS	S	S	S	S	S
					Pantothenate-Deficient Diet					
A	1.1±0.4	1.5±0.3	1.3±0.2	1.6±0.2	1.7±0.1	1.9±0.1	2.0±0.5	2.2±0.2	1.8±0.2	1.5±0.1
B	1.2±0.2	1.1±0.2	1.4±0.4	1.2±0.6	1.5±0.3	1.7±0.2	2.1±0.3	1.6±0.2	1.2±0.2	1.0±0.2
C	0	0.4	0	0.4	0.2	0.2	0	0.6	0.6	0.5
D	NS	NS	NS	NS	NS	NS	NS	S	S	S
					Supplemented Pantothenate-Deficient Diet					
E	1.0±0.2	1.1±0.1	1.1±0.5	1.4±0.3	1.5±0.2	1.9±0.3	2.1±0.2	2.0±0.2	2.5±0.2	2.4±0.5
F	1.2±0.2	1.1±0.2	1.4±0.4	1.2±0.6	1.5±0.3	1.7±0.2	2.1±0.3	1.6±0.2	1.2±0.2	1.0±0.2
C	0	0	0	0.2	0	0.2	0	0.4	1.3	1.4
D	NS	NS	NS	NS	NS	NS	NS	S	S	S

A = Infected
B = Noninfected
C = Increase over controls
D = Statistical significance at 1% level
E = Experimental (supplemented with pantothenate)
F = Control (pantothenate-deficient)
NS = Not statistically significant
S = Significant

TABLE 66. *Food intake of rats fed complete, pyridoxine-deficient, and supplemented pyridoxine-deficient diets and infected with* Trichinella spiralis *(data in grams per rat per day ±SD). (After Sen, 1969)*

	Day 15	*25*	*35*	*45*	*55*
			Control Diet		
Controls (uninfected)	11.5±0.7	13.0±1.0	14.4±0.5	15.6±1.0	17.3±1.0
Experimentals (infected)	11.6±1.1	13.6±1.0	15.2±1.0	16.1±0.9	18.0±0.4
Increase over controls	0.1	0.6	0.8	0.5	0.7
Statistical significance at 5% level	NS	NS	NS	NS	NS
			Pyridoxine-Deficient Diet		
Controls (uninfected)	5.9±1.4	5.1±0.7	4.4±0.5	3.9±0.7	3.6±0.7
Experimentals (infected)	6.3±1.8	5.6±0.6	6.2±0.4	7.4±1.4	9.1±0.8
Increase over controls	0.4	0.5	1.8	3.5	5.5
Statistical significance at 1% level	NS	NS	S	S	S
			Pyridoxine-Deficient Supplemented Diet		
Controls (unsupplemented)	5.9±1.4	5.1±0.7	4.4±0.5	3.9±0.7	3.6±0.7
Experimentals (supplemented)	5.9±1.3	5.8±0.4	7.2±0.5	9.1±1.0	10.4±0.8
Increase over controls	0	0.7	2.8	5.2	6.8
Statistical significance at 1% level	NS	NS	S	S	S

NS=Not statistically significant
S=Significant

The A isolate of *T. lewisi* was assayed in mice and rats after 39, 75, 95, 189, and 293 consecutive serial transfers in calorically restricted mice supplemented with a daily quantum of normal rat serum to determine whether: (1) the isolate would grow in starved and nonstarved mice without the aid of supplementary serum; (2) it grew better in serum-supported mice than trypanosomes originating from rats; and (3) it would reestablish itself in its normal rat host (Lincicome, 1959,b). The principal findings were that: (1) it could

TABLE 67. *Mortality of thiamine-deficient rats infected with* Trypanosoma lewisi. *(After Lincicome and Shepperson, 1965)*

Rat	*Day of Death* Uninfected	*Infected*
1	31	35
2	34	35
3	35	38
4	37	39
5	37	41
6	38	43
7	38	43
8	39	43
9	39	46
10	41	47
11	41	50
12	43	56
13	45	57
14	51	64
$\bar{X}\pm$SE	*39±1.3*	*46±2.3*

TABLE 68. *Average day of death ±SD of pantothenate-deficient mice inoculated with living* Trypanosoma duttoni, T. duttoni *cell homogenates and* T. duttoni *cell metabolic products. (After Lincicome and Lee, 1970a)*

Rat	*Controls*	*Infected*	*Homogenate*	*Metabolic Products*
1	25	65	66	33
2	37	72	72	35
3	53	81	73	47
4	59	89	76	50
5	59	91	78	52
6	61	96	80	58
$\bar{X}$	*49±15*	*82±12*	*74±5*	*46±10*
Statistical significance at 1% level		S	S	NS

NS=Not statistically significant
S=Significant

TABLE 69. *Microliters of oxygen (±SE) consumed by rat liver and kidney slices from* Trypanosoma lewisi-*infected rats fed thiamine-adequate but restricted diets, and thiamine-inadequate diets. (After Lincicome and Shepperson, 1965)*

Animal Group	*Day 21*	*28*	*35*	*42*
Liver				
Adequate diet				
uninfected	5.1±1.0	4.3±0.8	4.5±0.4	3.2±0.5
infected	4.8±0.6	4.2±0.3	3.2±0.2	2.5±0.8
Thiamine-def. diet				
uninfected	3.5±0.4	3.9±0.6	2.4±0.5	
infected	4.1±0.6	4.8±0.1	3.7±0.3	
Pairfed control diet				
uninfected	3.2±0.4	2±0.2	2.5±0.5	
infected	2±0.3	3±0.4	2.1±1.0	
Kidney				
Adequate diet				
uninfected	9.1±0.5	7.7±0.8	7.4±0.3	5.8±0.6
infected	11.2±0.3	6.7±0.3	9.5±0.3	9.8±0.2
Thiamine-def. diet				
uninfected	3.4±0	5.4±0.5	4.6±0.6	
infected	6.6±1.0	7.9±2.0	6.2±0.9	
Pairfed control diet				
uninfected	2.8±0.7	4.2±0.7	3.1±0.9	
infected	5.6±0.4	11.9±0.1	10.8±0.8	
Week of infection	*1*	2	*3*	*4*

not grow in calorically restricted mice without normal rat serum, which showed that any superficial adaptation it may have acquired would not allow it to grow unsupported in a heterologous environment; (2) it grew better in calorically restricted mice than trypanosomes which had never been passed through heterologous hosts—this was interpreted to mean that the A isolate became adapted to an heterologous environment only so long as rat serum was supplied as a supplement; (3) it had not lost its innate ability to reestablish itself in its natural homologous host, though there were indications it did not grow as well as did control trypanosomes.

At the 75th passage in calorically restricted albino mice the A isolate developed in albino mice whose dietary intake was normal but to which it was necessary to supply the rat serum supplement. This was the C isolate. Using the same criteria as above, the C isolate was maintained by serial transfer in mice for 220 passages over the period from April 1956 to August 1958. The length of the parasitemia in mice was stable, varying on the average per group of 10 transfer units from 5.2 to 6.5 days. Approximately 60 % of the 831 mice used in this particular aspect of the study developed maximal (4+)

TABLE 70. *Endogenous uptake (in μl O_2 per mg N per hour) of oxygen ± SD by liver slices from* Trypanosoma duttoni-*infected and noninfected mice fed complete, pantothenate-deficient, and pairfed control diets. (After Lee and Lincicome, 1970c)*

	Day 15		*23*		*31*		*39*		*47*		*55*		$\bar{X}$	
							Complete Diet							
A	19±3	15±6	26±2	23±2	20±4	38±5	39±2	33±2	34±4	39±1	32±3	34±2	*28±3*	*31±3*
B	14±6	20±3	21±4	15±2	18±2	21±5	26±3	19±4	19±5	21±5	17±5	24±3	*19±4*	*20±4*
C	5	0	5	8	2	17	13	14	15	18	15	10	*9*	*11*
D	36	0	24	53	11	80	50	74	79	86	88	42	*47*	*55*
E	NS	NS	NS	S	NS	S	S	S	S	S	S	S		
							Pantothenate-Deficient Diet							
A	18±4	18±4	31±3	19±2	28±2	21±5	27±4	31±6	36±5	29±2	31±2	34±4	*29±3*	*25±4*
B	20±7	18±5	27±6	17±3	19±3	21±2	14±3	17±8	18±4	22±2	19±5	24±1	*20±5*	*20±4*
C	0	0	4	2	9	0	13	14	18	7	12	10	*9*	*5*
D	0	0	15	12	47	0	93	80	100	31	68	40	*45*	*25*
E	NS	NS	NS	NS	S	NS	S	NS	S	S	S	S		
							Pairfed Control Diet							
A	24±5	21±5	37±2	25±2	36±3	31±4	40±2	40±5	35±4	37±8	27±3	39±2	*35±3*	*32±4*
B	18±2	26±2	26±1	18±3	21±3	20±2	22±4	27±1	24±2	14±3	19±1	21±5	*22±2*	*23±3*
C	5	0	11	7	15	11	18	13	11	13	8	18	*13*	*9*
D	28	0	42	39	71	55	81	48	46	93	42	86	*59*	*41*
E	NS	NS	S	S	S	S	S	S	S	S	S	S		

A = Infected
B = Uninfected
C = Increase over controls
D = Percent increase over controls
E = Statistical significance at 1% level
NS = Not statistically significant
S = Significant

TABLE 71. *Survival (in days) of* Trypanosoma lewisi *in albino mice receiving serum supplements and a restricted diet. (After Lincicome, 1958b)*

Treatment *Expt. No.*	*4*	*8*	*10*	*11*	*22*	*23*	*28*	*30*	*34*
No serum; diet unrestricted	1 (3)[a]	1 (3)	1 (6)		1 (3)	1 (3)			1 (3)
No serum; restricted diet				1 (3)	1 (3)	1 (3)			4 (3)
Serum; diet unrestricted	6[b] (3)	6 (3)	6(C_3H) 7(Beige) (6)		7–9 (3)	3–6 (3)			7 (3)
Serum; restricted diet				>9 (3)	10–17 (2)	>10 (3)	6–14 (6)	10–15 (8)	9–16 (3)

[a] Number of animals used are in parentheses
[b] 0.5 ml normal rat serum in this experiment. 1 ml used in all other experiments

infection. The first 100 passages showed evidences of major adjustments of the trypanosome to a heterologous environment. The second 100 passages showed relative stability of the numbers of host animals succumbing. An average of 3.7 days was required for the interval from inoculation to the observed maximal parasitemia, and there was little variation in the duration of this maximal population (1.7 days). An average of 3.9 days elapsed to the next passage. The trypanosome apparently made an adjustment to its new

TABLE 72. *Growth of* Trypanosoma lewisi *in the heterologous mouse host supplied with varying amounts of food and one milliliter normal rat serum per day. (After Lincicome, 1958b)*

Amount host food/day in g *Expt. No.*	*Length of parasitemia in days*[a] *5*	*7*[b]	*14*	*15*	*16*	*Maximum parasitemias expressed on an arbitrary scale* *5*	*7*[b]	*14*	*15*	*16*
1g			4 10 9	7	9			3+ 4+ 4+	4+	2+
			7 8 7	8	9			4+ 3+ 2+	4+	4+
			7 6 8		11			4+ 2+ 4+		3+
			13 8 9					4+ 3+ 4+		
2g					10					1+
					0					0
					8					2+
3g	6	5		8	7	2+	1+		1+	1+
	5	4		6	6	2+	1+		1+	2+
		2		0	4		1+		0	1+

[a] Each figure represents a single animal
[b] 0.5 ml serum given each animal in this experiment only

TABLE 73. *Summary of data collected from the first 100 serial passages of* Trypanosoma lewisi *("A" isolate) in the heterologous mouse host. The data are arranged as averages per group of 10 passages. Ranges of observations are in parentheses*

Passages	1–10	11–20	21–30	31–40	41–50	51–60	61–70	71–80	81–90	91–100	*Totals*
No. animals used	16	20	19	20	19	27	30	32	34	29	*246*
Interval to next passage,	3.9	5.2	3.9	4.2	4.1	4.0	4.1	3.9	4.2	3.5	*4.1*
days	(2–10)	(3–8)	(2–6)	(3–6)	(3–5)	(3–5)	(3–6)	(3–5)	(3–5)	(3–4)	*(2–10)*
Parasitemia											
% develop.											
3+Inf.	20	25	37	30	37	15	24	16	15	10	*21*
2+Inf.	27	5	0	0	5	18	13	9	12	4	*10*
1+Inf.	13	5	5	0	0	0	13	3	2	0	*4*
Interval (days) to maxi-	5	5	4.3	4.2	4.0	3.9	4.2	3.8	4.0	3.7	*4.1*
mal parasitemia	(2–9)	(3–7)	(1–6)	(3–7)	(2–7)	(2–7)	(2–6)	(2–6)	(2–6)	(3–5)	*(1–7)*
Duration of maximal	2.2	2.2	3.0	2.9	1.9	2.4	2.1	2.7	2.1	2.5	*2.1*
parasitemia (days)	(1–4)	(1–5)	(1–7)	(1–6)	(1–5)	(1–6)	(1–6)	(1–6)	(1–4)	(1–5)	*(1–7)*

TABLE 74. *Summary of data collected from serial passages 101 through 200 of* Trypanosoma lewisi *("A" isolate) in the heterologous mouse host. The data are arranged as averages per group of 10 passages. Ranges of observations in parentheses*

Passage	101–110	111–120	121–130	131–140	141–150	151–160	161–170	171–180	181–190	191–200	*Totals*
No. animals used	30	30	30	30	30	30	30	30	30	30	*300*
Interval to next passage, days	3.2 (3–4)	3.9 (3–4)	4.2 (4–5)	3.6 (3–4)	4.0 (3–5)	4.0 (3–5)	4.1 (3–5)	4.1 (3–5)	4.0 (3–5)	4.2 (3–5)	*3.9 (3–5)*
Parasitemia											
% develop.											
3+Inf.	17	20	20	4	23	3	13	13	20	13	*15*
2+Inf.	0	3	7	3	4	7	14	13	3	4	*5*
1+Inf.	0	0	0	0	0	0	0	7	0	0	1
Interval (days) to maximal parasitemia	3.5 (3–5)	3.8 (3–5)	4.3 (3–7)	3.7 (2–6)	3.9 (3–7)	4.0 (3–5)	4.0 (2–6)	3.9 (3–6)	3.8 (2–5)	4.4 (3–6)	*3.9 (2–7)*
Duration of maximal parasitemia (days)	2.8 (1–7)	2.3 (1–4)	2.2 (1–5)	2.8 (1–5)	2.5 (1–6)	2.2 (1–6)	1.9 (1–5)	1.7 (1–3)	2.1 (1–5)	1.9 (1–4)	*2.2 (1–7)*

TABLE 75. *Summary of data collected from serial passages 201 through 300 of* Trypanosoma lewisi *("A" isolate) in the heterologous mouse host. The data are arranged as averages per group of 10 passages. Ranges of observations in parentheses. (After Lincicome, 1959a)*

											Totals	*Grand total*
Passage	201–210	211–220	221–230	231–240	241–250	251–260	261–270	271–280	281–290	291–300	*201–300*	*1–300*
No. animals used	30	30	30	29	30	30	30	30	30	27	*296*	*842*
Interval to next	4.1	4.1	4.1	3.7	3.8	4.0	3.9	3.5	3.9	3.3	*3.8*	*3.9*
passage (days)	(4–5)	(3–5)	(3–5)	(3–4)	(3–4)	(4)	(3–4)	(3–4)	(3–5)	(3–4)	*(3–5)*	*(2–10)*
Parasitemia												
% develop.												
3+Inf.	10	20	13	7	17	13	10	3	6	7	*11*	*15*
2+Inf.	0	13	10	0	6	0	3	0	0	8	*4*	*6*
1+Inf.	0	0	0	0	0	0	0	0	0	0	*0*	*2*
Interval (days) to	4.2	3.9	4.3	3.7	3.5	3.8	3.7	3.5	3.2	3.3	*3.7*	*3.9*
maximal parasitemia	(2–6)	(3–5)	(3–7)	(3–5)	(3–5)	(2–4)	(3–6)	(3–5)	(2–4)	(3–5)	*(2–7)*	*(1–7)*
Duration of maximal	1.9	2.3	2.3	1.9	2.1	2.5	2.2	2.0	1.7	1.7	*2.1*	*2.2*
parasitemia (days)	(1–4)	(1–4)	(1–5)	(1–5)	(1–4)	(1–5)	(1–5)	(1–6)	(1–3)	(1–3)	*(1–6)*	*(1–7)*

TABLE 76. *Summary of data from 43 serial passages of* Trypanosoma lewisi *("B" isolate) in the heterologous mouse host. The data are arranged as averages per group of 10 passages. Ranges of observations in parentheses. (After Lincicome, 1959a)*

Passage	1–10	11–20	21–30	31–40	41–43	*Totals*
No. animals used	18	20	19	19	6	*82*
Parasitemia length in days	7.4	8.0	8.7	8.0	6.3	*7.9*
	(3–14)	(4–11)	(5–14)	(4–10)	(4–7)	*(3–14)*
Interval to next passage, days	4.9	5.4	4.6	4.7	5.0	4.9
	(3–6)	(3–8)	(3–6)	(4–6)	(4–6)	(3–8)
Death day after inoculation	7.5	8.5	8.7	8.5	6	*8.1*
	(5–12)	(5–12)	(6–12)	(5–11)	(4–7)	*(4–12)*
% animals dying	72	75	68	63	50	*68*
						(50–75)
Parasitemia—% develop. 4+Inf.	56	60	63	74	17	*60*
,, 3+Inf.	22	0	5	16	33	*12*
,, 2+Inf.	22	25	16	10	33	*20*
,, 1+Inf.	0	15	16	0	17	*8*
Interval (days) to max. parasitemia	4.6	4.3	4.2	4.8	3.6	*4.4*
	(2–9)	(2–8)	(2–6)	(3–7)	(3–5)	(2–9)
Duration of max. parasitemia (days)	2.1	3.7	4.3	2.7	2.2	3.1
	(1–5)	(1–7)	(1–8)	(1–6)	(1–3)	(1–8)

environment, as judged by the characteristic of the proportion of host animals dying during the first 100 transfers. There was no change in day of death of host animals. The rat serum supplement to these mice remained at 1 ml/mouse/day.

The C isolate was assayed in mice and rats after 28, 114, and 209 consecutive serial transfers in nonstarved mice given a quantum of normal rat serum daily to determine whether: (1) the isolate would grow in nonstarved mice without the aid of supplementary serum; (2) it grew better in serum-supported mice than trypanosomes originating from rats; and (3) it would reestablish itself in rats. The chief findings of this analysis were: (1) it could not grow in nonstarved mice without normal rat serum; (2) it grew better in nonstarved mice supplemented with normal rat serum than trypanosomes which had not previously been in mice—the interpretation of this finding was that the isolate had become superficially adapted to life in a heterologous environment, but that this adaptation did not enable it to exist without rat serum; (3) it had not lost its innate ability to reestablish itself in the rat. Its growth in rats was superior to that of control trypanosomes (Lincicome, 1959b).

The major result of these studies on continued development of *T. lewisi* in the heterologous host was that there was no important metabolic change in the trypanosome permitting it to develop in the heterologous host without rat serum, and that it was capable of growing in its normal, homologous host upon return to it. The implication of this finding was that rat serum contained a growth factor necessary for the development of the trypanosome and that the heterologous mouse host could be used to assay the factor.

Quantitative Studies on the Rat Serum Growth Factor

Preliminary titration of rat serum for its potency in supporting trypanosome growth in mice showed that it was active at volumes of 1.0, 0.5, 0.25, and 0.1 ml (Tables 77, 78, and 79) but not at 0.01 ml. In view of the inherent variability of rat serum among samples taken at different times, the unit of measurement was changed from a volume basis to one of protein equivalents. Table 80 shows that the titration endpoint of activity lies somewhere between 2.5 and 2.0 mg protein equivalent volumes. Assuming 5.0 g/100 ml of serum as the protein concentration in albino rats, a 50 mg protein equivalent volume would be 1.0 ml of serum; a 2.5 mg protein equivalent volume would be 0.05 ml.

Heterologous Sources of Growth Factor

It was of interest to determine the specificity of the growth-promoting properties of rat serum for *T. lewisi*. Accordingly, the sera of man, chicken, cow, dog, swine, chinchilla, and horse were all tested (Lincicome and Francis, 1960, 1961). None was found to have any activity. The sera of hamster,

TABLE 77. *Populations of* Trypanosoma lewisi *developing in C3H mice weighing 15 g or less. (After Lincicome and Francis, 1961)*

	Average hemacytometer count of trypanosomes Normal rat serum (ml)		
Days after inoculation	1 (3)[a]	0.5 (3)	0.1 (3)
3	+[b]	+	+
4	7	6	2
6	206	123	27
8	499	584	132
10	1069	807	277
12	Dead	Dead	Dead

[a] Indicates number of hosts used
[b] Indicates insufficient numbers to be recorded in hemacytometer

TABLE 78. *Populations of* Trypanosoma lewisi *developing in 10–15g mice supplemented with 1.0, 0.1, and 0.01 ml of normal rat serum. (After Lincicome and Francis, 1961)*

Days after inoculation	Normal rat serum (ml): 1, Exp. No. *1*	1, *2*	0.1, *1*	0.1, *2*	0.01, *1*	0.01, *2*
3	4	2	3	2	—[a]	—
4	7	7	5	9	—	—
5	11	33	5	15	—	—
6	17	95	9	19	—	—
7	16	94	8	28	—	—
8	15	—	8	—	—	—

Average hemacytometer count of trypanosomes

[a] Indicates no trypanosomes found

TABLE 79. *Populations of* Trypanosoma lewisi *developing in mice supplemented with varying volumes of normal rat serum. (After Lincicome and Francis, 1961)*

Days after inoculation	Normal rat serum (ml): 1, Exp. No. *3*	1, *4*	0.05, *3*	0.05, *4*	0.01, *3*	0.01, *4*
3	2	19	3	3	—[a]	—
4	7	18	6	7	—	—
5	32	78	11	7	—	—
6	34	101	8	12	—	—
7	—	98	—	10	—	—

Average hemacytometer count of trypanosomes

[a] Indicates no trypanosomes found

guinea pig, and rabbit were found to have activity, and, therefore, were quantitatively compared with rat serum. Table 81 shows that hamster serum is essentially three quarters as active as rat serum in promoting maximal growth of the trypanosome. Table 82 shows that guinea pig serum is about half as effective (Table 83).

TABLE 80. *Populations of* Trypanosoma lewisi *developing in mice supplemented with variable quantities of normal rat serum protein equivalents. (After Lincicome and Francis, 1961)*

Days after inoculation	*Average hemacytometer count of trypanosomes* — Normal rat serum (mg of protein equivalents) 50			2.5			2.0			0.5			0.1		
Exp. No.	*5*	*6*	*7*	*5*	*6*	*7*	*5*	*6*	*7*	*5*	*6*	*7*	*5*	*6*	*7*
3	8	—[a]	—	3		—		—	—	—				—	
4	62	4	6	9		4		—	—	—				—	
5	58	39	7	15		8		—	—	—				—	
6	75	49	22	15		8		—	—	—				—	
7	88	79	38	18		10		—	—	—				—	
8	87	78	43	12		8		—	—	—				—	

[a] Indicates no trypanosomes were found

TABLE 81. *Populations of* Trypanosoma lewisi *developing in 10–15 g mice supplemented with varying quantities of rat and hamster sera using protein equivalents as a basis of measurement. (After Lincicome and Francis, 1961)*

Days after inoculation	*Average hemacytometer trypanosome count* — Normal rat serum (mg of protein equivalents) 2.5		2.0		Normal hamster serum (mg of protein equivalents) 2.5		2.0	
Exp. No.	*9*	*10*	*9*	*10*	*9*	*10*	*9*	*10*
3	2	2	—[a]	—	1(50)	—(0)[b]	—	—
4	5	5	—	—	3(60)	3(60)	—	—
5	9	8	—	—	6(67)	8(100)	—	—
6	11	10	—	—	9(82)	5(50)	—	—
7	10	11	—	—	8(80)	8(73)	—	—
8		8				4(50)		

[a] Indicates that no trypanosomes were found
[b] Efficiency relative to Rat Serum for corresponding day in %
Effective Index = 9/11 for Exp. 9 = 82%
Effective Index = 8/11 for Exp. 10 = 73%
} *Av.* = *78%*

Identity of the Growth Factor

The ability of normal rat serum to support *T. lewisi* in heterologous mice can be altered by host starvation (Lincicome and Hinnant, 1962). Table 84 shows

TABLE 82. *Populations of* Trypanosoma lewisi *developing in 10–15 g mice supplemented with varying quantities of rat and guinea pig sera using protein equivalents as a basis of measurement. (After Lincicome and Francis, 1961)*

	Average hemacytometer trypanosome count							
	Normal rat serum (mg of protein equivalents)				Normal guinea pig serum (mg of protein equivalents)			
	2.5		2.0		2.5		2.0	
	Exp. No.							
Days after inoculation	*11*	*12*	*11*	*12*	*11*	*12*	*11*	*12*
3	3	1	—[a]	—	2(67)[b]	1(100)	—	—
4	5	5	—	—	3(60)	4(80)	—	—
5	8	9	—	—	5(63)	5(56)	—	—
6	9	10	—	—	5(56)	5(50)	—	—
7	9	10	—	—	5(56)	5(50)	—	—
8	6	7	—	—	5(83)	5(71)	—	—

[a] Indicates no trypanosomes were found

[b] Efficiency Index in percent relative to control of corresponding day

$$\text{Effective Index} = \frac{\text{Max. Parasitemia of Exp.}}{\text{Max. Parasitemia of Control}} \quad \begin{matrix} = 5/9 = 56\% \text{ for Exp. 11} \\ = 5/10 = 50\% \text{ for Exp. 12} \end{matrix}$$

TABLE 83. *Populations of* Trypanosoma lewisi *developing in 10–15 g mice supplemented with varying quantities of rat and rabbit serum using protein equivalents as a basis of measurement. (After Lincicome and Francis, 1961)*

	Average hemacytometer trypanosome count[a]							
	Normal rat serum (mg of protein equivalents)				Normal rabbit serum (mg of protein equivalents)			
	2.5		2.0		2.5		2.0	
	Exp. No.							
Days after inoculation	*13*	*14*	*13*	*14*	*13*	*14*	*13*	*14*
3	1	3	—[b]	—	—	—	—	—
4	4	4	—	—	1(25)[a]	2(50)	—	—
5	7	7	—	—	1(14)	1(14)	—	—
6	7	9	—	—	2(28)	4(44)	—	—
7	5	10	—	—	3(60)	4(40)	—	—
8	3	6	—	—	1(33)	2(33)	—	—

[a] Effective Index = 3/7 = 43% for Exp. 13 and 4/10 = 40% for Exp. 14. *Av. Index* = *42%*

[b] Indicates no trypanosomes were found

[c] Efficiency in percent relative to control of corresponding day

TABLE 84. *Populations of* Trypanosoma lewisi *in tail bloods of mice supplemented with sera of starved (5 days) and well-fed young rats. (After Lincicome and Hinnant, 1962)*

	Animal (all inoculate on day 1)	*Hemacytometer count Day of experiment* 3	5	7	9
	101–7	<1	<1	2	1
	101–8	2	6	12	8
	101–9	<1	2	12	D[c]
C[a]	101–10	<1	<1	1	2
	101–11	3	8	25	28
	101–12	2	8	25	D[c]
	101–1	<1	<1	<1	D[c]
	101–2	<1	2	3	2
	101–3	<1	<1	1	<1
X[b]	101–4	<1	<1	1	1
	101–5	D			
	101–6	<1	3	3	D[c]

[a] C=Control mice receiving sera of well-fed young rats
[b] X=Experimental mice receiving sera of starved young rats
[c] D=Died

that this ability is reduced at 5 days of starvation, more so after 7–8 days (Table 85), and drastically reduced by 12–14 days of starvation.

Greenblatt and Lincicome (1966) carried the analysis of the identity one step further by fractionation studies on serum. Through ultrafiltration, dialysis, and Sephadex G-25 separations they showed that it was the protein and not the crystalloid fraction that was the active component. Precipitation of the globulin fractions by $(NH_4)_2SO_4$ and Na_2SO_4 gave β and γ globulins which

TABLE 85. *Populations of* Trypanosoma lewisi *in tail bloods of mice supplemented with sera of starved (7–8 days) and well-fed old rats. (After Lincicome and Hinnant, 1962)*

	Animal (all incoulated on day 1)	*Hemacytometer count Day of experiment* 4	6	8	10
	65–1	3	72	300	D[c]
	65–2	6	24	107	D
C[a]	65–3	<1	17	25	D
	65–4	3	61	184	D
	65–5	2	47	99	D[c]
	65–6	3	13	49	D[c]
X[b]	65–7	4	60	88	D[c]
	65–8	<1	13	17	11

[a] C=Control mice receiving sera of normal old rats
[b] X=Experimental mice receiving sera of starved old rats
[c] D=Died

TABLE 86. *Populations of* Trypanosoma lewisi *in tail bloods of mice supplemented with sera of starved (12–14 days) and well-fed old rats. (After Lincicome and Hinnant, 1962)*

	Animal (all incoulated on day 1)	*Hemacytometer count Day of experiment* 4	7	9	11
	58–1	2	47	76	D[c]
	58–2	2	78	D	
C[a]	58–3	2	57	124	307
	58–4	1	36	26	290
	58–5	1	81	114	D[c]
	58–6	<1	2	4	D[c]
	58–7	<1	<1	3	D[c]
X[b]	58–8	<1	<1	<1	D[c]
	58–9	<1	<1	<1	<1
	58–10	<1	<1	4	4

[a] C = Control mice receiving sera of normal old rats
[b] X = Experimental mice receiving sera of starved old rats
[c] D = Died

TABLE 87. *Hemacytometer counts of trypanosome populations in peripheral tail bloods of normal and chimeric mice given rat serum. (After Cosgrove,* et al., *1969)*

Mice[a]		*Day of Experiment* 1[b]	2	3	4	5	6	7	8	9	10
	1	—[c]	—	D[d]							
	2	—	—	+[e]	3	6	11	D			
	3	—	—	9	59	158	578	D			
Chimeras	4	—	—	—	+	4	9	42	137	483	D
	5	—	—	—	21	D					
	6	—	—	8	110	D					
Normal	1	—	—	3	13	72	81	103	134	+	—
	2	—	—	1	D						

[a] Equal numbers of chimeras and normal animals were also inoculated with trypanosomes but given no rat serum supplements. Since in none of these could trypanosomes be detected even by the sensitive rat-inoculation test (Greenblatt and Lincicome, 1966) these animals have not been included in this table
[b] Day of injection
[c] "—" = No trypanosomes detected
[d] "D" = Died
[e] "+" = Trypanosomes seen but in numbers insufficient for enumeration by an hemacytometer

TABLE 88. *Aminotransferases (GPT and GOT) in homogenates and metabolic products of* Trichinella spiralis *(data in milliunits of enzyme activity per* 4×10^4 *larvae. (After Sen, 1969)*

	Homogenates				*Metabolic Products*			
Sample[a]	*GOT*		*GPT*		*GOT*		*GPT*	
1	4.0	6.0	5.0	5.0	4.0	4.0	5.0	2.5
2	8.2	6.2	5.3	6.0	4.5	4.5	5.0	3.0
3	5.1	5.0	6.0	7.0	5.0	4.0	5.0	3.5
4	5.0	5.0	6.5	7.0	5.0	3.0	4.0	4.0
5	4.5	4.5	7.0	7.0	4.5	3.5	4.0	4.0
6	6.0	5.2	7.0	7.0	5.0	4.0	4.0	3.0
$\bar{X}$	*5.1*	*5.3*	*6.1*	*6.5*	*4.6*	*3.8*	*4.5*	*3.3*
$\bar{X}\pm$SD	*5.2±0.7*		*6.3+0.7*		*4.2±0.6*		*3.9±0.7*	

[a] Each sample in duplicate

TABLE 89. *Pyridoxine content of larval* Trichinella spiralis *(data in micrograms per* 4×10^4 *larvae.) (After Sen, 1969)*

Sample	*Exp. 1* μg	*Exp. 2* μg
1	0.75	0.75
2	0.50	0.55
3	0.50	0.75
4	0.50	0.75
$\bar{X}\pm$SD	*0.56±0.07*	*0.70±0.1*

were active and an albumen which was not. Further analysis of globulins by DEAE and Sephadex G-200 column chromatography and ultracentrifugation indicated that the active growth-promoting component was in the gamma-two globulins and macroglobulins.

A Nutritional or Immunologic Problem?

From the very inception of this work, there was the question whether the phenomenon was mediated through nutritional or immunologic factors. Was the growth-promoting substance in rat serum (and to a lesser extent in hamster, guinea pig, and rabbit sera) a nutritional phenomenon, or was it due to immunologic blocking of heterologous tissue reactions, thus permitting the trypanosome to grow? Cosgrove *et al.* (1969) shed some light on this subject. Table 87 lists the results of inoculating chimeric mice (mice whose immunologic systems are destroyed by irradiation and replaced with rat hematopoietic tissue) with and without rat serum supplements. It is evident that *T. lewisi* still does not grow without the rat serum factor, even in chimeric mice.

DISCUSSION AND CONCLUSIONS

Parasitism was viewed by Lincicome (1963) as a great pattern of life on earth, and as a fundamental expression of a chemical (or molecular) relationship between two organic beings. The nature of this chemical basis determined "the degree of dependence, the harmoniousness of the relationship, the equalities of the exchanges, the antagonisms that may develop, etc." The brief review of parasitism, commensalism, phoresis, mutualism, and symbiosis in that paper (Lincicome 1963) held that all these phenomena referred to the same broad phenomenon of life. Commensalism and phoresis were regarded as ecologic descriptors identifying certain relationships in which there was no metabolic dependence by any participant. Little or no justification was expressed for retention of these terms in parasitology. By Smyth's (1962) definition, mutualism and symbiosis are variations of parasitism. Indeed, he says "both mutualism and symbiosis are merely recognized as special cases of parasitism in which some metabolic by-products of the parasite are of value to the host."

Read (1958) recognized the indefinable boundaries among mutualism, symbiosis, and parasitism and chose to regard them all, along with commensalism, as variations of a common theme of symbiosis. This position was seriously questioned (Lincicome 1963) on the basis that, by practice over the years, the term had become restrictively applied to mutually dependent organisms with an obligation attached lacking in a mutualist association. The concept of parasitism as an intimate association between two organisms in which the dependence of parasite is metabolic, expressed variously by Smyth (1962), Cameron (1958), Lewis (1953, 1957), and Lincicome (1953, 1963), applies no restrictions and provided for broad interpretations. It is, therefore, the umbrella over all previous terms attempting to define this fundamental way of life.

Under the "chemical basis of parasitism" (Lincicome, 1963) the various manifestations of the phenomenon could be accomodated if the whole were regarded as a "two-way affair," or in other words as "a going and a coming" or "a giving and a taking." There was something more than just this however: an element of goodness accompanying a parasitic relationship that was the structural foundation of this phenomenon. While the "chemical basis" idea explained the molecular structuring of this ecologic association, the fundamental function of this relationship was not so adequately or comprehensively explained, though the development of such terms as mutualism, commensalism, and symbiosis were logical attempts to do so. The concept of the goodness of parasitism provides an all-embracing basis for a rational philosophy and a more profound appreciation of the significance of this life expression. Goodness is, of course, an anthropometric descriptor but how else may we view the organic environment around us? It gives to us a sense of order and progression of nature. The descriptors I have used here to provide evidence for the idea are by no means limited to these forms. I have meant only that *Trypanosoma lewisi*, *Trypanosoma duttoni*, or *Trichinella spiralis* are but examples. These organic units were but convenient, handy experimental tools and were intended to represent two extremes in association. Such tools as DNA, poliomyelitis virus, or phytomonads might have been used had I had the technical expertise.

There is now a considerable mass of experimental evidence which has been growing in my laboratory for several years, supporting the concept of the goodness of parasitism. I have attempted to bring this evidence together in these pages. Two major lines of investigation were pursued: (1) the contributions of the parasite; (2) the contributions of the host.

The Contributions of the Parasite

The employment of the metabolic imbalance technique (Lincicome, 1953) has made possible many studies of the metabolic role of parasites in parasitism. This procedure has been useful in analyzing the following:

1. Host body weight gains
2. Host food consumption
3. Enzyme levels in host tissues
4. Pantothenate in host tissues and trypanosomes
5. Pyridoxine levels in host tissues, trypanosomes, and *Trichinella*
6. Host longevity
7. Respiration of host tissues

Host Body Weight Gains.—Assuming that the manner in which an animal grows is a good measure of its well-being, the growth of rats and mice was studied in detail. These rodents were placed on balanced diets and experimentally given appropriate inocula of *Trypanosoma lewisi, T. duttoni,* or *Trichinella spiralis*. The trypanosomes represented well-known nonpathogenic parasites of the blood stream of rats and mice, respectively, and the *Trichinella* (an oft-studied pathogen) was capable of development in either rats or mice. The first two organisms are host-specific while the latter infects a large number of vertebrates, including man. All experimentally-infected animals grew at a faster rate than did their control, uninoculated counterparts. In other words they were bigger and apparently healthier than their control mates (Tables 5, 6, 10, 11, 12, 15, 16, 21). As the study of body weights progressed, it was clear that statistically significant results were more likely if detailed attention were given to the animals' initial body weights. Test animals (rats) whose average weight had a standard deviation of 1 g or less were acceptable. Animals beyond this range were excluded from test.

Two metabolic-imbalance conditions were applied to rodents with the three organisms above for further study of body weight gains. The rationale supporting these studies was that, were the host experimentally deprived of a single metabolic factor, any restoration of function in the animals provided with a parasite would be good presumptive evidence of a contribution on the part of the parasite. A pantothenate imbalance in mice with *T. duttoni* clearly led to better weight gains than those observed in similarly imbalanced hosts without the trypanosome (Table 22). A thiamine imbalance in rats with *T. lewisi* likewise provided clear evidence that this trypanosome was providing thiamine or thiamine-like substance to its host (Table 19). Rats imbalanced in pyridoxine and inoculated with *T. lewisi* (Table 25) also grew better than did rats without the trypanosome. Finally, pyridoxine-imbalanced rats with *Trichinella* grew better than did animals without trichina worms (Table 26). All these experiments, of course, were fully controlled by incorporation of pair feeding and supplementation of the vitamins.

Growth stimulation in the above instances was related to the parasite. How was this brought about? Was it the presence of the foreign cell or organism? Was it some product secreted by the parasite? To answer these questions homogenates of whole trypanosome cells or whole bodies of larval *Trichinella* were washed free of contaminating host proteins, homogenized, and inoculated into rats and mice. Rats carrying *Trypanosoma lewisi*, whether they were well fed or pyridoxine deficient (Table 43), grew faster than did uninoculated hosts, in a manner similar to that produced by living trypanosomes. Comparable observations were made for mice infected with *T. duttoni* and for rats infected with *Trichinella* (Tables 49 and 51). This suggested that whatever factor was involved, it bore some relationship to the cell or organism. To determine this relationship, metabolic products of these parasites were prepared and inoculated. No stimulation of growth of mice or rats followed administration of metabolic products of both trypanosomes, whatever the metabolic balance or imbalance (Table 56 and 58). This fact emphasized that the factor was probably associated with the living cell or something within it, and not with something it secreted. Rats infected with *Trichinella*, on the other hand, showed growth stimulation only in states of pyridoxine imbalance, thus indicating that *Trichinella* probably secretes a substance having such properties.

The fact that these observations have been carried out under the well-controlled experimental conditions of the laboratory does not assure that the same results will obtain under natural conditions. We have, however, made an attempt to simulate nature in that the inocula of parasites were small, thus possibly reflecting exposure under natural conditions. To have administered large numbers of parasites would have defeated the purpose, and more than likely would have killed at least some of the hosts.

The general, uncontrolled daily observations in the laboratory have left the impression that infected rats and mice were more active physically, and were more responsive to the human presence. If this be true, then the heavier infected animal would likely in nature have an advantage over the uninfected animal, since by its kinetic movements and heavier weight it would be in a better competitive position. This is certainly an idea worthy of experimental analysis. I can anticipate that this sort of ecologic advantage would have significant implications for the survival of an individual and ultimately for survival of a particular species, provided the method of natural propagation of the parasite assured its transfer to subsequent generations. This idea then has considerable merit in the evolutionary development of species and faunas.

Host Food Consumption.—The amount of food eaten served to throw light on the stimulation of growth of infected hosts. Theoretically the greater kinetic activity of a host might be reflected in the overall amount of food eaten, possibly indicating the number of times the animal moved to the feeding station; or, as in metabolically imbalanced hosts, as a marker for the improvement in an anorexic state produced by the withdrawal of the metabolic factor.

The quantity of food consumed was measured for animals on a complete diet and under the several states of metabolic imbalance (pyridoxine, thiamine, and pantothenate). It was also measured for animals inoculated with homogenates and metabolic products. Both rats infected with *T. lewisi* and mice infected with *T. duttoni* ate more food (Tables 64 and 65), while no such differences were demonstrable in *Trichinella*-infected rats (Table 66). However, pyridoxine-deficient rats showed significant differences in food intake.

Food consumption of metabolically imbalanced rats receiving *T. lewisi* homogenate was greater than those not receiving homogenate, thus confirming the observation on body weights (Table 44). Significant relief from the anorexia was also shown by mice with a pantothenate imbalance (Table 50) and again the homogenate gave similar results. The pattern of *Trichinella* was similar to that for body weights in that there were no significant differences in rats on the complete diet (Table 53), but pyridoxine-imbalanced rats with this worm homogenate ate more food.

With respect to metabolic products, neither of the trypanosomes responded and it is therefore concluded that these trypanosomes do not relieve the anorexia through a secretory substance released to the rat via the blood. Following the same pattern observed for homogenates, *Trichinella*-infected rats on a complete diet showed no differences in food intake, but did so under pyridoxine-deficiency states. This is further confirming evidence that the *Trichinella* larva secretes products affecting the metabolic state of the rat. The isolation of such products might have important and interesting applications. A problem for the future might deal with the metabolic effects of secretory products (sexcretions) gathered *in vitro* on cell respiration or cell growth.

Enzyme levels in Host Tissues.—These were measured under several conditions of the host: in well-fed animals; in metabolically-imbalanced animals; in animals inoculated with homogenates; in animals inoculated with metabolic products.

The levels of transketolase in rat liver and kidney homogenates gave evidence that tissues of infected rats contained more thiamine than did tissues of noninfected rats. The effect of trypanosomes upon transketolation was more pronounced in thiamine-deficient rat tissues. Brin (1962) pointed out that the addition of thiamine to tissue in which apotransketolase was already saturated with the coenzyme, had little effect on further enhancement of transketolase activity. This is possibly the explanation for the small enhancement of enzyme noted in tissues of well-nourished, infected rats (Table 27 and 28). Tissues from thiamine-deficient rats with or without infections of *T. lewisi* showed increased transketolase activity upon addition of the homogenates.

Singer (1961) studied thiamine levels in liver, kidney, and blood of rats infected with *T. rhodesiense* and fed normal and deficient diets, and found enhanced levels of the vitamin in liver and blood throughout the infection in all animals. It is obvious that beneficial substances may pass from parasite to host even in host-parasite relationships traditionally regarded as pathogenic. Lincicome and Shepperson (1965) thought that substances from the cells of *T. lewisi* (they called them metabolic products) contributed to the increased metabolic activity of host cells, and mentioned carbon dioxide, ethanol, succinate, lactate, pyruvate, and acetate, all of which appear to be formed by this trypanosome (Ryley, 1951), as possible contributions to serve as metabolic sparkers. This is probably true in thiamine-deficient animals as shown subsequently in longevity studies.

Two aminotransferases (GPT and GOT) requiring pyridoxine as cofactor were also studied under several conditions of metabolic imbalance and administration of homogenates and metabolic products of *Trichinella spiralis* and *Trypanosoma lewisi*. All the evidence (Tables 29, 30, 31) indicated that the activities of these aminotransferases were reduced in deficiency states but were elevated when animals were infected. It is clear then that either the experimentally-eliminated factor was restored by the parasite or some substance having similar activity was supplied. Nelson and Lincicome (1966) studied serum transaminases in rats inoculated with *T. lewisi* and came to the conclusion that GOT and GPT levels were elevated during infection. Increased serum transaminase activity was believed to come from the host's tissues as well as from disintegrating parasite cells. More GPT than GOT (nine times more) was found in *T. lewisi* which is in contrast to most mammalian tissues. This fact correlated well with the finding that GPT levels are higher in the latter phases of the parasitemia because of immunologic disintegration of the trypanosomes. The elevation of aminotransferases in trypanosome infections would therefore seem to be caused by two mechanisms: one which activates the latent host enzyme systems through provision of the cofactor; the other which releases intact enzyme systems toward the latter phases of the events of the parasitemia.

Metabolism of Pantothenate in Host Tissues and Trypanosomes.—The finding that pantothenate levels in liver and plasma of mice infected with *T. duttoni* are all elevated (Tables 34 and 35) after the course of infection, correlates well with the natural history of the infection. The evidence suggests that pantothenate, like some aminotransferases (Nelson and Lincicome, 1966), is probably released into the circulation directly as the result of immunologic destruction of the *T. duttoni* cells. This comes from two to three weeks or more after introduction of the trypanosomes. Pantothenate is probably not directly secreted into the plasma flow since no pantothenate activity was observed for metabolic products of this trypanosome. This is indeed a convenient and natural contribution by the parasite and has ecologic significance to the individual host animal. Pantothenate was shown to be present in *T. duttoni* cell homogenates (Table 41).

The blood pyruvate levels of mice with *T. duttoni* are not easily interpretable, for in some instances these levels were elevated (Table 38) at times and reduced at others. Since the oxidation of pyruvate is dependent upon pantothenate's role in coenzyme A activity, the elevated levels may be a reflection of pyruvate as a metabolic end-product of the trypanosome (though this is not certain for *T. duttoni* as it is with *T. lewisi*). On the other hand, the reduced levels seen may be the result of additional pantothenate supplied through the destructive immunologic forces at work in this infectious process.

Pyruvate metabolism of host tissue slices clearly indicated that the trypanosome (*T. duttoni*) was responsible for the increased oxidation of pyruvate by liver slices of infected mice. This was probably, again, the result of release of pantothenate from immunologically destroyed cell bodies of the trypanosome.

Metabolism of Pyridoxine in Host Tissues, Trypanosomes, and Trichinella.—The increased pyridoxine content of livers of rats infected with *T. lewisi* (Table 36) reflects the natural history of events in this infectious process and is similar to that of *T. duttoni.* The vitamin appears to be released from the trypanosome not by secretory mechanisms but by disruption of the cell at the time when immunologic control forces reduction of the parasitemia. This phenomenon actually works for the benefit of the host, and is another example of metabolic exchanges between the components of this system. There appears to be about 0.55 microgram of pyridoxine bound in 2×10^8 trypanosomes. This figure provides some insight into the probable quantity of this vitamin that could be available to the organic environment.

A similar situation obtained for *Trichinella* infections (Table 37), because livers of infected rats contained more pyridoxine than did those of uninfected animals. Pyridoxine was also demonstrated to be present in larval homogenates (Table 89) as well as in larval metabolic products. The mechanism of pyridoxine availability in *Trichinella* infections, therefore, appears to be different than in trypanosome infections. *Trichinella* larvae appear to secrete pyridoxine or a substance having pyridoxine activity. This would appear logical since the worms are not immunologically destroyed as are trypanosomes, but are walled off so that their bodies are not actually in contact with host proteins. This is done by mineral deposits sufficient to isolate the larvae from physical contact but insufficient to prevent seepage of sexcretions into the host environment.

Host Longevity.—The significantly longer life of thiamine-deficient rats infected with *T. lewisi* (Table 57) is confirmation of previous observations that there is more thiamine available (reflected as more transketolase activity) to the tissues of infected animals than to uninfected ones. The additional thiamine probably permits more coenzyme formation, and this in turn means more oxidation of pyruvate and citric acid intermediates for greater energy supplies. Indeed, it is shown here that mice deficient in pantothenate also live longer if infected with *T. duttoni* (Table 58). This is a direct indicator of pantothenate in liver and plasma of mice infected with *T. duttoni.* More thiamine or more pantothenate serves to keep up the energy supply, thus extending the life of the organic environment. It would be instructive to design and execute an experiment, simulating nature, showing the energy balance accruing to trypanosome-infected rodents and how this affects the social behavior and success of these animals.

Respiration of Host Tissues.—The greater endogenous respiratory rates of tissues from mice and rats harboring trypanosomes generally confirm the conclusion already reached that thiamine and pyridoxine (or products having similar metabolic functions of these vitamins) are actually supplied by the trypanosomes. Endogenous rates reflect the greater metabolic activity and presumably a greater supply of utilizable substrates. Lincicome and Bruce

(1965) found greater metabolic activity in liver slices of mice infected with *Trypanosoma rhodesiense*, but not in heart slices. The presence of trypanosomes is definitely reflected in host metabolic behavior. It is certainly of much interest to determine the exact nature of this phenomenon in animals maintained on a complete diet. The mechanism in metabolic imbalance states may be readily explainable, as is possible from the above discussion, but not so in animals on well-balanced regimens.

The Contribution of the Host

The evidence of the host's (organic environment) metabolic contribution to the parasite accumulated from experimental studies of *Trypanosoma lewisi*. The basic tool developed for these studies was the heterologous host technique (Lincicome 1955, 1957, 1958a,b, 1959c). The philosophy responsible for the employment of heterologous hosts is given by Lincicome (1959c). A heterologous host is defined as one in which a parasite ordinarily does not grow. A parasite such as *Trichinella spiralis*, with natural ability to develop and grow in many kinds of hosts, is not host-specific, and is not a good candidate for transfer to a heterologous host. *Trypanosoma lewisi* is restricted to the small number of hosts in which it may grow or survive (Lincicome, 1958b) and is therefore suitable. It will grow well only in the albino rat, but may survive for a while in the rabbit, guinea pig, and hamster (see Lincicome, 1958b for review). The heterologous host technique has proved to be a valuable tool in experimental study of host metabolic contributions. The application of this technique made possible the study of trypanosome-environment relationships unobtainable in any other way, chiefly because the trypanosome must be studied in combination with its vertebrate host, otherwise it loses its metabolic identity and metamorphoses. The morphogenetic shift from trypanosome to crithidial (trypomastigote to epimastigote in the terminology of Hoare and Wallace, 1966) stage is not understood either functionally or metabolically.

Many years ago Coventry (1929) tried giving rat serum supplements to guinea pigs but could not detect any favorable effect upon the parasitemia. Perhaps this failure may be explained by the fact that the guinea pig permits some survival of *T. lewisi*, and its serum contains a survival factor not fully as potent as rat serum.

The development of *T. lewisi* in mice was shown first to be dependent upon rat serum supplements to the mouse (Table 71). The original idea in transmission of the trypanosome was that impairment of the mouse's immunologic defenses might favor the establishment of the trypanosome. Indeed this was true (Table 72), but after establishment of the cell in mice on a regular passage basis it was no longer necessary to starve the host prior to and during the parasitemia, but the rat serum supplementation was necessary. It was, therefore, apparent that rat serum contained a component or behaved in such a manner to make it possible for this trypanosome to live in a foreign environment without any morphogenetic change.

It next became a paramount question whether *T. lewisi* underwent any demonstrable metabolic change as it developed in and was passaged in supplemented mice. This problem had to be solved, for if the trypanosome underwent any significant metabolic change which made it no longer necessary to supplement the heterologous host, obviously the heterologous idea would not be workable.

The overall conclusion reached after more than three years of 300 serial subpassages of one isolate of *T. lewisi* was that it had not undergone sufficient detectable metabolic change to make it possible for it to grow in the mouse without the rat serum factor or to be unable to grow in its homologous host upon reintroduction (Lincicome, 1959a,b, 1960).

The minimal quantity of rat serum necessary on a daily basis was next considered (Tables 77, 78, 79). On the basis of volume alone the endpoint of activity lay between 0.01 and 0.10 ml of rat serum. Subsequent titration on a protein equivalent basis (Table 80) placed the endpoint between 2.5 and 2.0 mg protein equivalent volumes of rat serum.

The specificity of the factor was next considered. Tests of a large number of sera from common animals including man failed to detect any substances having the same function, except the sera of the hamster, guinea pig, and rabbit. Coventry (1929), as well as Kanthack *et al.* (1898), Laveran and Mesnil (1912), Laveran and Roudsky (1914), Nieschulz and Wawo-Roentoe (1929), observed survival in the guinea pig for up to 13 days. Other researchers had failed to do this. Using rat serum as the 100 % base, guinea pig serum was only about 50 % as effective (Table 82). Whether this means that only 50 % as much factor was present was not revealed in these experiments. Rabbit serum (Table 83) was less than 50 % as potent as rat serum, but hamster serum was 75 % as effective (Table 81).

Greenblatt and Shelton (1968) have done interesting work on mouse serum as an environment for *T. lewisi*. Trypanosomes contained in dialysis bags implanted in the mouse peritoneal cavity were found to undergo slight multiplication of the cells, but when contained in diffusion chambers, they grew well and were carried through seven weekly serial transplants. Rat serum supplements to mice with diffusion implants containing trypanosomes produced a great stimulus to further growth of the trypanosome.

Greenblatt and Shelton (1968) also tested human, monkey, dog, cow, newborn calf, horse, and chicken sera and nonspecific substances such as egg albumin, trypicase soy broth, dextran, zymosan, and polyvinyl pyrollidone for growth stimulation of *T. lewisi* in mice but without general success. However, human and monkey sera proved successful in supporting the trypanosome. They also were successful in getting *T. lewisi* to grow in gerbils. These results are important but require confirmation.

From the beginning of the work on establishment of *T. lewisi* in the heterologous mouse, there was the possibility that blockade or breakdown of the mouse's immunodefensive mechanisms might play a deciding role. Concerning the action of rat serum Lincicome (1958b) stated that it "may be simply protective as Pollack (1947) and Davis and Dubos (1946) have observed in cultures of the pertussis organism and the tubercle bacillus," and "It is conceivable that the relatively large amounts of rat serum employed effectively block the reticulo-endothelial system of the mouse. . . ." Recently it was possible to check the possibility of immunologic breakdown by using chimeric mice whose natural immunogenic mechanisms — having been destroyed by irradiation — were replaced by a functional rat system. In such chimeric mice, Cosgrove *et al.* (1969) found that rat serum was still a requirement for development, thus adding additional support for the view of Lincicome (1958b) that the phenomenon was nutritional in nature.

The identity of the factor in rat serum has been studied in part. It was found initially that starvation of the rat host reduced the potency of the factor (Tables 84 and 85). Starvation of the rat for 12–14 days drastically reduced the potency of serum, while 5–7 days of starvation had a lesser effect. The effects of starvation have not been studied further, but they should be because of modern, rapid means of electrophoretic and chromatographic separations. Greenblatt and Lincicome (1966) showed that the factor was associated with the protein part of the serum, and that the active component was associated with the α-2 globulins and macroglobulins. Recently Greenblatt *et al.* (1969) have extended the analytic procedure for the separation of the active components. The 7S α-2 globulins were separated chromatographically to yield β globulins which were active in supporting growth. This fraction was electrophoretically related to other similar fractions shown to stimulate cells in culture, or to support *in vitro* culture of *Plasmodium knowlesi* (Greenblatt *et al.*, 1969). Once this factor is fully identified and separated into pure form, it will be exciting to determine how specific it is for *Trypanosoma lewisi*. It could be the key to why *T. lewisi* is a *dependent* cell—a parasite.

ACKNOWLEDGMENTS

I am grateful to my students, without whose zealous companionship in research and without whose friendship and steadiness of purpose the work summarized here could not and would not have been done. Of these I want particularly to mention Dr. A. A. Warsi, Dr. Jacqueline Shepperson, Dr. John I. Bruce, Dr. Dilip Sen, Dr. Alfred Smith, Dr. Clarence M. Lee, Dr. Juanita Hinnant, Dr. Earlie Francis, Mr. Mohammed ElHelu, Mr. Joseph Adaramola, Mr. Sidney Draggan, Miss Gladys Wells, Miss Miriam McLean, Dr. George C. Hill, Dr. Emmanuel E. Eni, and Dr. Roy C. Watkins. To my colleague Dr. Charles Greenblatt I express appreciation for many fine hours of work and discussion together when I spent a sabbatical year at the National Institute for Arthritis and Metabolic Diseases in Bethesda, Maryland.

Most of the research reported here was carried out in my laboratory and was supported by grants AI-03409 (E-3409) and 2Tl AI 40 (2E-40) from the National Institutes of Health, Bethesda, Maryland, during the period from 1958 to 1968.

SUMMARY

The lines of evidence supporting a hypothesis of the goodness of parasitism are reviewed. Parasitism is viewed as a metabolic ecological association of two organisms, the basis of which is chemical and the function of which is fundamentally one of molecular exchanges of social, ecological, and evolutionary values.

Evidences of the contributions of the parasite came from the following parameters:

1. Body weight gains of well-fed rats.
2. Body weight gains of metabolically-imbalanced hosts.
3. Measurements of transketolases and aminotransferases.
4. Pantothenate levels in host tissues.
5. Pantothenate content of *Trypanosoma duttoni* cells.
6. Pyridoxine content of *T. lewisi* cells and larvae of *Trichinella.*
7. Pyridoxine levels in host tissues.
8. Pyruvate levels and utilization by host tissues.
9. Homogenates of parasites.
10. Metabolic products of parasites.
11. Host food consumption.
12. Host longevity.
13. Respiration of host tissues.

Evidences of the host's contributions to the parasite came from studies on:

1. Growth of *T. lewisi* in heterologous hosts.
2. Effect of continued development of *T. lewisi* in heterologous hosts.
3. Quantitative analyses of the rat serum factor.
4. Heterologous sources of growth factor.
5. Identity of the growth factor.

The technical procedures of metabolic-imbalance and the heterologous host were basic tools developed for these studies.

REFERENCES

Agosin, M. and Aravena, L. C. 1959. Anaerobic glycolysis in homogenates of *Trichinella spiralis* larvae. Exp. Parasitol. 8:10–30.

Baker, H., Oscar, F., Pasher, I., Dinnerstein, A., and Sobotka, H. 1960. An assay for pantothenic acid in biological fluids. Clin. Chem. 6:36–42.

Brin, J. 1962. Effects of thiamine deficiency and of oxythiamine on rat tissue transketolase. J. Nutr. 78:179–183.

Brin, M., Tai, M., Ostashever, A. S., and Kalinsky, H. 1960. The effect of thiamine deficiency on the activity of erythrocyte hemolysate transketolase. J. Nutr. 71:273–281.

Cameron, T. W. M. 1958. Parasites and parasitism. John Wiley and Sons, New York.

Cosgrove, G. E., Lincicome, D. R. and Warsi, A. A. 1969. Development of *Trypanosoma lewisi* in experimental chimaeras. J. Protozool. 16:47–49.

Coventry, F. A. 1929. Experimental infections with *Trypanosoma lewisi* in the guinea pig. Amer. J. Hyg. 9:247–259.

Davis, B. D., and Dubos, R. J. 1946. Serum albumin as a protective rather than nutritional growth factor in bacteriological media. Fed. Proc. 5:246.

Freed, M. 1966. Methods of vitamin assay. Interscience Publishers, New York.

Greenblatt, C. L., Jori, L. A. and Cahnmann, H. J. 1969. Chromatographic separation of rat serum growth factor required by *Trypanosoma lewisi*. Exp. Parasitol. 24:228–242.

Greenblatt, C. L. and Lincicome, D. R. 1966. Identity of trypanosome growth factors in serum. II. Active globulin components. Exp. Parasitol. 19:139–150.

Greenblatt, C. L. and Shelton, E. 1968. Mouse serum as an environment for the growth of *Trypanosoma lewisi*. Exp. Parasitol. 22:187–200.

Hoare, C. A. and Wallace, F. G. 1966. Developmental stages of trypanosomatid flagellates: a new terminology. Nature 212:1385–1386.

Kagan, I. G. 1960. Trichinosis: a review of biologic, serologic and immunologic aspects. J. Infect. Dis. 107:65–93.

Kanthack, A. A., Durham, H. E. and Blandford, W. F. H. 1898. On Nagana, or Tsetse fly disease. Roy. Soc. (London), Proc., B. 64:100–118.

Lang, C. A. 1958. Simple microdetermination of Kjeldahl nitrogen in biological materials. Anal. Chem. 30:1692–1694.

Laveran, A., and Mesnil, F. 1912. Trypanosomes et Trypanosomiases. 2nd Edition, Paris

Laveran, A. and Roudsky, D. 1914. Contribution a l'étude de la virulence du *Trypanosoma lewisi* et du *Tr. duttoni* pour quelques especes animales. Bull. Soc. Pathol. Exot. 7:528–535.

Lee, C. M. 1969. Pantothenic acid metabolism in mice infected with *Trypanosoma duttoni*. Ph.D. Thesis, Howard Univ.

Lee, C. M. and Lincicome, D. R. 1970a. *Trypanosoma duttoni*: oxygen uptake of liver slices of pantothenate-deficient mice. (MS submitted for publication.)

Lee, C. M. and Lincicome, D. R. 1970b. *Trypanosoma duttoni*: cell populations and antibody formation in pantothenate-deficient mice. (MS submitted for publication.)

Lee, C. M. and Lincicome, D. R. 1970c. *Trypanosoma duttoni*: pyruvate and pantothenate levels in plasma and liver tissue of normal and pantothenic acid-deficient mice. MS submitted for publication.

Lewis, R. W. 1953. An outline of the balance hypothesis of parasitism. Amer. Natur. 87:273–281.

Lewis, R. W. 1957. A graphic presentation of the balance hypothesis of parasitism. Acta Bot. 3:27–29.

Lincicome, D. R. 1953. The nutrition of parasitic protozoa. I. Preliminary experiment. utilizing *Trypanosoma lewisi*, pp. 173–184. *In* Thapar Commemorative Volume, Univ. Lucknow, India.

Lincicome, D. R. 1955. Growth factor in normal rat serum for *Trypanosoma lewisi*. J. Parasitol. 4 (Sec. 2):15.

Lincicome, D. R. 1957. Growth of *Trypanosoma lewisi* in the heterologous mouse host. Amer. J. Trop. Med. Hyg. 6:392.
Lincicome, D. R. 1958a. Normal rat serum as a growth factor for *Trypanosoma lewisi.* Proc. Sixth Int. Congr. Trop. Med. Malaria 3:71–76.
Lincicome, D. R. 1958b. Growth of *Trypanosoma lewisi* in the heterologous mouse host. Exp. Parasitol. 7:1–13.
Lincicome, D. R. 1958c. Normal rat serum as a growth factor for *Trypanosoma lewisi.* U.S. Naval Med. Res. Inst. Lect. Rev. Ser. 58–7.
Lincicome, D. R. 1959a. Serial passage of *Trypanosoma lewisi* in the heterologous mouse host. I. Development in calorically-restricted hosts. J. Protozool. 6:310–315.
Lincicome, D. R. 1959b. Observations on changes in *Trypanosoma lewisi* after growth in calorically restricted and in normal mice. Ann. Trop. Med. Parasitol. 53:274–287.
Lincicome, D. R. 1959c. The heterologous host as a research tool in nutrition studies on parasitic protozoa. U.S. Naval Med. Res. Inst. Lect. Rev. Ser. 59–4.
Lincicome, D. R. 1960. Serial passage of *Trypanosoma lewisi* in the heterologous mouse host. II. Developmental history during transfer in adequately-fed hosts. Ann. Parasitol. Hum. Comp. 35:457–468.
Lincicome, D. R. 1963. Chemical basis of parasitism. Ann. N. Y. Acad. Sci. 113:360–380.
Lincicome, D. R. and Bruce, J. I. 1965. Oxygen uptake of liver and heart slices of *Trypanosoma rhodesiense*-infected mice. Exp. Parasitol. 17:332–339.
Lincicome, D. R. and Fergusson, K. A. 1964. Serum protein changes in rats experimentally infected with *Trichinella spiralis.* Parasitol. 50(Sec. 2):28.
Lincicome, D. R. and Francis, E. H. 1960. Induction of development of *Trypanosoma lewisi* in the mouse by heterologous sera. J. Parasitol. 46(Sec. 2):42.
Lincicome, D. R. and Francis, E. H. 1961. Quantitative studies on heterologous sera inducing development of *Trypanosoma lewisi* in mice. Exp. Parasitol. 11:68–76.
Lincicome, D. R. and Hinnant, J. A. 1962. Identity of trypanosome growth factors in serum. I. Alteration of factors in rat serum by host starvation. Exp. Parasitol. 12:128–133.
Lincicome, D. R. and Lee, C. M. 1970a. *Trypanosoma duttoni*: body weight gains, food consumption and longevity of pantothenate-deficient mice given living cells, cell homogenates and cell metabolic products. (MS submitted for publication.)
Lincicome, D. R. and Lee, C. M. 1970b. *Trypanosoma duttoni*: tissue utilization of pyruvate in pantothenate-deficient mice. MS submitted for publication.
Lincicome, D. R. Rossan, R. N. and Jones, W. C. 1960. Rate of body weight gain of rats infected with *Trypanosoma lewisi.* J. Parasitol. 46(Sec. 2):42.
Lincicome, D. R., Rossan, R. N. and Jones, W. C. 1963. Growth of rats infected with *Trypanosoma lewisi.* Exp. Parasitol. 14:54–65.
Lincicome, D. R., Sen, D. K. and Cambosus, B. 1965. Development of *Trypanosoma duttoni* in two strains of mice. J. Parasitol. 5(Sec. 2):27.
Lincicome, D. R. and Shepperson, J. 1961. Increased growth of experimental hosts associated with foreign autonomous cells. The Physiologist 4(3).
Lincicome, D. R. and Shepperson, J. 1963. Increased rate of growth of mice infected with *Trypanosoma duttoni.* J. Parasitol. 49:31–34.
Lincicome, D. R. and Shepperson, J. R. 1965. Experimental evidence for molecular exchanges between a dependent trypanosome cell and its host. Exp. Parasitol. 17:148–167.
Lincicome, D. R. and Watkins, R. C. 1963. Method of preparing pure cell suspensions of *Trypanosoma lewisi.* Bull. Amer. Inst. Biol. Sci. 13:53–54.
Lincicome, D. R. and Watkins, R. C. 1965. Antigenic relationships among *Trypanosoma lewisi*-complex cells. I. Agglutinins in antisera. Parasitol. 55:365–373.
Majno, G. and Bunker, W. E. 1957. Preparation of tissue slices for metabolic studies: a hand microtome especially suitable for brain. J. Neurochem. 2:11–14.

Mills, C. K. and Kent, N. H. 1965. Excretions and secretions of *Trichinella spiralis* and their role in immunity. Exp. Parasitol. 16:300–310.

Nelson, B. D. and Lincicome, D. R. 1966. Serum transaminases and aldolase in rats inoculated with *Trypanosoma lewisi*. Proc. Soc. Exp. Biol. Med. 121:566–569.

Nieschulz, O. and Wawo-Roentoe, F. K. 1929. Infektionsversuche von Meerschweinchen mit *Trypanosoma lewisi*. Z. Parasitenk. 2:294–296.

Pollack, M. R. 1947. The growth of *H. pertussis* on media without blood. Brit. J. Exp. Pathol. 28:295–307.

Read, C. P. 1958. A science of symbiosis. AIBS (Amer. Inst. Biol. Sci.) Bull. 8:16–17.

Riley, J. F. 1951. Studies on the metabolism of protozoa. I. Metabolism of the parasitic flagellate, *Trypanosoma lewisi*. Biochem. J. 49:577–585.

Sadun, E. H. and Norman, L. 1955. The use of an acid soluble protein fraction in the flocculation test. J. Parasitol. 41:477–482.

Segal, L., Alberta, E. and Wyngaarden, J. B. 1956. An enzymatic spectrophotometric method for the determination of pyruvic acid in blood. J. Lab. Clin. Chem. 48:137–143.

Sen, D. K. 1969. *Trichinella spiralis*: metabolic exchange of pyridoxine with the rat host. Ph.D. Thesis, Howard Univ.

Singer, I. 1961. Tissue thiamine changes in rats with experimental trypanosomiasis or malaria. Exp. Parasitol. 11:391–401.

Smyth, J. D., 1962. Introduction to animal parasitology. Charles C Thomas, Springfield, Ill.

Stirewalt, M. A., Shepperson, J. R. and Lincicome, D. R. 1965. Comparison of penetration and maturation of *Schistosoma mansoni* in four strains of mice. Parasitology 55: 227–235.

Taliaferro, W. H. and Taliaferro, L. G. 1922. The resistance of different hosts to experimental trypanosome infections with special reference to a new method of measuring this resistance. Amer. J. Hyg. 2:264–319.

Thillet, C. J., and Chandler, A. C. 1957. Immunization against *Trypanosoma lewisi* in rats by injection of metabolic products. Science 125:346–347.

Umbreit, W. W., Burris, R. H. and Stauffer, J. F. 1957. Manometric techniques. Burgess Publishing Co., Minneapolis, Minn.

Warsi, A. A. 1968. Metabolic exchange of pyridoxine between a dependent cell and its organic environment. Ph.D. Thesis, Howard Univ.

Nutritional Aspects of the Symbiosis Between Echinoecus pentagonus and its Host in Hawaii, Echinothrix calamaris

PETER CASTRO

Department of Biological Sciences
University of Puerto Rico
Río Piedras, Puerto Rico

INTRODUCTION AND REVIEW OF THE LITERATURE

Echinoecus pentagonus (Crustacea, Brachyura), a member of the brachyuran subfamily Eumedoninae of the family Parthenopidae, is a symbiont of several Indo-West Pacific sea urchins (Serène *et al.*, 1958). Most of the known species assigned to the Eumedoninae are associated with crinoids and echinoids. *E. pentagonus* is found in Hawaii associated with the diadematid sea urchin, *Echinothrix calamaris* (Echinoidea), a common littoral species. Male and juvenile female crabs live on the host's peristome. Adult females are confined to the rectum, where calcification of the periproct produces a gall-like structure (Fig. 1). Males occasionally move in and out of rectums.

Feeding habits and nutrition of the symbiont were investigated by analyzing stomach contents, feeding behavior, and the ingestion and assimilation of host tissues by use of radioactive tracer techniques. The dynamic equilibrium in existence between the symbiont and its host was analyzed, in part, by estimating the energetic relationships between the partners.

Symbiotic forms are common among the decapod crustaceans (see reviews by Balss, 1956; Patton, 1967a) but surprisingly little is known about the nutritional aspects of these symbioses. Knowledge on the subject is generally restricted to incidental notes on feeding habits. Potts (1915) has described the feeding habits and the functional morphology of the buccal parts and stomach of *Hapalocarcinus marsupialis*, a brachyuran crab that induces the formation of galls in madreporarian corals. The same features have been studied in a few xanthid crabs symbiotic in madreporarian corals, namely, *Domecia acanthophora* (see Patton, 1967b) and species of *Trapezia* and *Tetralia* (see

Contribution No. 348 of the Hawaii Institute of Marine Biology, University of Hawaii, Honolulu, Hawaii 96822

Fig. 1. Rectum and calcified periproct of *Echinothrix calamaris* cut open in one side to show a female *Echinoecus pentagonus* in its normal microhabitat. (×2.3)

Knudsen, 1967). The scant information available on the feeding habits of pinnotherid crabs associated with bivalves has been summarized by Cheng (1967). Observations on the feeding habits of other pinnotherids have been given by Rathbun (1918) and MacGinitie and MacGinitie (1968).

MATERIALS AND METHODS

Ingestion and Assimilation of Host Tissues

The stomach contents of numerous specimens of *Echinoecus pentagonus* were analyzed in order to estimate their diet. Stomach contents of crabs kept isolated from their host were also examined.

Ingestion of host tissues by the symbiont was determined by injecting uniformly labeled d-glucose-^{14}C (Volk Radiochemical Co., Burbank, California; specific activity of 210 mc/mM) into the perivisceral coelomic cavity of sea urchins and determining the amount of labeled material present in the tissues of crabs found living in the rectum, peristome, and test of the host. Six sea urchins, each with a male or female in the rectum and another crab on the peristome or test, were injected through the peristome with 100 μc of the glucose solution (equivalent to 84 μg of d-glucose) using a disposable 1.0 ml syringe. The injected animals were kept in a 115-gallon fiberglass tank provided with running seawater. Two additional sea urchins, each with two crabs, were kept in the same tank as controls. All sea urchins had been starved for 2–3 days and few or no fecal pellets were being voided. One sea urchin was removed at each of the following time intervals: 3, 6, 12, 18, 24, and 48 hours after the injection of glucose. A control sea urchin was removed 24 hours after being in contact with the injected individuals; the other was removed after 48 hours. Crabs were frozen after being thoroughly washed with seawater. After weighing (frozen dry weight), each crab was sonicated (using a Biosonik oscillator) over ice in 10 % methanolic trichloroacetic acid (TCA). The samples were subsequently placed in a water bath-shaker at 48°C for 2 hours, centrifuged, and the supernatant solution was adjusted to 4.0 ml with anhydrous methyl alcohol before counting. The TCA-insoluble material was digested in 4.0 ml of 2N methanolic potassium hydroxide, placed in a water bath-shaker at 48°C for 18 hours, centrifuged, and the supernatant solution was adjusted to 4.0 ml. The insoluble material that remained was separated by centrifugation into skeletal fragments and a white precipitate that appeared to be insoluble protein. Both residues were washed with absolute methyl alcohol and digested in an ice bath with 4.0 ml of concentrated perchloric acid. All samples were decolorized overnight with 5–10 drops of 30 % hydrogen peroxide and were then counted in a Beckman LS-100 liquid scintillator. Beckman's "Fluoralloy" (a mixture of naphthalene, butyl PBD, "Cab-O-Sil," and PBBO) dissolved in distilled dioxane was used as the scintillant mixture. Reagent blanks were used for background corrections. Quenching corrections were made from quench ratio curves prepared for each of the solvents employed. A ^{14}C-labeled bicarbonate solution was used as the external standard.

The distribution of ^{14}C-labeled material in crabs was analyzed by placing individuals in contact with sea urchins similarly injected with labeled glucose. The stomach, hepatopancreas-gonads, gills, and the remaining tissues (including the skeletal parts) of each crab were individually digested with 1.0 ml of concentrated formic acid, decolorized, and counted as before.

The chemical composition of the ^{14}C-labeled material in the coelomocyte aggregations of the rectum was determined in samples taken from a sea urchin injected 48 hours before with 100 μc of labeled glucose. The cells were extracted with cold 10 % aqueous TCA and were then centrifuged. The supernatant solution was decolorized with 30 % hydrogen peroxide and one portion was counted. Two volumes of 95 % ethyl alcohol were added to the remaining supernatant solution for the precipitation of glycogen. The TCA-alcohol supernatant solution was counted after centrifugation. The insoluble material was digested in concentrated formic acid, decolorized, and counted.

The possible accumulation in the symbiont of the naphthoquinone pigments characteristic of sea urchins was also investigated. Pigments were extracted from the peristome epithelium, coelomocyte aggregations of the rectum, tube feet, and spines of sea urchins, as well as from crabs from which the digestive tract had been removed. Absolute methyl alcohol was used in the extractions of the pigments. Acetic acid (5 % of the total volume) was added to the alcohol as a stabilizing agent (Spruit, 1949). Spectra were determined in 1.0cm cells on a Beckman DB-G spectrophotometer.

Host Fecal Pellets

Samples of the fecal pellets of the host were analyzed in order to estimate their nutritive value. Pellets were obtained directly from the rectum or immediately after being voided. All samples were kept frozen before the analyses.

Alcohol-soluble carbohydrates were determined by extracting the fecal samples three times with hot 80 % ethyl alcohol. After filtering through glass fiber, the alcohol was evaporated under reduced pressure, the residue was dissolved in distilled water, and the sugars were determined spectrophotometrically by the phenol-sulfuric acid method of Dubois *et al.* (1956). TCA-soluble carbohydrates in the alcohol-washed feces were extracted by slowly heating the samples with 5 % aqueous TCA, centrifuging, and then measuring the sugars in the supernatant solution by the phenol-sulfuric acid method as before. A Beckman Spectronic 20 spectrophotometer was used in both analyses. Determinations were carried out in triplicate and a correction was made for the presence of pigments. D-glucose standards were used.

Total protein was measured by extracting the alcohol-washed and TCA-washed samples with hot 1.0N sodium hydroxide, centrifuging, and then analyzing triplicates of the supernatant solution for proteins by employing the Folin-phenol method of Lowry *et al.* (1951). The preparation of reagents followed the modifications given by Price (1965). Standards were prepared by precipitating a standard solution of crystalline bovine serum (Armour Pharmaceutical Company) with 5 % TCA followed by an extraction with hot 1.0N sodium hydroxide. A Beckman Spectronic 20 spectrophotometer was used in the determinations.

Total lipid content of a different set of samples was determined by the method of Mukerjee (1956) as outlined by Strickland and Parsons (1968). Their procedure was modified by the use of a small water condenser attached to the test tubes during saponification. A Beckman DB-G spectrophotometer with 1.0cm cells was used in the determinations.

The quantitative determination of organic carbon was carried out by employing the acid dichromate digestion method of Johnson (1949) as adapted to spectrophotometry by Strickland and Parsons (1968). The sodium sulfate washing suggested by these authors was omitted. Absorbance was determined in 1.0cm cells on a Beckman DB-G spectrophotometer.

For the histochemical analysis of the fecal pellets envelope (peritrophic membrane), whole, undecalcified pellets were fixed in 70 % ethyl alcohol, embedded in paraffin, and sectioned. Sections were stained using the periodic acid-Schiff (PAS) reaction for polysaccharides, pH 2.6 alcian blue counterstained with metanil yellow, and Lillie's toluidine blue O technique, all outlined by Humason (1967). In order to maintain the metachromatic properties of toluidine blue, the stained sections were not dehydrated and glycerin was used as the mounting medium. The last two reactions are specific for acid mucopolysaccharides.

The presence of enzymes in the symbiont capable of hydrolyzing algal polysaccharides was investigated by incubating sugar-free algal homogenates with stomach and hepatopancreas extracts. Fresh *Ulva fasciata*, a green alga, was homogenized and repeatedly washed in hot 80 % ethyl alcohol and cold distilled water. The washed tissues were dried in vacuum and 20.0 mg samples were incubated with filtered extracts of the stomach and hepatopancreas of individual crabs in 0.1N sodium phosphate buffer (pH 6.8). The pH of the digestive organs of crabs was found to vary between 6.8 and 7.0. The crab tissues were previously washed in filtered seawater, homogenized in cold buffer, and centrifuged. The crab tissue-algae preparations, as well as digestive system and algae controls, were kept in test tubes maintained in a water bath at 30°C. One drop of toluene was added to each test tube as a bacteriostatic agent. The rate of release of reducing sugars in all of the preparations was measured spectrophotometrically as assayed by the 3,5-dinitrosalicylate reduction (Clark, 1964). Absorbance was determined in a Beckman DB-G spectrophotometer.

Energetic Relationships

The energy budget of the host was estimated by measuring food intake, production of feces, and oxygen consumption in five adult individuals, including two with rectums previously occupied by crabs. The sea urchins had been kept for three weeks with the red alga *Laurencia* sp. as the only food source. After one day of starvation for the voiding of feces from previously ingested algae, all sea urchins were placed in individual 10-gallon tanks provided with air stones. The amounts of *Laurencia* ingested during 24-hour periods was determined by weighing (wet weight) the algae present in each tank at the beginning and at the end of each interval of time. The fecal pellets voided during the same 24-hour periods were removed with a large volumetric pipette and were preserved by freezing. Daily ingestion of food was recorded in each of the five individuals for a total of five days. Similar wet-weighed samples of the alga were dried in an oven, weighed, and a correction factor was determined in order to express the alga ingested in terms of dried weight. Fecal pellets were dried in an oven and their dry weight was recorded. Even when the animals were not provided with food for one day before quantifying the food intake, feces from the food ingested prior to the experiment were still being voided by two individuals. As a consequence, no records were taken of feces voided during a period of time equal to that during which feces were still produced after no more food was provided at the end of the five day period. Caloric values of the alga and feces were determined by using a Phillipson oxygen microbomb calorimeter (Phillipson, 1964) connected to a Beckman potentiometric recorder. Benzoic acid was used in the calibration of the instrument.

Oxygen consumption in each of the five sea urchins was determined by using an 8-liter, wood and plexiglass respirometer provided with a stirrer and a lower compartment through which tap water flowed to maintain constant temperature. The total volume of the respirometer and the connecting tubes was approximately 8,478 ml. Absolute volumes, that is, total volume corrected for the volume of each sea urchin, varied between 8,196 and 8,364 ml. The temperature of the water in the chamber varied between 26° and 27°C. The oxygen content of the water in the respirometer was measured at the beginning of the experiment (after a 30 minute period of acclimation in which running sea water was allowed to flow through) and one hour after the chamber was sealed. Oxygen concentrations were determined by using a Winkler technique as outlined by Strickland and Parsons (1968). A control determination was carried out with an empty respirometer and all values were corrected for it.

Oxygen consumption in the symbiont was measured in a respirometer built from a 500 ml glass jar and a rubber stopper provided with two sealable openings. Absolute volume of the jar and connecting tubes varied between 402 and 403 ml. The respirometer was placed in a large pan with running tap water. The temperature of the water in the jar varied between 26° and 27°C. The Winkler technique was used to measure the oxygen content of the water at the beginning of the experiment and one hour after the sealing of the jar. Each determination was preceded by a 30 minute period of acclimation. Oxygen consumption was measured in trials involving three males and two adult females. Corrections were made from a control determination.

Caloric values were obtained from samples of coelomocyte aggregations taken from the rectum of three sea urchins and from samples of peristome epithelium from four sea urchins. A Phillipson oxygen microbomb calorimeter was used in the determinations.

RESULTS

Feeding Habits of the Symbiont

Tissue and material from the fecal pellets of the host appear to be the only sources of nutrition to the symbiont. The stomach contents of female and male crabs removed from the host's rectum consisted mostly or entirely of host tissues. Partly digested material appeared as a reddish-brown, amorphous mass, but spherical cells were observed occasionally (Fig. 2). These cells are

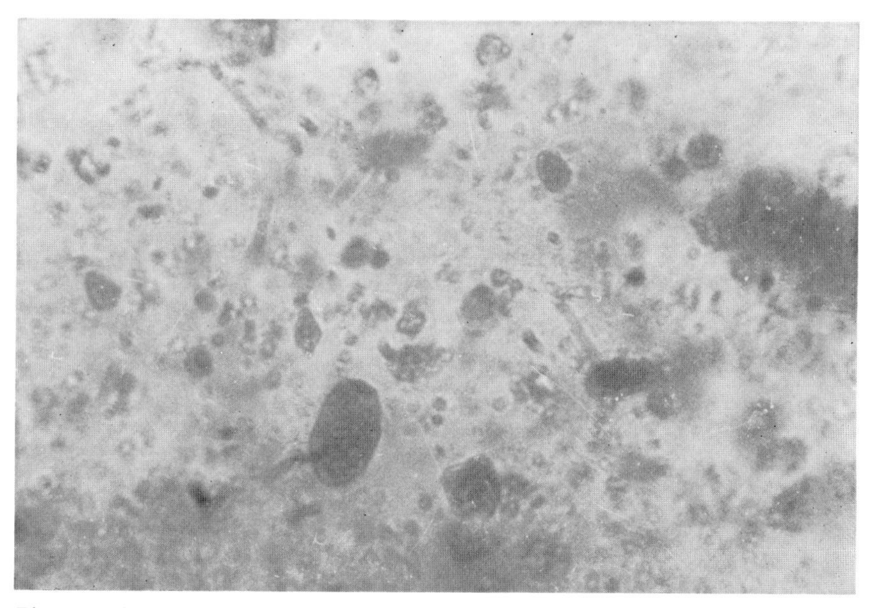

Fig. 2. Portion of the stomach contents of a female *Echinoecus pentagonus* previously found living in the rectum of *Echinothrix calamaris*. The oval-shaped cell appears to be a pigmented eleocyte from the host. (×600)

pigmented eleocytes, coelomocytes that migrate across the inner epithelium of the host's rectum (Fig. 3) and accumulate in abundant, mucus-like aggregations along the lumen (Fig. 4). Material from the fecal pellets was found to be present in roughly 75 % of the individuals that were examined. Most of this material was in the form of fragments and filaments of algae; but sponge spicules, diatom frustules, bristle-like skeletal fragments, and unidentified calcareous sediment were frequently observed. The contents of the stomach of males and juvenile females taken from the peristome and test of the host consisted of partly digested host tissues. Peristome epithelium and tube feet have been observed being ingested by crabs living outside the rectum.

Crabs were kept alive in isolation from their host for as long as 36 days. A drastic reduction in the size of the hepatopancreas was observed in these crabs. Their stomachs were usually empty.

Host tissues and fecal pellets are handled by the chelipeds before they are moved to the mouth region. Scooping movements by the chelipeds are involved in the gathering of the coelomocyte aggregations in the rectum. The third maxillipeds are involved in the sorting of material from the fecal pellets. They are characterized by a well developed palp which bears long setae. Two rows of short teeth are found along the inner side of each seta. The remain-

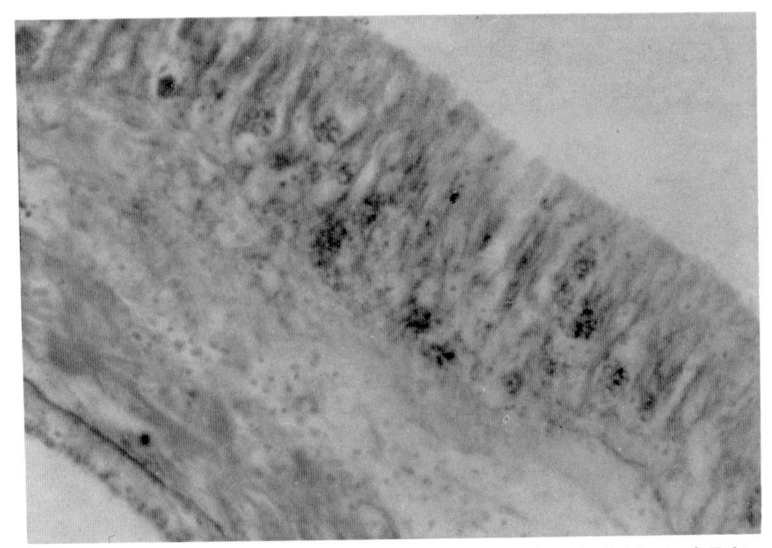

Fig. 3. Cross-section of the medial portion of the rectum of an individual of *Echinothrix calamaris* previously occupied by a female *Echinoecus pentagonus*. Eleocytes can be seen migrating across the inner epithelium. (PAS reaction; ×400)

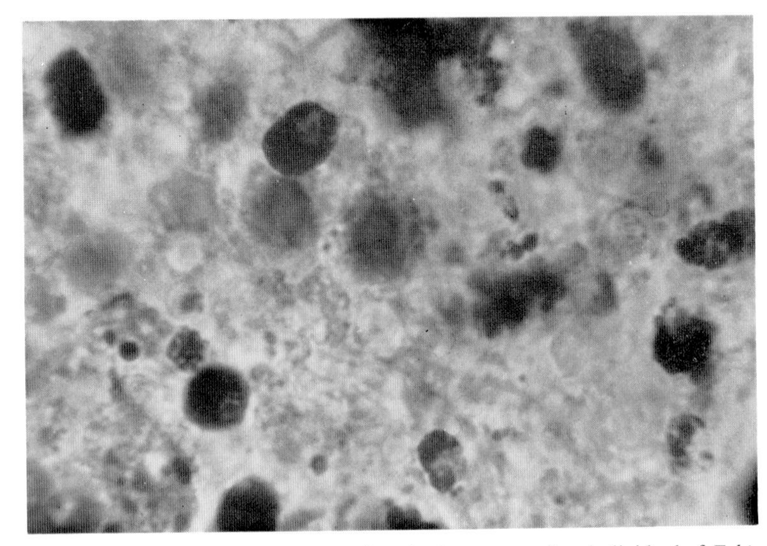

Fig. 4. Smear of the coelomocyte aggregations in the rectum of an individual of *Echinothrix calamaris* occupied by a female *Echinoecus pentagonus*. The material is composed of whole and fragmented eleocytes. (×600)

ing buccal appendages and the ossicles of the stomach are not modified as in other symbiotic brachyurans (Potts, 1915; Patton 1967b; Knudsen, 1967).

Ingestion of the host's epithelial tissue and tube feet from the peristomial region by the postlarval stages, juvenile females, and males is in an apparent balance with the remarkable healing and regenerating abilities of the sea urchin. These crabs are small (the largest male measured was 9.9 mm in carapace width) and their ability to move between sea urchins across the spines does not restrict them to a single animal. Females living in the rectum (8.0–17.0 mm in carapace width) are constantly being supplied by food in the form of coelomocytes and fecal pellets. There is no evidence for the ingestion of tissue from the rectum. Serious damage, however, is observed when females are placed on sea urchins having rectums already occupied by another female or when placed on *Echinothrix diadema* or *Diadema paucispinum*, two diadematid sea urchins in which the rectum is not sufficiently large to accomodate the adult females. In either case, spines are removed and test epithelium is ingested (Fig. 5), a situation which leads to the eventual death of the sea urchins after 2–4 days of contact.

Fig. 5. An adult female of *Echinoecus pentagonus* after being left overnight on the test of an individual of *Echinothrix diadema*. (×1.8)

Ingestion and Assimilation of Host Tissues

Relatively high amounts of ^{14}C-labeled material were found in the tissues of crabs which had been in contact with sea urchins injected with labeled glucose (Table 1; the location indicated for each crab is that at the end of the period of contact). All females were fully grown; thus their presence outside the rectum represents an abnormal situation. At least two males moved from their original hosts and it is possible that some moved in and out of rectums. Significant concentrations of labeled material started appearing at least 6 hours after a sea urchin was injected. Labeling took place faster and in larger amounts in crabs found in the rectum. The specific activity of the alcoholic potassium hydroxide fraction (assumed to be composed mostly of proteins) was found to be higher (except in one case) than the other. The distribution

TABLE 1. *^{14}C-Labeled material (cpm/mg of tissue) from individuals of* Echinoecus pentagonus *living in contact with* Echinothrix calamaris *(one sea urchin for each of the time intervals) injected with 100 μc of D-glucose-UL-^{14}C*

Hours in contact with sea urchin	*Sex*	*Location in sea urchin*	*Alcoholic TCA fraction*[1]	*Alcoholic KOH fraction*[2]	*$HClO_4$ fractions*[3]	*Total specific activity*
3	female	test	0	56.8	0.1	56.9
3	male	test	0	0	0.3	0.3
6	female	rectum	31.2	1,061.4	1.0	1,093.6
6	female	test	16.1	181.8	0.2	198.1
12	male	rectum	201.3	692.5	1.7	895.5
12	male	peristome	2.4	0	0.2	2.6
18	female	rectum	714.7	1,248.7	2.0	1,965.4
18	male	peristome	64.1	381.1	0.4	445.6
24	male	rectum	118.4	1,433.4	2.1	1,553.9
24	male	peristome	407.0	1,030.8	1.4	1,439.2
48	male	rectum	13.4	171.4	0.8	185.6
Controls						
24	female	rectum	0	0	—	0
24	male	peristome	0	0	—	0
48	female	rectum	2.4	0	—	2.4
48	female	test	2.6	0	—	2.6

[a] Low molecular weight molecules
[b] Mostly proteins
[c] Skeletal fragments

of ^{14}C-labeled material in crabs kept in contact for 24 hours with similarly injected sea urchins is summarized in Table 2. The presence of radiocarbon outside the stomach is an indication that host tissues were assimilated. Most of the label was found in the tissues which remained after the major organ systems were removed. Nevertheless, the gills showed the highest specific activity in most individuals.

The composition of the ^{14}C-labeled material in the coelomocyte aggregations of a sea urchin is detailed in Table 3. The sea urchin was injected with labeled glucose 48 hours before sampling. Most of the label appeared to be incorporated into proteins.

The labeled glucose injected into the sea urchins was found to be rapidly removed from the coelomic fluid. It appeared as little as one hour after injection in the coelomocytes of the perivisceral fluid, in the coelomocyte aggregations of the rectum, and in the peristome epithelium (Castro, 1969). Evidence indicates that the coelomic fluid, rather than the coelomocytes, serves as the most important vehicle for nutrient transport in the host.

TABLE 2. *Distribution of ^{14}C-labeled material in four individuals of* Echinoecus pentagonus *in contact for 48 hours with an individual of* Echinothrix calamaris *injected with 100 μc of D-glucose-UL-^{14}C*

Crab No. and location in host	*cpm per mg tissue weight*	*cpm per mg total weight*	*% Total activity*
1. *female* (outside rectum)			
stomach	2.0	<0.1	11.7
hepatopancreas-gonads	0.6	<0.1	12.8
gills	4.3	<0.1	16.0
remaining tissues	0.2	0.2	59.5
whole animal	—	0.3	—
2. *female* (outside rectum)			
stomach	6.3	0.1	8.9
hepatopancreas-gonads	2.2	0.1	5.7
gills	18.7	0.3	21.3
remaining tissues	0.8	0.8	64.1
whole animal	—	1.2	—
3. *female* (test)			
stomach	16.5	0.3	24.0
hepatopancreas-gonads	1.7	0.1	6.0
gills	39.7	0.3	28.4
remaining tissues	0.5	0.5	41.6
whole animal	—	1.1	—
4. *male* (peristome)			
stomach	69.9	0.9	10.1
hepatopancreas-gonads	62.5	1.5	18.1
remaining tissues	6.4	6.1	71.8
whole animal	—	8.5	—

TABLE 3. *Composition of the ^{14}C-labeled material in the coelomocyte aggregations in the rectum of an individual of* Echinothrix calamaris *injected with 100μc of D-glucose-UL-^{14}C 48 hours before sampling*

Fraction	*cpm/mg*	*% Total activity*
TCA (cold, aqueous)	1,622.8	—
TCA-ethyl alcohol[a]	651.0	12.2
Glycogen (TCA fraction–TCA-alcohol fraction)	971.8	18.3
Residue[b]	3,692.1	69.5
Total	*5,314.9*	—

[a] Low molecular weight molecules
[b] Mostly proteins

The alcohol-soluble pigments from the peristome epithelium and the coelomocyte aggregations of the host show the typical absorption spectrum for naphthoquinones (Millot, 1957). These bright orange-red pigments appear not to be retained in the skeleton or tissues of the symbiont. Nevertheless, crabs show the same dark reddish brown color of the host's tissues.

Composition and Nutritive Value of the Fecal Pellets of the Host

The shape and composition of the fecal pellets of *Echinothrix calamaris* vary considerably. They are typically spherical, measuring 3–5 mm in diameter, and are sand-like in color. These pellets are mostly composed of sediment and small fragments of coral and coralline algae. Filamentous algae are found in considerable amounts. Large fragments of brown algae are sometimes present, giving the pellets a dark coloration. Ciliates, common inhabitants of the gut of sea urchins, are very abundant. Sponge tissue and spicules, diatoms, living nematodes, and the shells of foraminiferans have also been observed. The presence of bacteria has been demonstrated (L. H. Di Salvo, per. comm). Most of the fecal pellets of individuals found living in the shallow water reefs of Kaneohe Bay, Oahu, Hawaii are composed of large fragments of a green alga, *Dictyosphaeria cavernosa*. The pellets are elongated, 5–10 mm in lengh, and are covered with a thicker peritrophic membrane.

Carbohydrate, protein, organic carbon, and lipid contents of fecal pellet samples are listed in Table 4 and 5. Most of the organic carbon in the samples was evidently in the form of the higher polysaccharides of the algae which were not measured due to their insolubility in either alcohol or TCA. Lipids appeared to be the second most important major constituent, representing

TABLE 4. *Carbohydrate and protein contents of fecal pellet samples of* Echinothrix calamaris *(samples nos. 1–5 were composed primarily of sand and sediment, no. 6 of sponge tissue)*

	1	*2*	*3*	*4*	*5*	*6*	*Mean Values*
Alcohol-soluble carbohydrate[a]							
μg/mg feces	0.22	0.37	0.38	0.51	0.60	0.16	*0.37*
% dry weight	0.02	0.04	0.04	0.05	0.06	0.02	*0.04*
TCA-soluble carbohydrate[b]							
μg/mg feces	0.69	0.72	1.25	5.40	0.48	1.21	*1.63*
% dry weight	0.07	0.07	0.13	0.54	0.05	0.12	*0.16*
Protein nitrogen							
μg/mg feces	0.63	0.78	1.07	1.78	0.51	1.29	*1.01*
% dry weight	0.06	0.08	0.11	0.18	0.05	0.13	*0.10*

[a] Mostly monosaccharides and derivates
[b] Fragments of polysaccharides, starches, glycogen

an average of 1.5 % of the organic carbon. The total organic carbon was not measured in the samples where carbohydrates and proteins were measured, but it appears from their concentrations that they account for less than 1.0 % of the organic carbon.

The peritrophic membrane of the fecal pellets gave a positive reaction with PAS, pH 2.6 alcian blue, and toluidine blue O, indicating the presence of acid mucopolysaccharides. Fecal pellets of the diadematic *Diadema antillarum* have been described by Lewis (1964) as covered with a thin "mucous secretion secreted by the oesophagus."

There is no evidence for the occurence of enzymes in the symbiont capable of hydrolyzing the alcohol-insoluble and water-insoluble carbohydrates of the green alga, *Ulva fasciata*.

TABLE 5. *Organic carbon and lipid contents of fecal pellet samples of* Echinothrix calamaris *(samples nos. 7–9 were composed primarily of algae, nos. 10–12 of sand and sediment)*

	7	*8*	*9*	*10*	*11*	*12*	*Mean Values*
Organic carbon							
μg/mg feces	650.18	494.00	—	236.05	285.03	133.59	*359.77*
% dry weight	65.01	49.40	—	23.60	28.50	13.35	*35.97*
Total lipid							
μg/mg feces	3.76	3.45	12.73	4.32	2.63	3.98	*5.14*
% dry weight	0.38	0.34	1.27	0.43	0.26	0.40	*0.51*
% organic carbon	0.58	0.69	—	1.82	0.91	3.00	*1.40*

Energetic Relationships

Mean values for the energy input, production of feces, and respiration as measured in the five individuals that were experimentally fed with *Laurencia* are shown in Fig. 6. Energy equivalents for the mean volume of oxygen consumed assume a nonprotein respiratory quotient (RQ) value of 0.80, halfway point between the extreme carbohydrate and lipid oxidations (West *et al.*, 1966). These measurements were taken from individuals which had been acclimated to somewhat atypical conditions. The shallow water alga is probably never ingested in nature and some elements of the normal diet of the sea urchin were possibly absent. Nevertheless, sea urchins with crabs living on them were kept under the same conditions for more than six months. A starvation period after quantifying food intake probably affected the assimilation of the food still present in the gut. Oxygen consumption was possibly overestimated since it was measured immediately after the resumption of normal feeding following the starvation period.

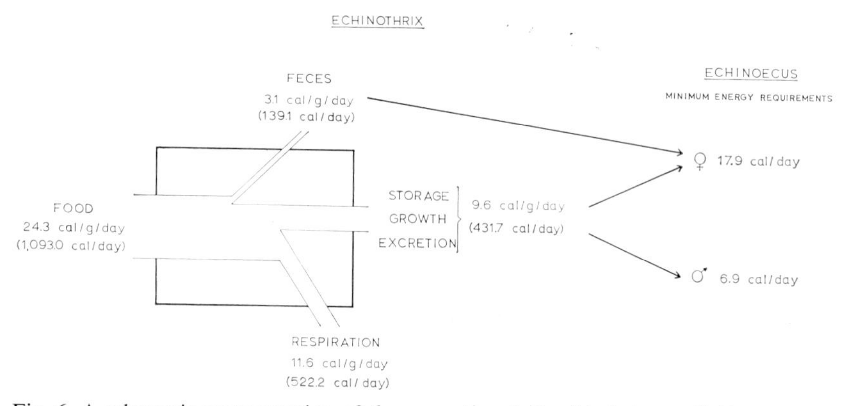

Fig. 6. A schematic representation of the energetic relationship between *Echinoecus pentagonus* and its host, *Echinothrix calamaris*.

The "assimilation efficiency" (AE) calculated as:

$$AE = \frac{(\text{weight of alga injested}) - (\text{weight of feces})}{(\text{weight of alga injested})} \times 100$$

varied between 77.0 and 89.0 % (mean value of 83.8 %). Values ranged from 43 to 93 % in the herbivorous sea urchin *Strongylocentrotus purpuratus* (Lasker and Boolootian, 1960; Boolootian and Lasker, 1964) and from 32.4 to 83.4 % in *S. intermedius* (Fuji, 1962, 1967). The presence of previously assimilated material in the fecal pellets (the peritrophic membrane, mucus) and the possibility of the release of unassimilated food in the form of dissolved material makes this measurement useful only for comparative purposes (Johannes and Satomi, 1967).

The minimum energy requirements of male and female crabs were estimated from oxygen consumption measurements (Figure 6). As in the host, the energy equivalents of the mean volumes of oxygen consumed assume a respiratory quotient value of 0.80.

DISCUSSION AND CONCLUSIONS

Symbioses involve the establishment of a dynamic equilibrium as a result of adaptive interaction between their members. Physiological, behavioral, and morphological adaptations in numerous symbiotic organisms enable them to utilize host material in order to meet at least part of their nutritional requirements. Metabolic dependency, being in the form of nutrients, enzymes, or development-inducing factors (Cheng, 1967), can be considered as a normal outcome of close symbioses.

The utilization of host material as a food source by *Echinoecus pentagonus* is directly related to its microhabitat in the host. The ingestion of coelomocytes and fecal material by adult females, as well as of peristome epithelium and tube feet by males and juvenile females, is in apparent equilibrium with the processes involved in the production or regeneration of these materials by the host.

The migration and accumulation of a large number of pigmented eleocytes in rectums occupied by the symbiont appear to take place as a response to the exposure of the rectum's epithelium to light. Few or no eleocytes are present in the inner epithelium and lumen of unoccupied rectums, where the tissues are covered by a prominent periproct. Furthermore, the accumulation of pigmented eleocytes in the lumen of the rectum was observed in sea urchins with excised periprocts. Migration of coelomocytes in response to mechanical irritation may help explain their accumulation in the rectum. Eleocytes, however, have not been recorded as being involved in phagocytosis or in any similar cellular defense mechanism.

Numerous functions have been attributed to the considerable variety of coelomocytes that are present in the coelomic fluid and tissues of echinoderms (Boolootian and Giese, 1958; Endean, 1966). Eleocytes (referred to as morula cells or spherulocytes by some authors) have been suggested as having a role in the translocation of nutrients (Boolootian and Lasker, 1964) and even in epidermal digestion and absorption (Pequignat, 1966). Echinochrome, a red naphthoquinone pigment characteristic of sea urchins, has been detected in eleocytes (MacMunn, 1885). Another orange-red naphthoquinone has been found in eleocyte-like coelomocytes in *Diadema antillarum* (Millot, 1957). The pigment, also found in colorless coelomocytes, changed to red and even to brown and black pigments when oxidized on exposure to air. An orange-red naphthoquinone was found in the coelomocyte aggregations of *Echinothrix calamaris*. Naphthoquinone pigments have been suggested as being involved in photoreception (Millot and Yoshida, 1957) and even to function as excretory products (Fox, 1953). The possible connection between echinochrome and the transport and storage of oxygen has been refuted (Farmanfarmaian, 1966).

Material from the fecal pellets of the host provides an additional food source to the adult females and to males present in the rectum. Considerable amounts of nutrients remain in the partly digested algae and encrusting organisms which constitute the diet of *Echinothrix*. Organisms associated with the pellets (ciliates, nematodes, and bacteria) are probably also ingested. The peritrophic membrane and any mucus present are other possible sources of nutrition. With the apparent exception of the higher polysaccharides of the algae, the remaining nutrients are assumed to be digested by the symbiont.

The organic constituents of the fecal pellets of some marine invertebrates have been analyzed by several workers (see Frankenberg *et al.*, 1967). In addition, rough estimates of the organic material in the gut contents of *Diadema antillarum* have been given by Lewis (1964). Studies by Johannes and Satomi (1966), Frankenberg *et al.*, (1967), and Frankenberg and Smith (1967) suggest that fecal pellets may have an important role as a nutrient source. It can be suggested that feces are probably ingested by the wide variety of crustaceans living in association with the rectum, cloaca, or hind gut of echinoderms (see review by Hyman, 1955).

The relationship between energy flow in *Echinothrix calamaris* and the average minimal energy requirements of male and female specimens of *Echinoecus pentagonus* is illustrated in Fig. 6. An average of 431.7 cal/day (equivalent to 9.6 cal/g/day or 39.5 % of the average energy input) is left available in the host for the storage of food energy and for the growth and repair of tissues. This figure also includes the energy that is lost in the form of excretory products. The amount of energy utilized in growth and storage depends mostly on the age, sex, and gonadal stage of each individual. Storage of food energy in *Echinothrix* and other sea urchins appears to be considerable since individuals can be starved for relatively long periods of time.

It was not possible to quantify the ingestion of host material by crabs. Data on their average minimal energy requirements at least indicate that it is theoretically possible for crabs to meet these requirements from host material. From the mean caloric value of peristome epithelium (1.75 cal/mg dry weight) it is calculated that a total of 3.9 mg of tissue would be required to be ingested per day in order to meet the minimum energy requirements of male crabs

(6.9 cal/day, 19.0 cal/g/day). Similarly, a total of 5.0 mg of coelomocyte aggregations from the rectum (mean caloric value of 3.57 cal/mg dry weight) would be required to be ingested per day to meet the minimum energy requirements of adult females (17.9 cal/day, 19.4 cal/g/day). A total of 13.5 mg of whole fecal pellets from *Laurencia*-fed sea urchins (mean caloric value of 1.33 cal/mg dry weight) would be needed to meet these requirements.

The estimated minimum amount of peristome tissue that is theoretically ingested by male crabs agrees with the limited degree of damage that is observed in sea urchins occupied by males or juvenile females. On the other hand, a total of 10.3 mg of peristome tissue (almost three times as much as in the males) would be required to be ingested in order to meet the minimum energy requirements of adult females. This is evidenced by the heavy damage that is observed when these females are kept outside the rectum (Fig. 5). The burden of adult females on the economy of the host is thus minimized by the utilization of the coelomocytes and fecal pellets that are almost continuously being emptied into the rectum.

Several schemes have been suggested in the attempt to classify symbioses (see review by Cheng, 1967). This has resulted in a number of meaningless categories and definitions. The use of the presence or absence of metabolic dependency in categorizing symbioses appears to be the most sensible. It is rather unfair, however, when used in dealing with cases where behavioral adaptations are predominantly involved (cleaning symbiosis and others). The metabolic dependency of *Echinoecus pentagonus* on its host would place the association under the category of parasitism.

SUMMARY

Feeding habits and nutrition of the symbiont were investigated by analyzing feeding behavior, stomach contents, and the ingestion and assimilation of host material using tracer techniques. The males and small immature females meet their energy requirements by ingesting epithelial tissue and tube feet from the peristomial region, damage being in equilibrium with the remarkable regenerating capacities of the host. The large females live in the rectum, where abundant aggregations of coelomocytes and material from the fecal pellets of the host are ingested. Coelomocytes, mostly pigmented eleocytes, migrate across the inner epithelium of the rectum and appear to be involved in the deposition of pigments in the normally unexposed tissue. The fecal pellets are mostly composed of sediment, but there is a relatively large amount of partly-digested algae and encrusting animals, as well as ciliates, nematodes, and bacteria. Dry feces average 0.51% lipid, 0.20% alcohol-soluble and TCA-soluble carbohydrates, and 0.10% protein. Organic material from the feces is selectively removed by the modified maxillipeds of the symbiont. The presence of large females outside the rectum results in heavy damage and in the eventual death of the host. Minimal energy requirements of males are supplied by a relatively small amount of epithelial tissue (3.9 mg/day), whereas almost three times this amount (10.3mg/day) would be required for adult females placed outside the rectum.

ACKNOWLEDGMENTS

This report represents a part of a dissertation submitted to the University of Hawaii in partial fulfillment of the requirements for the Ph. D. degree. I am deeply indebted to Dr. Thomas C. Cheng for his encouragement and inspiring enthusiasm. This work was supported by a scholarship from the Economic Development Administration of the Commonwealth of Puerto Rico.

REFERENCES

Balss, H. 1956. Decapoda. VI. Okologie (Fortsetzung). *In* Dr. H. G. Bronn's Klassen und Ordnungen des Tierreichs, Vol. 5, Part 1, Buch 7. Akadem. Verlag., Leipzig.

Boolootian, R. A. and Giese, A. C. 1958. Coelomic corpuscles of echinoderms. Biol. Bull. 115:53–63.

Boolootian, R. A. and Lasker, R. 1964. Digestion of brown algae and the distribution of nutrients in the purple sea urchin *Strongylocentrotus purpuratus*. Comp. Biochem. Physiol. 11:273–289.

Castro, P. 1969. Symbiosis between *Echinoecus pentagonus* (Crustacea, Brachyura) and its host in Hawaii, *Echinothrix calamaris* (Echinoidea). Ph.D. Thesis, Univ. Hawaii.

Cheng, T. C. 1967. Marine molluscs as hosts for symbioses (Adv. Marine Biol. Vol. 5). Academic Press, London. 424 p.

Clark, J. M. 1964. Experimental biochemistry. Freeman, San Francisco. 228 p.

Dubois, M., Gilles, K. A., Hamilton, J. K., Rebers, P. A., and Smith, F. 1956. Colorimetric method for determination of sugars and related substances. Anal. Chem. 28:350 356.

Endean, R. 1966. The coelomocytes and coelomic fluid. *In* Physiology of echinodermata (R. A. Boolootian, ed.). Interscience, New York. 822 p.

Farmanfarmaian, A. 1966. The respiratory physiology of echinoderms. *In* Physiology of echinodermata (R. A. Boolootian, ed.). Interscience, New York. 822 p.

Fox, D. L. 1953. Animal biochromes and structural colours. Cambridge Univ. Press, Cambridge. 378 p.

Frankenberg, D. and Smith, K. L. 1967. Caprophagy in marine animals. Limnol. Oceanogr. 12:443–450.

Frankenberg, D., Coles, S. L., and Johannes, R. E. 1967. The potential trophic significance of *Callinassa major* fecal pellets. Limnol. Oceanogr. 12:113–120.

Fuji, A. 1962. Studies on the biology of the sea urchin. V. Food consumption of *Strongylocentrotus intermedius*. Jap. J. Ecol. 12:181–186.

Fuji, A. 1967. Ecological studies on the growth and food consumption of Japanese common littoral sea urchin, *Strongylocentrotus intermedius* (A. Agassiz). Mem. Fac. Fish. Hokkaido Univ. 15:83–160.

Humason, G. L. 1967. Animal tissue techniques. Freeman, San Francisco. 569 p.

Hyman, L. H. 1955. The invertebrates, Vol. 4: Echinodermata. The Coelomate Bilateria. McGraw-Hill, New York. 763 p.

Johannes, R. E. and Satomi, M. 1966. Composition and nutritive value of fecal pellets of a marine crustacean. Limnol. Oceanogr., 11:191–197.
Johannes, R. E. and Satomi, M. 1967. Measuring organic matter retained by aquatic invertebrates. J. Fish. Res. Bd. Can. 24:2467–2471.
Johnson, M. J. 1949. A rapid micromethod for estimation of non-volatile organic matter. J. Biol. Chem. 181:707–711.
Knudsen, J. W. 1967. *Trapezia* and *Tetralia* (Decapoda, Brachyura, Xanthidae) as obligate ectoparasites of pocilloporid and acroporid corals. Pacif. Sci. 21:51–57.
Lasker, R. and Boolootian, R. A. 1960. Digestion of the alga, *Macrocystis pyrifera*, by the sea urchin, *Strongylocentrotus purpuratus*. Nature 188:1130.
Lewis, J. B. 1964. Feeding and digestion in the tropical sea urchin *Diadema antillarum* Philippi. Can. J. Zool. 42:549–557.
Lowry, O. H., Rosebrough, N. J., Farr, A. L., and Randall, R. J. 1951. Protein measurement with the Folin phenol reagent. J. Biol. Chem. 193:265–275.
MacGinitie, G. E. and MacGinitie, N. 1968. Natural history of marine animals. McGraw-Hill, New York. 523 p.
MacMunn, C. A. 1885. On the chromatology of the blood of some invertebrates. Quart. J. Microscop. Soc. 25:469–490.
Millot, N. 1957. Naphthoquinone pigment in the tropical sea urchin *Diadema antillarum* Philippi. Proc. Zool. Soc. London, 129:263–272.
Millot, N. and Yoshida, M. 1957. The spectral sensitivity of the echinoid *Diadema antillarum* Philippine J. Exp. Biol. 34:394–401.
Mukerjee, P. 1956. Use of ionic dyes in the analysis of ionic surfactants and other ionic organic compounds. Anal. Chem. 28:870–873.
Patton, W. K. 1967a. Commensal crustacea. Proc. Symp. Crustacea [Marine Ass. India] 3:1228–1244.
Patton, W. K. 1967b. Studies on *Domecia acanthophora*, a commensal crab from Puerto Rico, with particular reference to modifications of the coral host and feeding habits. Biol. Bull. 132: 56–67.
Pequignat, E. 1966. Skin digestion and epidermal absorption in irregular and regular urchins and their probable relation to the outflow of spherule-coelomocytes. Nature 210:397–399.
Phillipson, E. 1964. A miniature bomb calorimeter for small biological samples. Oikos 15:130–139.
Potts, F. A. 1915. *Hapalocarcinus*, the gall-forming crab, with some notes on the related genus *Cryptochirus*. Papers Dep. Marine Biol. Carnegie Inst. Wash. 8:33–69.
Price, C. A. 1965. A membrane method for determination of total protein in dilute algal suspensions. Anal. Biochem. 12:213–218.
Rathbun, M. J. 1918. The graspoid crabs of America. Bull. US Nat. Mus. 97:1–461.
Serène, R., Duc, T. V., and Luom, N. V. 1958. Eumedoninae du Viet-Nam (Crustacea). Treubia 24:135–242.
Spruit, C. J. P. 1949. Absorption spectra of quinones. I. Naphthoquinones and naphthohydroquinones. Rec. Trav. Chim. Pays Bas 68:309–324.
Strickland, J. D. H., and Parsons, T. R. 1968. A practical handbook of seawater analysis. Bull. Fish. Res. Bd. Can. 167:1–311.
West, E. S., Todd, W. R., Mason, H. S., Van Bruggen, J. T. 1966. Textbook of biochemistry. Macmillan, New York. 1595 p.

Effects of Light and Other Factors on Host-Symbiont Interactions in Green Hydra

CHARLES F. LYTLE, LEONARD G. EPP
and
GEORGE T. BARTHALMUS

Department of Biology
The Pennsylvania State University
and
Department of Zoology
North Carolina State University

INTRODUCTION AND REVIEW OF THE LITERATURE

The association between certain types of hydra and green algae is one of the classical examples of symbiosis, and numerous papers have been published on various aspects of this association and on the interrelationship between the partners. These papers range from early speculations on the nature of the green color and the possibility of an "animal chlorophyll," to recent investigations utilizing sophisticated biophysical and biochemical techniques to reveal subtle interactions between the symbiont and its host.

Systematics

The systematics of green hydras were summarized by Ewer (1948) and Forrest (1959). Two morphological species are now recognized, the European *Chlorohydra viridissima* (Pallas 1766) and the American *Chlorohydra hadleyi* Forrest 1959. The natural occurrence of *C. viridissima* in the United States has not been definitely established, although some recent experimental studies in this country have been conducted with green hydra derived from European stocks.

The genus *Chlorohydra* is of questionable validity, particularly since the work of Forrest, but was retained by her pending further study. No serious taxonomic study of green hydras has appeared since her report. The determination of species in hydras is difficult and in most cases requires culturing to obtain sexual forms. Unfortunately, some experimental studies have been conducted with hydra inadequately described and arbitrarily named.

Some early workers, including Marshall (1882) and Schulze (1917), noted morphological differences in green hydras obtained from various sources and suggested the existence of distinct races or possibly additional species. Investigators in recent years have employed several strains (isolates) of green hydra obtained from various sources for experimental studies. At least some of these strains appear to exhibit distinctive physiological and morphological features (cf. Muscatine, 1965; Oschman, 1967; Cernichiari *et al.*, 1969), although there has as yet been no extensive study of the comparative morphology and physiology of various strains. Such studies become of greater importance as increasingly sensitive techniques are employed in the investigations of symbiont interactions and other aspects of the biology of green hydra.

Algal Morphology

The algal symbionts of green hydra are generally found in membrane-bounded vacuoles within the digestive cells of the gastrodermis (Fig. 1). They are generally assumed to be members of the genus *Chlorella*, which they

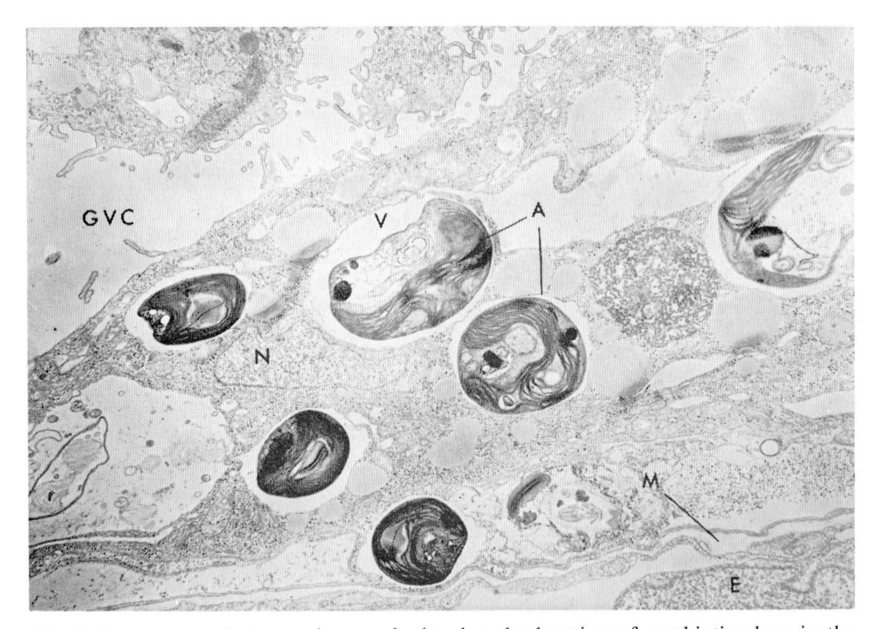

Fig. 1. Low-power electron micrograph showing the location of symbiotic algae in the digestive cells of *Chlorohydra hadleyi*. The algal cells (*A*) occur in distinct vacuoles (*V*) within the digestive cell. Part of the nucleus (*N*) of a digestive cell is seen at the left. Below the mesoglea (*M*) a portion of the epidermis (*E*) is visible. Part of the gastrovascular cavity (*GVC*) can be seen at the top left. Material in Figure 1–3 was fixed in glutaraldehyde, postfixed in osmium, embedded in Maraglas, and stained with lead citrate. (X 4,500. Micrograph by Carl F. T. Mattern.)

resemble in general morphology. Beyerinck (1890) reported successful culture of algae isolated from a green hydra and identified the alga as *Chlorella vulgaris*. Recent workers, however, have been unable to culture the algal symbionts removed from hydra (Park *et al.*, 1967; Cernichiari *et al.*, 1969).

Information on the ultrastructure of the symbiotic algae has been provided by Wood (1959), Oschman (1967), and by Park *et al.* (1967). The algae generally resemble the *Chlorella* symbionts found in the peripheral cytoplasm of *Paramecium bursaria* described by Karakashian *et al.* (1968).

The algal symbionts of *Chlorohydra viridissima* (Burnett strain) are ellipsoidal in shape and are surrounded by a distinct cell wall (Fig. 2). A large, cup-shaped chloroplast surrounds the centrally-located nucleus, mitochondria, Golgi apparatus, and numerous ribosomes. The chloroplast consists of a series of lamellar stacks made up of several thylacoids. The lamellar stacks are separated by a granular stroma. A large pyrenoid typically is found near

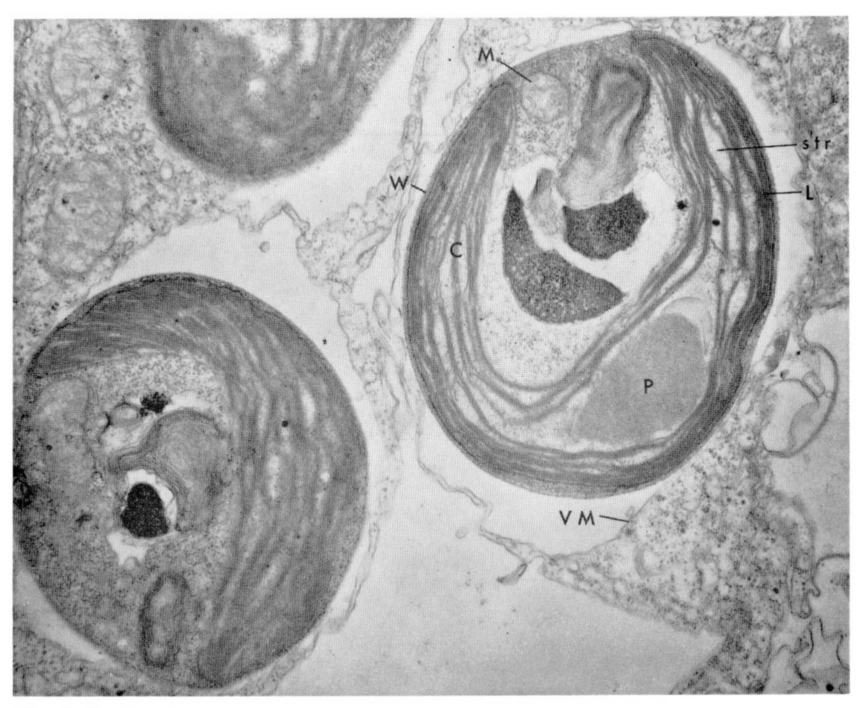

Fig. 2. Section through an algal cell in *Chlorohydra viridissima* showing the cup-shaped chloroplast (*C*) with its lamellar stacks (*L*), granular stroma (*str*) and large pyrenoid (*P*). A single mitochondrion (*M*) is visible near the top. The thickened algal cell wall (*W*) and the membrane surrounding the vacuole (*VM*) can also be seen. (X 17,250. Micrograph by Carl F. T. Mattern.)

the base of the chloroplast, and several ellipsoidal starch grains are found between the thylacoids.

A shell of starch grains surrounds the pyrenoid, and the granular matrix of the pyrenoid is traversed by a single thylacoid extending from the chloroplast lamellae (Fig. 3).

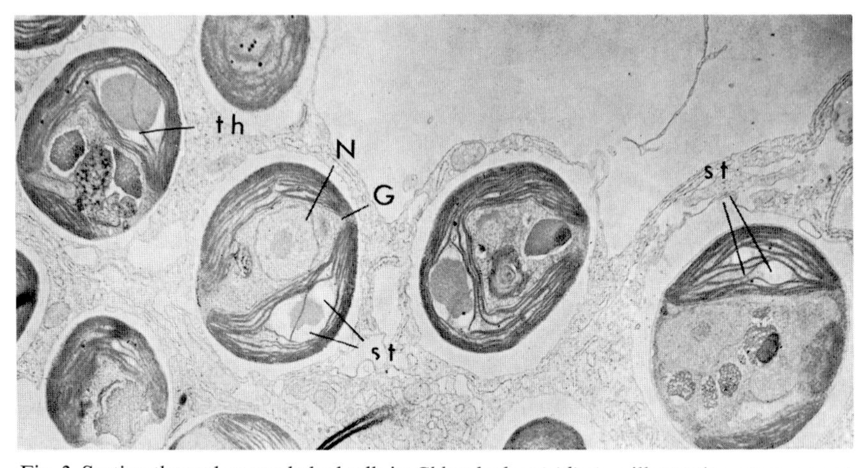

Fig. 3. Section through several algal cells in *Chlorohydra viridissima* illustrating arrangement of starch grains (*st*) around the pyrenoid (*P*) and the single thylacoid traversing the pyrenoid. One algal cell also shows ellipsoidal starch grains (*st*) between the chloro plast lamellae. The nucleus (*N*) and Golgi apparatus (*G*) are also shown in one of the algal cells. (X 6,500. Micrograph by Carl F. T. Mattern.)

The symbiotic algae in the Kenilworth strain of *Chlorohydra hadleyi* appear to be closely similar to those of the Burnett strain of *Chlorohydra viridissima* (see Park *et al.*, 1967). The algal cells in the former strain are generally smaller in size, however, than those in the Burnett strain.

Oschman (1967) described algae from another strain of green hydra (California strain) which lacks a pyrenoid, and which also appears to have a thinner cell wall than the algae in the Burnett strain. In the same paper Oschman reported preliminary observations on algae from two other strains of green hydra (Carolina and European strains) which appear to be morphologically distinct. Thus it seems likely that careful studies of the comparative morphology of algae from various strains of green hydra may reveal still further evidence of diversity in the symbiotic algae found associated with the various strains.

Dissociation and Reassociation of Symbionts

Several investigators have demonstrated that green hydra can be freed of their algae and that the resulting aposymbiotic (albino) hydra survive and reproduce for extended periods of time. The most widely employed technique for bleaching green hydra is that introduced by Whitney (1907, 1908), who employed 0.5–1.5 % glycerine solutions to induce expulsion of the algae from the hydra. This technique has been employed to produce aposymbiotic animals for numerous experimental studies on green hydras, including those of Burnett and Garofalo (1960), Muscatine (1961), Muscatine and Lenhoff (1965a; 1965b), Stiven (1965a,b), Park *et al.* (1967), and Epp and Lytle (1969). Fulton (see discussion following Muscatine, 1961), reported successful bleaching of green hydra with chloramphenicol solutions. Also, Goetsch (1924) produced aposymbiotic hydra by growing green hydra at low temperatures in the dark in a calcium-deficient medium for several weeks.

Park *et al.* (1967) reported a difference in the susceptibility to bleaching treatments between two types of green hydras. These workers were able to bleach the Kenilworth strain of *Chlorohydra hadleyi* with glycerine treatment and established aposymbiotic clones of this strain. In contrast, they found the Burnett strain of *Chlorohydra viridissima* refractory even to prolonged glycerine treatment. Chloramphenicol treatment was also found ineffective in bleaching the Burnett strain of *C. viridissima.*

We have obtained similar results in experiments with the same two types of hydra in our laboratory. Bleached specimens of the Kenilworth strain of *C. hadleyi* were obtained in each of several trials after 10–21 days of exposure to 0.5 % glycerine solutions. We have also obtained bleached hydra of this strain after 7–14 days of culture in solutions containing 200 mg per liter of chloramphenicol. The Burnett strain of *C. viridissima*, however, retained its algae even after 6 weeks of exposure to glycerine solutions (0.5–0.8 %). We also found chloramphenicol solutions (50–400 mg/liter) to be ineffective in bleaching the Burnett strain.

Several investigators have demonstrated successful reinfection of aposymbiotic hydra with symbiotic algae. Goetsch (1924) was able to reestablish symbioses by grafting green halves to nongreen halves and also by feeding aposymbiotic hydras on *Daphnia* containing small pieces of green hydra.

Park *et al.* (1967) provided the first experimental evidence of specificity between *C. hadleyi* and its algal symbionts. Successful infections were estabished after injections of algal suspensions prepared from *C. hadleyi* into aposymbiotic *C. hadleyi*, but not following injections of algae from *C. viridissima*. Interspecific grafts between the two hydras were also ineffective in reestablishing infections in the aposymbiotic *C. hadleyi*. Karakashian (1963) and Karakashian and Karakashian (1965), however, succeeded in establishing algae from a green hydra in bleached *Paramecium bursaria*. Subsequent experiments with this artificially infected stock of *P. bursaria* demonstrated growth rates higher than those in the same stock reinfected with its normal symbiont.

Physiology and Biochemistry

Several recent experimental studies have provided biochemical evidence that, at least under certain conditions, the symbiotic algae contribute to the growth and survival of green hydra. Muscatine and Lenhoff (1963) demonstrated that significant amounts of the CO_2 photosynthetically fixed by symbiotic algae are later transferred to the tissues of the hydra. It was later shown by Muscatine *et al.* (1967) that algae isolated from green hydra release large amounts of their photosynthetically fixed carbon into the incubation medium. Hence these results suggested that the algae *in vivo* provide maltose to the hydra. Recently, Roffman and Lenhoff (1969) have provided further evidence for such a transfer of carbohydrates to the hydra tissues. Their data appear to show transfer of labelled carbon from the algae to the hydra tissues in the form of some small molecule or molecules and later indicated relatively high concentrations of the label (32–44 %) in the polysaccharide-containing fraction.

Maltose release by symbiotic algae *in vitro* has been shown by Ciernichiari *et al.* (1969) to be very sensitive to the pH of the incubating medium; therefore, these authors have suggested that changes in intracellular pH may provide a mechanism by which the hydra may be able to regulate maltose excretion by the symbiotic algae. Evidence was also obtained that the synthesis of maltose occurs at or near the surface of the algal cells, and these authors have proposed new biochemical mechanisms for maltose synthesis and excretion by the algal symbionts. Comparative studies of symbiotic and free-living strains of *Chlorella* have revealed pronounced differences in the patterns of carbohydrate metabolism between the two types. An extensive review of carbohydrate exchange in various types of symbioses, including green hydra, has been provided by Smith *et al.* (1969).

The occurrence and distribution of enzymes of carbohydrate metabolism in green hydra and other coelenterates has been studied by Powers *et al.* (1969) and by Rutherford and Lenhoff (1969). Their work has demonstrated the presence of an active glucose-6-phosphate dehydrogenase in both green and aposymbiotic hydra but an apparent absence of the companion enzyme, 6-phosphogluconate dehydrogenase, usually present in plant and animal tissues with an active hexose monophosphate pathway. The results of these studies, therefore, indicate some apparent peculiarities of carbohydrate metabolism in coelenterates which may be of importance in the interactions between green hydra and their algal symbionts.

INFLUENCE OF ENVIRONMENTAL FACTORS

Several previous investigations have provided evidence of a contribution of the symbiotic algae to the growth and survival of green hydra under certain conditions. This contribution appears to be most readily demonstrable when food is limited (Muscatine and Lenhoff, 1965b) or under some stress as in a parasitic infection (Stiven, 1965b). Other studies have indicated distinct physiological differences between different types of green hydras. We, therefore, undertook a study of the interactions between the symbiotic algae and their hosts through an analysis of the effects of light, nutrition, and population density on reproduction in green and sapoymbiotic hydra (Epp and Lytle, 1969).

Materials and Methods

Clone cultures of the Burnett strain of *Chlorohydra viridissima* and the Kenilworth strain of *Chlorohydra hadleyi* were established as described previously (Epp and Lytle, 1969). A new clone of aposymbiotic *C. hadleyi* was established by use of the glycerine technique. The animals were reared in modified spring water (100 mg NaH_2CO_3 + 50 mg Na-EDTA per liter) and maintained at 20 ± 1°C with controlled light intensity and photoperiodicity. Cultures were fed excess *Artemia nauplii* on alternate days and the culture solution was replaced daily.

Buds were counted and removed each day during an experiment. Statistical analysis of the data was performed using a single classification analysis of variance followed by Duncan's multiple range test (Duncan, 1955, modified by Kramer, 1956) using Harter's critical values. Data processing was done on an IBM 360 computer at The Pennsylvania State University Computation Center.

Results

Hydra cultures grown under a 12-hour photoperiod demonstrated a significant difference in budding between the Burnett Green hydra and the Kenilworth Green and Albino hydras. The time course of budding for a 21-day period is illustrated in Figure 4. We found no significant difference, however, in bud production between Kenilworth Green and Albino hydra under a 12-hour light-dark illumination cycle. A comparison of animals reared under different illumination conditions revealed a significant influence of light on

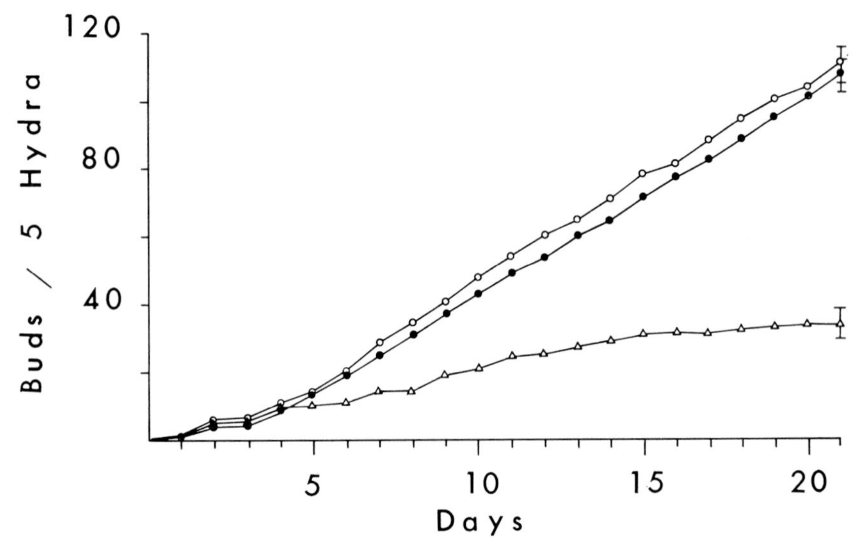

Fig. 4. Budding of Burnett Green (*triangles*), Kenilworth Green (*closed circles*), and Kenilworth Albino (*open circles*) hydras with a 12 hour light-dark cycle (After Epp and Lytle, 1969).

bud production in the three types of hydra (Tables 1 and 2). Statistical analysis of the data showed that the Kenilworth Green hydra produced significantly more buds under constant light or under a 12-hour light-dark cycle than in darkness. Perhaps more surprisingly, the illuminated cultures of Kenilworth Albino hydra exhibited a similar increase in bud production, despite the absence of symbiotic algae. These results clearly suggest that the increased budding of the Kenilworth Green hydra under illumination may be due at least in part to some factor other than the symbiotic algae.

TABLE 1. *Buds produced by three types of hydra during a 21-day experiment under three different light regimes. Each value represents the mean and standard error of three replicate cultures. Hydra were fed on alternate days*

Hydra	*Continuous dark*	*Continuous light*	*12 hours dark/ 12 hours light*
Burnett Green	51.7±4.62	34.7±3.81	33.0±1.66
Kenilworth Green	92.3±4.24	107.7±5.54	107.3±3.37
Kenilworth Albino	89.3±2.55	112.3±5.67	112.0±5.81

TABLE 2. *Summary of Duncan's Multiple Range Test comparing the budding of three types of hydra exposed to different light regimes*

Regime		*12 hours light/ 12 hours dark*			*Continuous light*			*Continuous dark*		
		B	G	A	B	G	A	B	G	A
12 hours light/ 12 hours dark	B	—	1	1	NS	1	1	5	1	1
	G	1	—	NS	1	NS	NS	1	5	5
	A	1	NS	—	1	NS	NS	1	5	5
Continuous light	B	NS	1	1	—	1	1	5	1	1
	G	1	NS	NS	1	—	NS	1	5	5
	A	1	NS	NS	1	NS	—	1	1	1
Continuous dark	B	5	1	1	5	1	1	—	1	1
	G	1	5	1	1	5	1	1	—	NS
	A	1	5	5	1	5	1	1	NS	—

B=Burnett Green
G=Kenilworth Green
A=Kenilworth Albino
5=$P<0.05$
1=$P<0.01$
NS=Not statistically significant

Burnett Green hydra exhibited a reverse pattern. Fewer buds were produced under either lighted regime than in darkness. Thus there was an apparent difference in the influence of light on budding in the two types of green hydra. Investigation of the effects of various photoperiods on budding provided further evidence of this difference between the Burnett Green hydra and the Kenilworth Green hydra. When Kenilworth Albino and Burnett Green hydra were reared under several different photoperiods, they were found to exhibit opposite trends in regard to the influence of light on their reproduction (Fig. 5). Budding tended to increase with increasing photoperiod in the Kenilworth Albino hydra but tended to decrease with increasing photoperiod in the Burnett Green hydra. Significant differences ($P<0.01$) were demonstrated in the number of buds produced under zero and two hours of light and between six and eight hours of light in the Kenilworth Albino hydra. The Burnett Green hydra, however, demonstrated significant differences in the number of buds produced between zero and eight hours of light.

Although the Kenilworth Green hydra did not demonstrate a significant contribution of the symbiotic algae as reflected by bud production, experiments with starved hydras have provided such evidence. The results of an experiment with Kenilworth Green, Kenilworth Albino, and Burnett Green hydra are shown in Figure 6. Prior to the period of starvation, the animals were maintained on a normal feeding schedule (fed on alternate days) under each of the light regimes indicated. Feeding was stopped at the beginning of the experiment and records were kept of the number of buds produced each

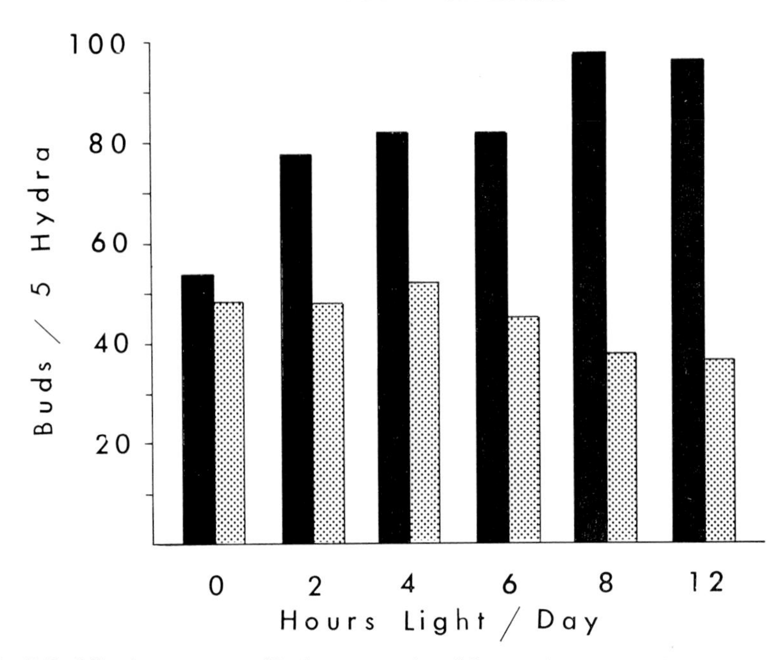

Fig. 5. Budding in two types of hydras exposed to different photoperiods. *Solid bar s*represent Kenilworth Green hydra; *stippled bars* represent Burnett Green hydra. (After Epp and Lytle, 1969.)

day until budding appeared to have ceased. The increased bud production by the Kenilworth Green hydra (47.0±2.20 buds) over that by the Kenilworth Albino (29.0±2.99 buds) suggests a contribution of the symbiotic algae present in the green form to the budding of the green hydra. Both types of Kenilworth hydra produced significantly more buds than did the Burnett Green hydra (6.6±1.65 buds; $P<0.01$), further indicating the difference in energy metabolism and reproduction between the Kenilworth and Burnett strains.

Similar evidence of a contribution of the symbiotic algae to the reproduction has also been obtained by several previous workers, including Muscatine and Lenhoff (1965b), Stiven (1965b), and Park *et al.* (1967).

Altering the light regimes of the starving animals also resulted in differences in the number of buds produced by the three types of hydra. The greatest number of buds was produced by the Burnett Green and Albino hydra under a 12-hour light-dark cycle (Table 3). Both types produced significantly fewer buds under constant light or darkness (Table 4). These results again indicated the operation of a light-mediated reaction in aposymbiotic hydra independent

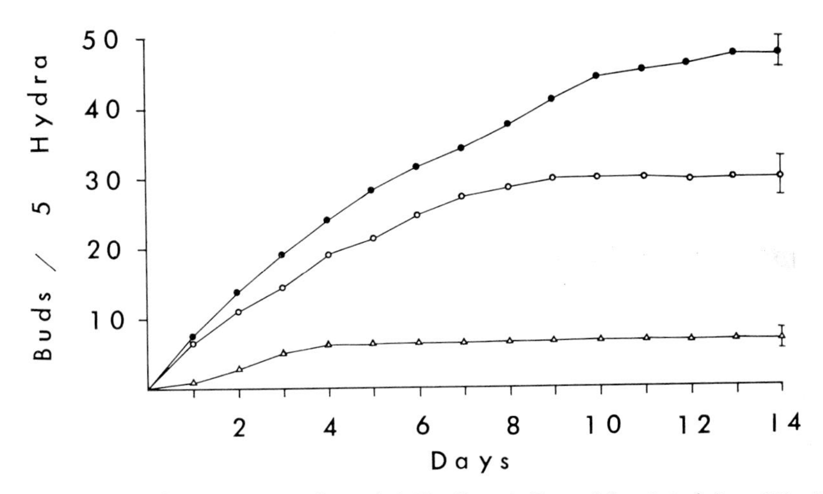

Fig. 6. Budding of Burnett Green (*triangles*), Kenilworth Green (*closed circles*), and Kenilworth Albino (*open circles*) hydras during starvation. Twelve-hour light dark cycle. (After Epp and Lytle, 1969.)

TABLE 3. *Buds produced by three types of hydra under three different light regimes during a 14-day starvation period. Each value represents the mean and standard error of three replicate cultures*

Hydra	*Continuous dark*	*Continuous light*	*12 hours light/ 12 hours dark*
Burnett Green	19.0±2.99	1.6±0.33	6.6±1.65
Kenilworth Green	28.6±2.69	37.6±2.69	47.0±2.20
Kenilworth Albino	21.0±2.69	26.3±2.43	29.0±2.99

TABLE 4. *Summary of Duncan's Multiple Range Test comparing the budding of three types of hydra during a 14-day starvation period*

Regime		*12 hours light/ 12 hours dark*			*Continuous light*			*Continuous dark*		
		B	G	A	B	G	A	B	G	A
	B	—	1	1	5	1	1	1	1	1
12 hours light/	G	1	—	1	1	1	1	1	1	1
12 hours dark	A	1	1	—	1	1	NS	1	NS	1
	B	5	1	1	—	1	1	1	1	1
Continuous	G	1	1	1	1	—	1	1	1	1
light	A	1	1	NS	1	1	—	5	NS	5
	B	1	1	1	1	1	5	—	1	NS
Continuous	G	1	1	NS	1	1	NS	1	—	1
dark	A	1	1	1	1	1	5	NS	1	—

B = Burnett Green
G = Kenilworth Green
A = Kenilworth Albino

5 = P < 0.05
1 = P < 0.05
NS = Not statistically significant

of the symbiotic algae. The greater number of buds produced by the Kenilworth Green, and the larger relative differences between the different light regimes, presumably reflect the combined effects of the symbiotic algae and of the hypothesized nonalgal light-mediated reaction.

The Burnett Green hydra, in contrast to the Kenilworth Green hydra, produced fewer buds under constant light than under a 12-hour light-dark cycle. This result is in agreement with our previous observations on the inhibitory influence of increased light exposure on the budding of Burnett Green hydra. The Burnett Green hydra again demonstrated greater bud production under the dark regime than under either lighted regime.

An effect of population density on the reproduction of green hydra was demonstrated by exposure of experimental animals to crowded conditions prior to a period of starvation. Hydras were removed from stock cultures containing several hundred hydra in 50 ml of culture medium, and were exposed to different light regimes without feeding. With a 12-hour light-dark cycle, the Kenilworth Green hydra produced more buds than did either the Kenilworth Albino or the Burnett Green hydra (Fig. 7). Exposure to different light regimes resulted in significant differences in the reproduction of both types of green hydra, but not in the Kenilworth Albino hydra (Tables 5 and 6). In contrast to the experiments discussed previously (Table 3), the Burnett Green hydra produced significantly more buds under both constant illumination and 12 hours of illumination than it did in darkness.

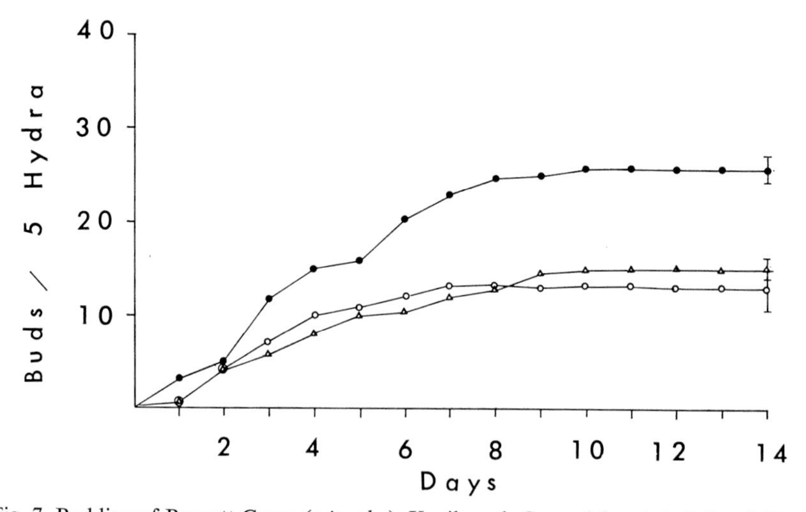

Fig. 7. Budding of Burnett Green (*triangles*), Kenilworth Green (*closed circles*) and Kenilworth Albino (*open circles*) hydras during starvation after previous exposure to crowded conditions. Twelve hour light-dark cycle. (After Epp and Lytle, 1969.)

TABLE 5. *Buds produced by three types of hydra during starvation under three different light regimes after previous exposure to crowded conditions*

Hydra	*Continuous dark*	*Continuous light*	*12 hours light/ 12 hours dark*
Burnett Green	8.3±0.6	13.0±1.43	15.3±1.00
Kenilworth Green	17.3±2.69	30.0±2.40	26.6±1.43
Kenilworth Albino	14.6±1.31	15.3±1.85	13.0±1.43

TABLE 6. *Summary of Duncan's Multiple Range Test comparing the budding of three types of hydra during starvation under three different light regimes after previous exposure to crowded conditions*

Regime		*12 hours light/ 12 hours dark*			*Continuous light*			*Continuous dark*		
		B	G	A	B	G	A	B	G	A
	B	—	1	NS	NS	1	NS	1	NS	NS
12 hours light/	G	1	—	1	1	NS	1	1	1	1
12 hours dark	A	NS	1	—	NS	1	NS	5	5	NS
	B	NS	1	NS	—	1	NS	5	5	NS
Continuous	G	1	NS	1	1	—	1	1	1	1
light	A	NS	1	NS	NS	1	—	1	NS	NS
	B	5	1	5	5	1	5	—	1	1
Continuous	G	NS	1	5	5	1	NS	1	—	NS
dark	A	NS	1	NS	NS	1	NS	1	NS	—

B=Burnett Green
G=Kenilworth Green
A=Kenilworth Albino
5=$P<0.05$
1=$P<0.01$
NS=Not statistically significant

Starved Kenilworth Green and Albino hydras produced fewer buds under all three light regimes after exposure to crowding conditions than did corresponding hydra in low density cultures (5 hydras in 50ml of culture medium), thus indicating an inhibitory effect of increased population density on budding in the Kenilworth hydra. The increased bud production of the Kenilworth Green hydra under constant light, or with 12 hours of illumination, thus appears to reflect the contribution of the symbiotic algae to budding. Burnett Green hydra, however, exhibited an opposite pattern. Fewer buds were produced in starved animals in darkness after exposure to crowding, but more buds were produced by similar Burnett Green hydra under both lighted regimes.

DISCUSSION AND CONCLUSIONS

Our experiments on the effects of light, population density, and nutrition have thus provided further evidence of a complex series of host-symbiont interactions in green hydra. As we suggested earlier (Epp and Lytle, 1969), our results appear to indicate that budding in the Burnett and Kenilworth green hydras is regulated by the supply of some critical metabolite or metabolites in the hydra tissues which can be derived both from food and from the symbiotic algae. The observed differences in the budding of the Burnett Green hydra from that of the Kenilworth Green hydra appears explainable on the basis of smaller metabolic pools of the metabolite present in the Burnett Green hydra and a sensitivity of this hydra to increased levels of the metabolite which tend to inhibit budding.

Our results on the effects of environmental factors are therefore compatible with the model of carbohydrate release proposed by Muscatine and his coworkers, although we have not undertaken a study of the chemical nature of the host-symbiont interactions.

Several of the experimental studies reviewed earlier provide evidence of other types of interactions between the algae and hydra as well as for a considerable degree of diversity in the host-symbiont relationships in this association; this merits further investigation. The algal injection experiments of Park *et al.* (1967) have indicated the existence of host specificity in at least one case. Different degrees of affinity between the symbionts have been shown by the differential responses of various types of green hydra to glycerine and chloramphenicol treatments. Furthermore, the inability of recent workers to successfully culture algal symbionts *in vitro* suggests the possibility of special nutritional requirements of the symbiotic algae. A recent report by Cook (1969) appears to provide the first experimental evidence for transfer of materials from a hydra to its symbiotic alga.

ACKNOWLEDGMENTS

We are indebted to Dr. Carl F. T. Mattern for allowing us to study his electron micrographs of the algal symbionts of *Chlorohydra viridissima* and *C. hadleyi* and for the use of Figures 1–3. We are also grateful to Dr. Helen D. Park for her interest and for her sharing of unpublished data. Figures 4–7 are reprinted through the courtesy of the editor of the *Biological Bulletin*.

REFERENCES

Beyerinck, M. W. 1890. Kulturversuche mit Zoochlorellen, Lichengonidien und anderen niederen Algen. Bot. Zeit. 48:741–754.

Burnett, A L. and Garofalo, M. 1960. Growth pattern in the green hydra, *Chlorohydra viridissima*. Science 131:160–161.

Cernichiari, E., Muscatine, L. and Smith, D. C. 1969. Maltose excretion by the symbiotic algae of *Hydra viridis*. Roy. Soc. (London) Proc., B. 173:557–576.

Cook, C. B. 1969. Benefit to algal symbionts from feeding of green hydra. Amer. Zool. 9:1138–1139.

Davis, L. V. 1966. Inhibition of growth and regeneration in *Hydra* by crowded culture water. Nature 212:1215–1217.

Duncan, D. B. 1955. Multiple range and multiple F tests. Biometrics 11:9–42.

Ewer, R. F. 1948. A review of the Hydridae and two new species of *Hydra* from Natal. Proc. Zool. Soc. London 118:228–244.

Epp, L. G. and Lytle, C. F. 1969. The influence of light on asexual reproduction in green and aposymbiotic hydra. Biol. Bull. 137:79–94.

Forrest, H. B. 1959. Taxonomic studies on the hydras of North America VII. Description of *Chlorohydra hadleyi*, new species, with a key to the North American species of hydras. Amer. Midland Natur. 62:440–448.

Goetsch, W. 1924. Die Symbiose der Süsswasser-Hydroiden und ihre kunstliche Beeinflussung. Z. Morphol. Ökol. Tiere 1:660–731.

Karakashian, S. J. 1963. Growth of *Paramecium bursaria* as influenced by the presence of algal symbionts. Physiol. Zool. 36:52–68.

Karakashian, S. J. and Karakashian, M. W. 1965. Evolution and symbiosis in the genus *Chlorella* and related algae. Evolution 19:368–377.

Karakashian, S. J., Karakashian, M. W., and Rudzinska, M. A. 1968. Electron microscopic observations on the symbiosis of *Paramecium bursaria* and its intracellular algae. J. Protozool. 15:113–128.

Marshall, W. 1882. Über einige Lebenserscheinungen der Süsswasserpolypen und über eine neue Form von *Hydra viridis*. Z. Wiss. Zool. 37:664–702.

Muscatine, L. 1961. Symbiosis in marine and fresh water coelenterates, p. 225–268 *In* H. M. Lenhoff and W. F. Loomis (eds.) The biology of hydra. Univ. Miami Press, Miami, Fla.

Muscatine, L. 1965. Symbiosis of hydra and algae—III. Extracellular products of the algae. Comp. Biochem. Physiol. 16:77–92.

Muscatine, L., Karakashian, S. J. and Karakashian, M. W. 1967. Soluble extracellular products of algae symbiotic with a ciliate, a sponge, and a mutant hydra. Comp. Biochem. Physiol. 20:1–12.

Muscatine, L. and H. M. Lenhoff. 1963. Symbiosis: on the role of algae symbiotic with hydra. Science 142:956–958.

Muscatine, L. and Lenhoff, H. M. 1965a. Symbiosis of hydra and algae. I. Effect of some environmental cations on growth of symbiotic and aposymbiotic hydra. Biol. Bull. 128:415–424.

Muscatine, L. and Lenhoff, H. M. 1965b. Symbiosis of hydra and algae. II. Effects of limited food and starvation on growth of symbiotic and aposymbiotic hydra. Biol. Bull. 129:316–328.

Ochsman, J. L. 1967. Structure and reproduction of the algal symbionts of *Hydra viridis*. J. Phycol. 3:221–228.

Park, H. D., Greenblatt, C. L., Mattern, C. F. T. and Merrill, C. R. 1967. Some relationships between *Chlorohydra*, its symbionts, and some other chlorophyllous forms. J. Exp. Zool. 165:141–162.

Powers, D. A., Lenhoff, H. M. and Leone, C. A. 1969. Glucose 6 phosphate dehydrogenase and 6 phosphogluconate dehydrogenase activities in coelenterates. Comp. Biochem. Physiol. 27:139–144.

Roffman, B. and Lenhoff, H. M. 1960. Formation of polysaccarides by hydra from substrates produced by their endosymbiotic algae. Nature 221:381–382.

Rutherford, C. L. and Lenhoff, H. M. 1969. Enzymes of glucose catabolism in hydra. I. Relative activities of enzymes and absence of 6 phosphogluconate dehydrogenase. Arch. Biochem. Biophys. 133:119–127.

Schulze, P. 1917. Neue Beitrage zu einer Monographie der Gattung *Hydra*. Arch. Biontologie 4:39–119.

Smith, D., Muscatine, L. and Lewis, D. 1969. Carbohydrate movement from autotrophs to heterotrophs in parasitic and mutualistic symbiosis. Biol. Rev. 44:17–90.

Stiven, A. L. 1965a. The relationship between size, budding rate, and growth efficiency in three species of hydra. Res. Population Ecol. 7:1–15.

Stiven, A. L. 1965b. The association of symbiotic algae with the resistance of *Chlorohydra viridissima* (Pallas) to *Hydramoeba hydroxena* (Entz). J. Invert. Pathol. 7:356–367.

Whitney, D. D. 1907. Artificial removal of the green bodies of *Hydra viridis*. Biol. Bull. 13:291–299.

Whitney, D. D. 1908. Further studies on the elimination of the green bodies from the entoderm cells of *Hydra viridis*. Biol. Bull. 15:241–246.

Wood, R. L. 1959. Intracellular attachment in the epithelium of hydra as revealed by electron microscopy. J. Biophys. Biochem. Cytol. 6:343–351.

Blue-Green Algae in Echiuroid Worms

SIRO KAWAGUTI

Department of Biology
Faculty of Science
Okayama University
Okayama, Japan

INTRODUCTION

The echiuroid worm, *Bonellia viridis*, is well known to zoologists because of its remarkable sexual dimorphism; that is, the degenerated male is found attached symbiotically to the female body. It is also well known for its green coloration which is caused by the presence of a pigment called bonellin (Fox, 1953; Lederer, 1939).

Another species of echiuroid, *Ikedosoma gogoshimense*, has small green spots on the whole body surface, including the proboscis. Each of these spots is composed of an aggregation of cells with many green particles in the connective tissue of the subepidermal layer. An electron microscopic study has revealed blue-green algae in these green cells (Kawaguti, 1968).

Bonellia fuliginosa, a species closely related to *B. viridis*, bears a bluish green color almost evenly over the whole body to the naked eye (Ikeda, 1904; Brafield, 1968). At a low magnification it shows rather diffuse patches. These patches are composed of distributions of small green particles which closely resemble those of *Ikedosoma gogoshimense*. An electron microscopic study of this worm has been anticipated for some time; but there was practically no exact information available concerning the ecology and distribution of this worm until we found it on a reef of Kikaishima, one of the Amami Islands of Japan.

MATERIALS AND METHODS

The specimens of *Bonellia fuliginosa* used in this experiment were collected on the reefs of Kikaishima and Amamioshima in the Amami Islands. This species is rather abundant in this region as was described by Ikeda (1904) for the Ryukyu Islands. However, Brafield (1968) recently reported that it was rare at Banyuls-sur-Mer, France. He collected it at a depth of 12 m, probably in daytime. One wonders if this difference reflects ecological differences between the two localities.

The worm lives in a burrow in coral heads of various sizes. It is most abundant a little above the lowest tidemark. It expands its proboscis from the burrow in the dark or in a dim light, but not when directly exposed to the sun (Fig. 1). It is very sensitive to strong light. The burrow has a small opening and a wide space at the end where the main body of the worm is situated. When the worm is taken out by splitting the coral, it contracts its proboscis and shows a continuous peristaltic movement in the main body. But, if placed in the dark, it rests quietly with its proboscis extended as it does in its habitat during the night. A successful collection can be made at low tide during the night or when the tide is rising in the evening.

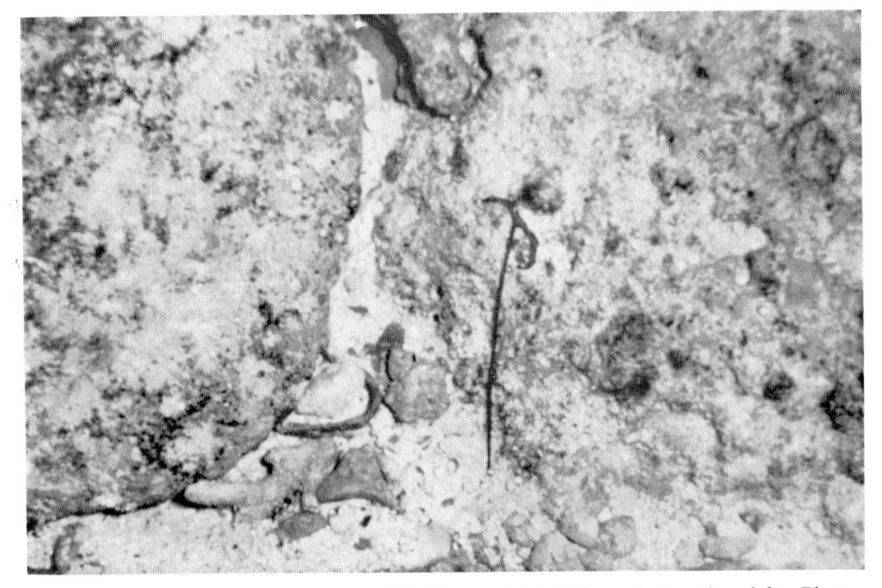

Fig. 1. *Bonellia fuliginosa* in a natural habitat, at Kikaishima, during the night. Photographed with a flash light, printed from a color slide.

The second species of echiuroid studied was *Ikedosoma gogoshimense*, collected in the vicinity of the Tamano Marine Laboratory of Okayama University. This worm lives in a burrow on the muddy sandbottom in the inter-tidal zone (Kawaguti, 1968).

For light microscope observations, a piece of the body wall was dissected and observed directly or the piece was fixed in Bouin's solution for histological sections.

For electron microscope observations, part of the body wall was pinned on a wooden plate and exposed to a fixative. After a few minutes when the tissue turned black, the portion was dissected out. The tissue was cut into small pieces and refixed for one hour. The fixative employed was 1 % osmium tetroxide buffered at pH 7.2 with phosphate buffer. Five per cent glutaraldehyde with cacodylate buffer was also used, followed by postosminification. The pieces were dehydrated in a series of graded alcohol solutions and embedded in epoxy resin. They were sectioned by use of a glass knife on an LKB ultramicrotome. The sections were mounted on formvar coated grids and observed under a Hitachi 11A or 11E electron microscope. Thick sections were also stained in toluidine blue for light microscope observations.

RESULTS

Light Microscope Observations

The outer surface of the body of *Bonellia fuliginosa* is covered with a single layer of epithelial cells as described for *Ikedosoma* in a previous paper (Kawaguti, 1968). In the intact epidermis, green cells are found evenly distributed (Fig. 2). Each of these cells is composed of an assembly of green particles. In the proboscis, the green cells are more densely distributed on the side opposite to the ciliated groove.

Differences between these green cells and those of *Ikedosoma* are clear even in light microscope observations.

Electron Microscope Observations

Epidermal Cells.—As is shown in Figure 3, the epidermal layer of the proboscis is composed of a single layer of rather tall cells with frequent insertions of mucous cells. The outer surface of each cell is covered with a row of long microvilli. However, the embedding mucus of these microvilli is very diffuse and almost transparent, and thus is strikingly different from that of *Ikedosoma* (Kawaguti, 1968). Each epidermal cell has its nucleus at its proximal end where it faces the underlying connective tissue.

In the ciliated groove, the surface of the epidermal cell is covered with a compact arrangement of long cilia. In the underlying connective tissue there are pigment cells and several other kinds of cells, such as muscles, nerves, and wandering cells. The muscle cells show a tendency to oblique striation with a few J-granules. Nerve-muscle junctions are observed frequently. Each of these is very interesting and will be dealt with separately.

Pigment Cells.—Pigment cells are found scattered independently in the subepidermal connective tissue along with other types of cells. This is quite different from the condition in *Ikedosoma* where these cells appear in groups (Kawaguti, 1968). Each pigment cell contains some granular bodies of various shapes as is shown in Figures 4–6. These granular bodies usually measure about 1 μ in diameter although a few may measure more than 2 μ. They show

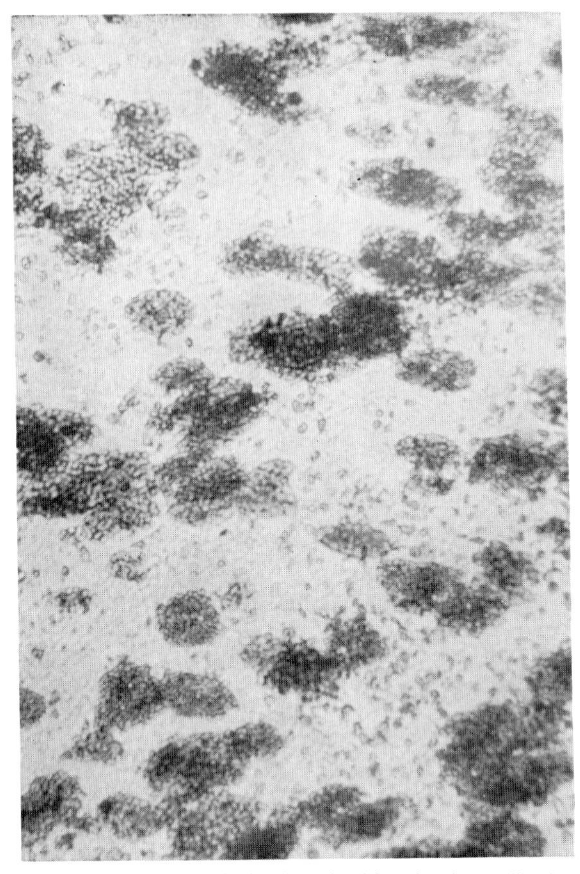

Fig. 2. Light micrograph of green cells in the subepidermis of *Bonellia fuliginosa*.

various internal structures such as lamellae and myelinated bodies. Sometimes dark particles also occur. These structures closely resemble those found in *Ikedosoma* and are similar to those frequently depicted in electron micrographs of bacteria and blue-green algae.

In some cases, as is shown in Figure 4, there are some inclusions which are easily identified as bacteria. One of these bacteria bears lamellar structures. Besides these particles there are many mitochondria, endoplasmic reticula, and vesicular bodies in the cytoplasm of the pigment cell.

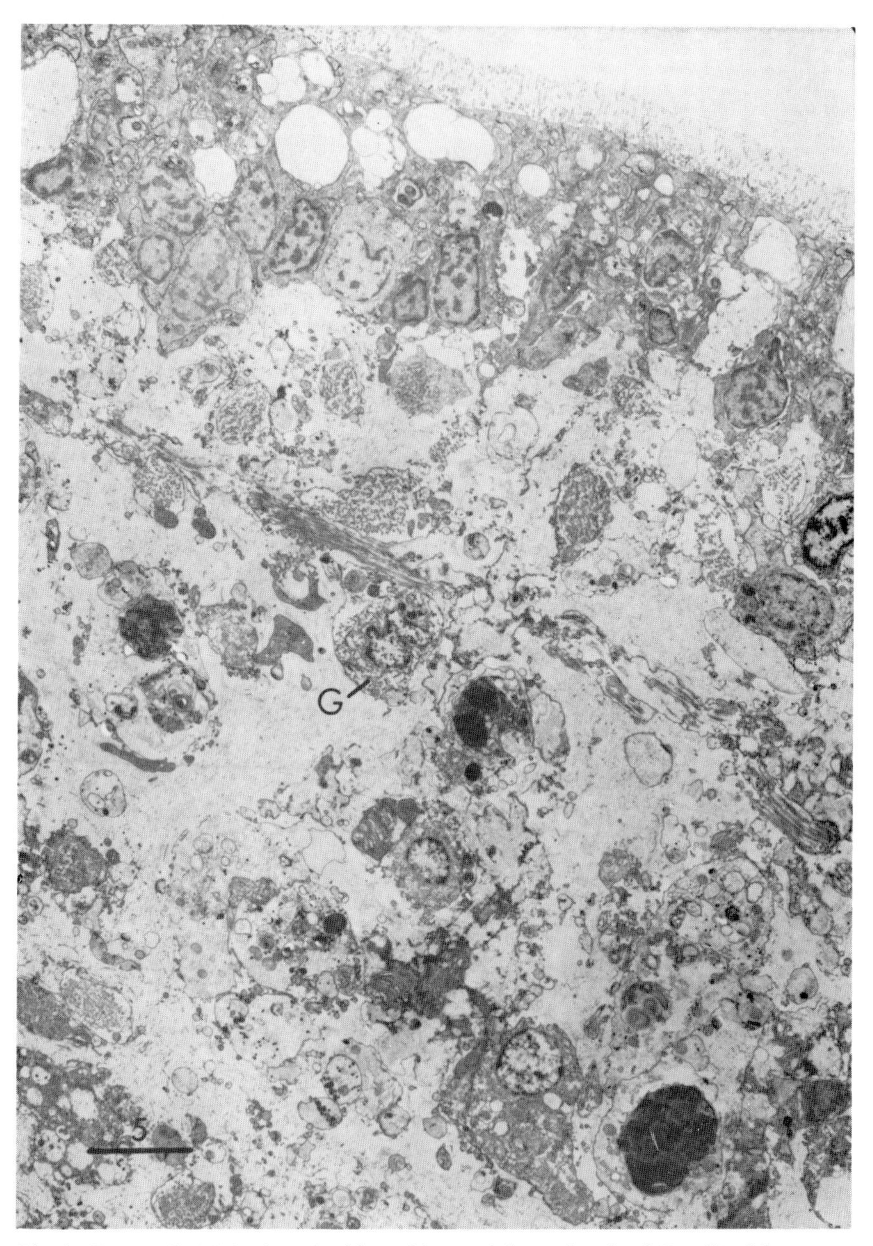

Fig. 3. Green cells (*G*) in the subepidermal layer of the proboscis of *Bonellia fuliginosa* at low magnification. *Bar* represents 5μ. (X 2,800.)

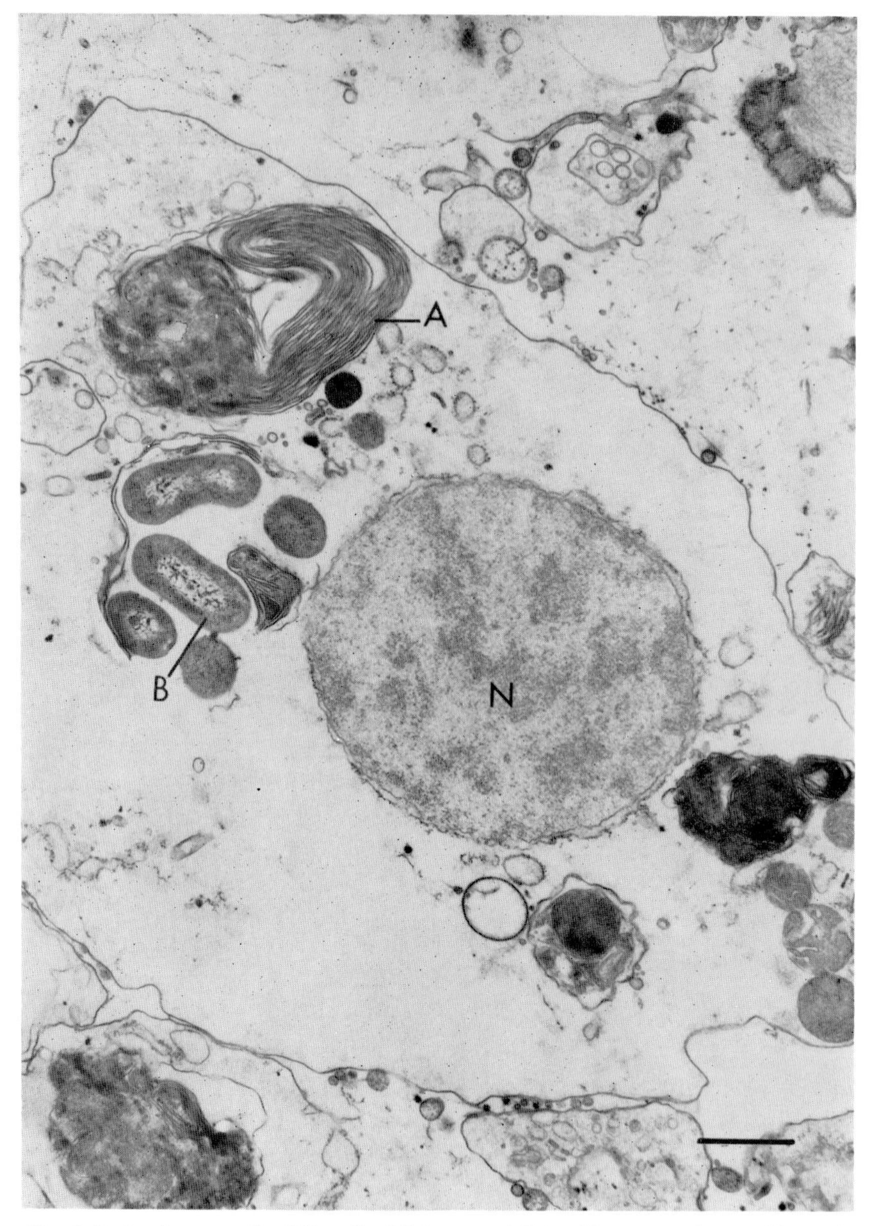

Fig. 4. Parts of green cells of *Bonellia fuliginosa* containing blue-green algae and bacteria. *N*, nucleus; *B*, bacteria; *A*, algae. *Bar* represents 1μ. (X 13,000.)

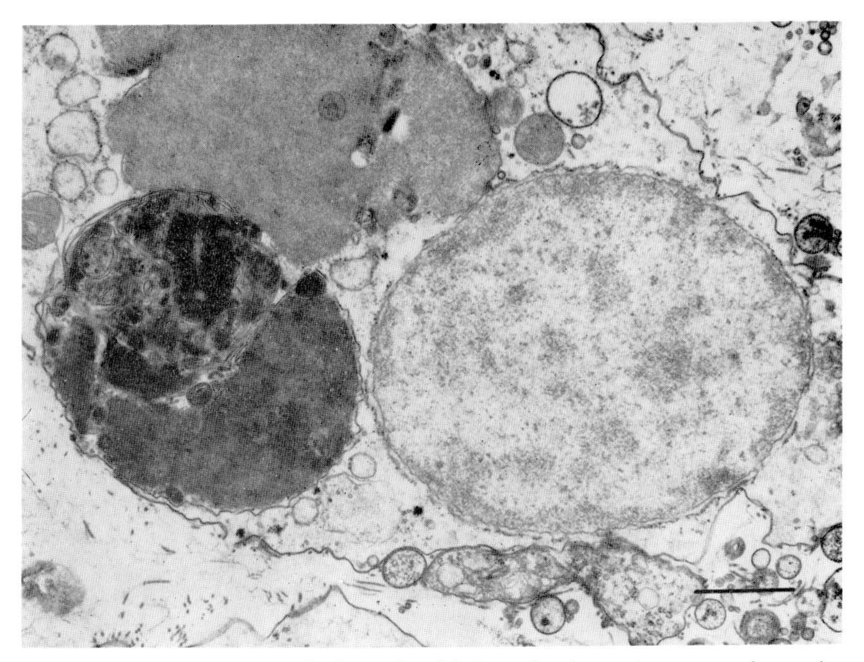

Fig. 5. A part of a pigment cell of *Bonellia fuliginosa* showing various types of granules. *Bar* represents 1 μ. (X 13,750.)

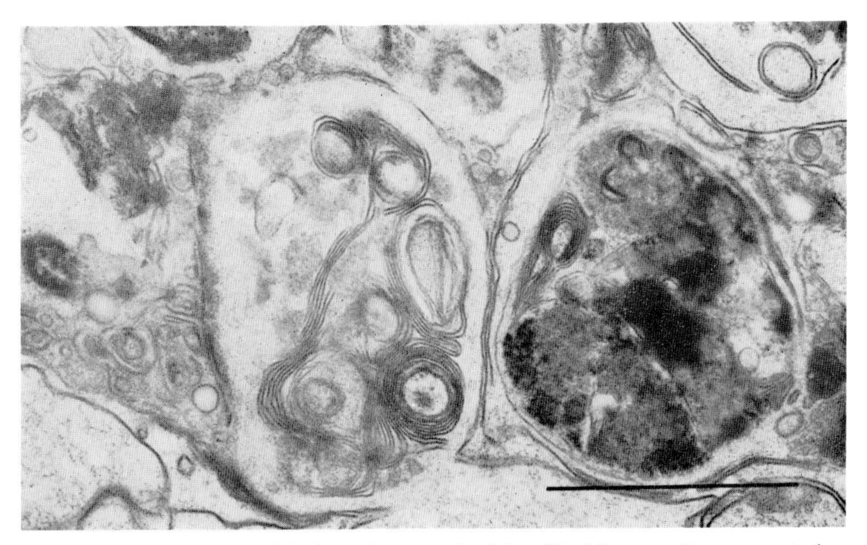

Fig. 6. Details of algal particles in a pigment cell of *Bonellia fuliginosa*. *Bar* represents 1 μ. (X 35,000.)

DISCUSSION AND CONCLUSIONS

Blue-green algae are found in two species of echiuroids, *Ikedosoma gogoshimense* and *Bonellia fuliginosa*. The algal symbionts occur in their epidermal connective tissue. These echiuroid species live in burrows in the muddy sand-bottom and in coral crevices, respectively. They are sensitive to light and are nocturnal. But they are shallow-water inhabitants and can receive dim light on their body surface.

It is probable that the same circumstances will be found in similar species, for example, *Bonellia viridis* and *Thalassema lankesteri*, which have bonellin, and also in other unexamined echiuroids which bear green spots.

The pigments of both species show practically identical absorption bands which are similar to those found for bonellin by Lederer (1939). However, their absorption spectra are quite different from those of chlorophyll *a* and bacterio-chlorophyll.

Brafield (1968) reported a reduced oxygen consumption in the proboscis of *Bonellia*. This was true in our experiments also. *Ikedosoma* can consume ammonium. These facts, along with the electron micrographs presented, support the view that the green spots in the pigment cells of echiuroids are really blue-green algae.

SUMMARY

Electron Microscope Observations

Echiuroid worms frequently bear green color on their body surfaces. These green worms are shallow-water inhabitants. *Ikedosoma gogoshimense* is found in the muddy sand near the low tide line; but *Bonellia fuliginosa* lives in a burrow with a small opening in coral heads of tropical and subtropical reefs. This worm expands its proboscis from the burrow in the dark or in a dim light, but not under direct sunlight.

When these worms are treated with alcohol, each of these solutions shows a green color having a bonellin-like absorption band. Histological observations indicate that they have a large number of green cells in their sub-epidermal connective tissue. These cells are very large and are composed of an aggregation of green particles. These particles can be cultured *in vitro*. Electron microscopic study has revealed that these are blue-green algae. These algal symbionts seem to have metabolic relations with their hosts.

REFERENCES

Brafield, A. E. 1969. Oxygen consumption of an echiuroid, *Bonellia viridis* Roland. J. Exp. Biol., 48:427–434.

Fox, D. L. 1953. Animal biochromes and structural colours. Cambridge Univ. Press, Cambridge. 379 p.

Ikeda, I. 1904. The Gephyrea of Japan. J. Coll. Sci. Univ. Tokyo, 20:1–87.

Kawaguti, S. 1968. Electron microscopy on blue-green algae in the body-wall of an echiuroid, *Ikedosoma gogoshimense*. Biol. J. Okayama Univ., 14:67–74.

Lederer, E. 1939. L'isolement et la constitution chimique de la bonelline, pigment vert de *Bonellia viridis*. C. R. Acad. Sci. Paris, 209:528–530.

Immunity in Invertebrates

M. R. TRIPP

Department of Biological Science
University of Delaware
Newark, Delaware

INTRODUCTION

Interest in mechanisms by which invertebrates react to foreign materials in their tissues has increased rapidly in recent years and several comprehensive reviews are available (Salt, 1961; Stephens, 1964; Cheng, 1967; Tripp, 1969). In general these studies show that susceptibility to infection is determined by genetic factors and by anatomical and physiological characters of the species that may be modified by age, diet, and environment. These principles have been reviewed in detail elsewhere (Tripp, 1969) and need not be considered here.

The purpose of this article is to examine some details of what is now known of invertebrate internal defense mechanisms, to see what generalizations can be made, to identify critical information that is lacking, and to examine some possible experimental approaches to problems of immunity in invertebrates.

CELLULAR RESPONSES

Phagocytosis

It is well established that foreign material in invertebrate tissues commonly elicits a marked cellular response. Phagocytosis of particles has been reported from most invertebrate metazoan phyla including sponges (Cheng *et al.*, 1968), earthworms (Cameron, 1932), arthropods, especially insects (Jones, 1962), and molluscs (Tripp, 1963). Amoebocytes engulf small particles and may either digest them or carry them to the exterior. Injected inert particles (e.g., India ink) form extracellular clumps, amoebocytes detach and engulf small particles, and often migrate through tissues until they are eliminated from the body. This progression has been demonstrated in such diverse groups as sponges (Cheng *et al.*, 1968) and molluscs (Stauber, 1950; Feng, 1967). A similar phenomenon has been observed recently in tunicates (Anderson, 1969), except that phagocytes transfer their particles to epithelial cells and do not migrate through epithelial layers to the exterior. This resembles the mechanism by which particles may be removed from vertebrate tissues via gut epithelia. Biological particles (e.g., bacteria, yeast, red blood cells) may be digested intracellularly by phagocytes (Tripp, 1960). The time required for such degradation varies with experimental conditions and the nature of the particle. Submicroscopic particles (e.g., viruses) or soluble proteins may be pinocytosed by amoebocytes (Fries and Tripp, 1970; Feng, 1968; Phillips, 1966) and eliminated from tissues but the precise mechanisms involved are not known.

Encapsulation

Capsule formation is a common reaction of invertebrates to foreign material too large for phagocytosis in their tissues. The structure of such capsules in insects has been reviewed in detail by Salt (1963) and more recently electron micrographs of capsules in insects (Salt, 1967) and molluscs (Rifkin *et al.*, 1969) have been published. Blood cells adhere to the surface of the foreign material and flatten out. Additional cells cohere to a depth of several layers and, at least in the oyster, extracellular fibers may be deposited (Rifkin *et al.*, 1969). Although these structures have several common morphological features it is likely that subtle, but significant, differences will be found with further study (Cheng, per. comm.). It is not known whether amoebocytes are actively attracted to the foreign material or whether they are carried passively by hemolymph and selectively adhere to a foreign body after contacting it.

"SELF" VERSUS "NOT SELF"

It is not surprising that grossly foreign substances elicit cellular responses. However, there are two other cases where more subtle distinctions between "self" and "not self" are required. These are "natural" host-parasite combinations and tissue grafting.

There are numerous reported examples of minimal or no cellular response of a host to a "normal" parasite. For example, Salt (1961) has shown that insect parasitoids deposited in "proper" insect hosts rarely elicit any cellular response as long as the parasites are healthy; Pan (1965) has shown a similar lack of response of *Biomphalaria glabrata* to the presence of *Schistosoma mansoni* in its tissues. These same animals infected with other parasitic species (or even with different strains of the normal parasitic species) respond vigorously and the infection may be terminated. Thus these seem to be situations where certain parasites have evolved external surfaces that host hemocytes cannot distinguish from the hosts' own internal surface.

Transplantation experiments show the same differential response. Sipunculids (*Dendrostomum zostericolum*) encapsulate stained or damaged homologous eggs or normal heterologous eggs, but normal homologous eggs are not encapsulated even in male sipunculids where it might be thought they would be "foreign" (Cushing *et al.*, 1965). Insects often, but not always, encapsulate interspecific transplants whereas intraspecific transplants usually elicit minimal responses (Salt, 1961). Among molluscs (*B. glabrata*), there is some evidence that heterologous tissue is rejected while homologous tissue is tolerated (Tripp, 1961). Starfish (*Patiria miniata* and *Asterias forbesi*) distinguish between coelomic implants of homologous and heterologous tissues (Ghiradella, 1965). Undoubtedly the clearest demonstration of transplantation immunity among invertebrates has been furnished by Cooper (1968, 1969) in his study of skin grafts of earthworms. These annelids (*Lumbricus terrestris* and *Eisenia foetida*) distinguished clearly between xenografts and autografts. Both types of grafts healed during the first 24 hours. The xenografts were always rejected subsequently, but autografts were not. Acidophilic cells were prominent in the rejection reaction. These studies take on additional significance because they are the only reported instances in which the grafted tissues clearly healed in (that is, circulation was reestablished). This is an important factor in deciding whether the graft was truly rejected or whether it was never accepted.

It seems clear from these findings that hemocytes of many invertebrates have the ability to distinguish "self" from "not self." Extensive studies of vertebrate macrophages show that "Clearly, a variety of mechanisms can operate to allow the macrophage to discriminate between material to be phagocytosed and material to be left alone. These vary with the nature of the particle but in most cases probably involve antibody — natural or immune, cytophilic or (in the absence of antigen) non-cytophilic — and possibly, in some cases, complement." (Nelson, 1969).

Whether humoral factors serve as "recognition substances" in invertebrates is not known. There is a consensus that materials comparable to vertebrate immunoglobulins are not produced in invertebrates (Grey, 1969). However, it is also clear that there are soluble substances in the hemolymph of many invertebrate species that possess biological activity. Tyler (1946) described many of those that agglutinate sperm or blood cells. Some of these substances possess remarkable specificity (Boyd *et al.*, 1966; Johnson, 1964) while others lack a high degree of specificity (Tripp, 1966). What little is known of the molecular basis of these interactions suggests that they are susceptible to analysis by well known biochemical techniques and that hemolymph proteins interact with "antigenic" sites much as do vertebrate antibodies (McDade and Tripp, 1967). The evidence that hemolymph may have an opsonic effect *in vitro* (Tripp, 1966) and possibly *in vivo* (Stauber, 1961) is inconclusive. It seems unlikely that factors in invertebrate hemolymph can play a crucial role in recognition and cellular responses because phagocytosis in the absence of hemolymph is so vigorous.

Thus the basic question becomes, "What properties of individual hemocytes allow recognition of foreignness?" There is some evidence that such specificity is associated with cells. Cooper (1969) has shown clearly that earthworms specifically reject skin grafts and that this phenomenon may be cell-mediated. Cushing *et al.* (1970) have recently described what may be specific inhibition ("tolerance"?) induced in sipunculid phagocytes. Obviously the site of particle recognition must be at the cell surface and here is where the search for specific receptors must lead.

Cell Surface Receptors

Specific cell surface receptors have been postulated as the mechanism which mediates certain well known specific cell-cell interactions. An "aggregating factor" (100S particle, 100–300 Å in diameter) has been shown capable of directing species-specific aggregations of cells of marine sponges (Humphreys, 1966). Specific cell surface components of different mating types of yeast (*Hansenula wingei*) have been isolated and characterized as glycoproteins of low molecular weight (2.9 S) (Crandall and Brock, 1968). Rubin (1966) has characterized a "conditioning factor" from chick cells in culture as a lipoprotein probably derived from the cell membrane. Surfaces of such cells are disrupted by surfactants, malignancy is enhanced, and this is thought to be associated with loss of contact inhibition, the ability to invade tissues, and to metastasize (Rubin, 1966). Surface coats ("fuzz" or hirsute coat) of cells are now known to be very widespread, perhaps universal, and they are complex chemically and functionally (Revel and Ito, 1966). Usually it is not possible to separate the receptor from the cell membrane so that biological activity is retained (Warren *et al.*, 1966). The technical problems associated with the preparation of cell membranes are formidable and they have been reviewed by Wallach (1966). One must conclude that progress in this area will be slow.

The powerful genetic and immunologic techniques available for the study of mammalian tumor antigens (Klein, 1966) are not available for studying the surfaces of invertebrate cells, although partially analogous techniques might be developed. Electrophoresis of viable cells has some usefulness (Weiss, 1967) but it does not tell much about specific cell surface components. Undoubtedly there are several reasons for this: a constant structural flux at the cell surface with old material being destroyed and new material synthesized; the delicate nature of some components does not permit isolation by relatively destructive techniques now available; spatial relationships, and therefore some functional relationships, are destroyed by current isolation techniques.

Model systems that have proved useful in studies of vertebrate cell surfaces suggest new approaches. The destruction of specific cell receptors by enzymes might prove useful. For example, the neuraminidase of *Vibrio cholerae*

(Burnet and Stone, 1947) is known to split off terminal N-acetylneuraminic acid from mucoproteins that normally serve as sites of attachment for myxoviruses and paramyxoviruses on red blood cells (Davis *et al.*, 1968). It would be interesting to know whether red cells treated with this enzyme attach to phagocytes as readily as do untreated cells; and whether phagocytes treated with such an enzyme are inhibited in their ability to attach to foreign particles. Variations of this approach, such as those developed by Rabinovitch (1969), offer a powerful experimental tool. This and other approaches would be useful if reliable invertebrate cell culture techniques were available. Unfortunately, except for a few types of insect cells, such techniques are not available.

Some of the more recent immunologic techniques developed to study cell-mediated responses might well be adapted to invertebrate systems. Cooper (1968, 1969) has already established a prototype system of skin grafting in earthworms for studying cell-mediated responses. Immune adherence would seem to be another obvious approach to measurement of cell-cell interaction (LoBuglio *et al.*, 1967). Induction of hypersensitive states might be assayed by the leukocyte inhibition tests that have proved so useful in the study of cell-mediated hypersensitivity of vertebrates (Davis *et al.*, 1964).

SUMMARY

Invertebrates react selectively to the presence of unusual substances in their tissues. Although hemolymph components may interact with such substances, the basic reactions seem to be cell-mediated. Nearly nothing is known about how cells distinguish between "self" and "not self" materials, but understanding of the mechanisms involved in cell-particle recognition is an important goal. It seems obvious that no single experimental technique will explain how invertebrate cells detect foreignness, but the application of a variety of techniques may do so. If these questions can be answered all of biology will be greatly benefited.

REFERENCES

Anderson, R. S. 1969. Cellular responses of tunicates to foreign bodies. Amer. Zool. 9:1116. (Abstr.)

Boyd, W. C., Brown, R. and Boyd, L. G. 1966. Agglutinins for human erythrocytes in mollusks. J. Immunol. 96:301–303.

Burnet, F. M. and Stone, J. D. 1947. The receptor-destroying enzyme of *Vibrio cholerae*. Austral. J. Exp. Biol. Med. Sci. 25:227–238.

Cameron, G. R. 1932. Inflammation in earthworms. J. Pathol. Bacteriol. 35:933–972.

Cheng, T. C. 1967. Marine molluscs as hosts for symbioses. (Adv. Marine Biol., Vol. 5) Academic Press, London. 424 p.

Cheng, T. C., Rifkin, E. R., and Yee, H. W. F. 1968. Studies on the internal defense mechanisms of sponges. II. Phagocytosis and elimination of India ink and carmine particles by certain parenchymal cells of *Terpios zeteki*. J. Invert. Pathol. 11:302–309.

Cooper, E. L. 1968. Transplantation immunity in annelids. I. Rejection of xenografts exchanged between *Lumbricus terrestris* and *Eisenia foetida*. Transplantation 6:322–337.

Cooper, E. L. 1969. Specific tissue graft rejection in earthworms. Science 166:1414–1415.

Crandall, M. A. and Brock, T. D. 1968. Molecular aspects of specific cell contact. Science 161:473–475.

Cushing, J., Boraker, D. and Keough, E. 1965. Reactions of sipunculid worms to intracoelomic injections of homologous eggs. Fed. Proc. 24:504. (Abstr.)

Cushing, J. E., Tripp, M. R. and Fuzessery, S. 1970. Adaptation of sipunculid amoebocytes to phagocytic inhibition by bovine serum. Fed. Proc. 29:771.

David, J. R., Al-Askari, S., Lawrence, H. S. and Thomas, L. 1964. Delayed hypersensitivity *in vitro*. I. The specificity of inhibition of cell migration by antigens. Immunology 93:264–282.

Davis, B. D., Dulbecco, R., Eisen, H. H., Ginsberg, H. S. and Wood, W. B., Jr. 1968. Principles of microbiology and immunology. Harper and Row, New York.

Feng, J. S. 1966. The fate of a virus, *Staphylococcus aureus* phage 80, injected into the oyster, *Crassostrea virginica*. J. Invert. Pathol. 8:496–504.

Feng, S. Y. 1965. Pinocytosis of proteins by oyster leukocytes. Biol. Bull. 129:95–105.

Feng, S. Y. 1967. Responses of molluscs to foreign bodies, with special reference to the oyster. Fed. Proc. 26:1685–1692.

Fries, C. R. and Tripp, M. R. 1970. Uptake of viral particles by oyster leukocytes *in vitro*. J. Invert. Pathol. 15:136-137.

Ghiradella, H. T. 1965. The reaction of two starfishes, *Patiria miniata* and *Asterias forbesi* to foreign tissue in the coelom. Biol. Bull. 128:77–89.

Grey, H. M. 1969. Phylogeny of immunoglobulins, p. 51–104. *In* F. J. Dickson and J. H. Humphrey (eds.), Advances in Immunology, Vol. 10. Academic Press, New York.

Humphreys, T. 1966. The cell surface and specific cell aggregation. *In* B. D. Davis and L. Warren (eds.) The specificity of cell surfaces. Prentice-Hall, Englewood Cliffs, N. J. 290 p.

Johnson, H. 1964. Human blood group A, specific agglutinin in the butter clam *Saxidomas giganteus*. Science 146:548–549.

Jones, J. C. 1962. Current concepts concerning insect hemocytes. Amer. Zool. 2:209–246.

Klein, G. 1966. Tumor antigens. *In* B. D. Davis and L. Warren (eds.) The specificity of cell surfaces. Prentice-Hall, Englewood Cliffs, N.J. 290 p.

LoBuglio, A. F., Cotran, R. S. and Jandl, J. H. 1967. Red cells coated with immunoglobulin G: binding and sphering by mononuclear cells in man. Science 158:1582–1585.

McDade, J. E. and Tripp, M. R. 1967. Mechanism of agglutination of red blood cells by oyster hemolymph. J. Invert. Pathol. 9:523–530.

Nelson, D. S. 1969. Macrophages and immunity. John Wiley and Sons, New York. 335 p.

Pan, C. T. 1965. Studies on the host-parasite relationship between *Schistosoma masoni* and the snail *Australorbis glabratus*. Amer. J. Trop. Med. Hyg. 14:931–976.

Phillips, J. H. 1966. Immunological processes and recognition of foreignness in the Invertebrates. *In* R. T. Smith, P. A. Miescher and R. A. Good (eds.) Phylogeny of immunity. Univ. Florida Press. 276 p.

Rabinovitch, M. 1969. Phagocytosis of modified erythrocytes by macrophages and L 2 cells. Exp. Cell Res. 59:326–332.

Revel, J. P. and Ito, S. 1966. The surface components of cells. *In* B. D. Davis and L. Warren (eds.) The specificity of cell surfaces. Prentice-Hall, Englewood Cliffs, N.J. 290 p.

Rifkin, E., Cheng, T. C. and Hohl, H. R. 1969. An electron-microscope study of the constituents of encapsulating cysts in the American oyster, *Crassostrea virginica*, formed in response to *Tylocephalum* metacestodes. J. Invert. Pathol. 14:211–226.

Rubin, H. 1966. The behavior of normal and malignant cells in tissue culture. *In* B. D. Davis and L. Warren (eds.) The specificity of cell surfaces. Prentice-Hall, Englewood Cliffs, N.J. 290 p.

Salt, G. 1961. The haemocytic reaction of insects to foreign bodies. *In* J. A. Ramsay and V. B. Wigglesworth (eds.) The cell and the organism. Cambridge Univ. Press, Cambridge. 350 p.

Salt, G. 1963. The defense reactions of insects to metazoan parasites. Parasitology 53:527–642.

Salt, G. 1967. Cellular defense mechanisms in insects. Fed. Proc. 26:1671–1674.

Stauber, L. A. 1950. The fate of India ink injected intracardially into the oyster, *Ostrea virginica* Gmelin. Biol. Bull. 98:227–241.

Stauber, L. A. 1961. Immunity in invertebrates, with special reference to the oyster. Proc. Nat. Shellfisheries Ass. 50:7–20.

Stephens, J. M. 1964. Immunity in insects. *In* E. A. Steinhaus (ed.) Insect pathology: An advanced treatise. Vol. 1 Academic Press, New York. 661 p.

Tripp, M. R. 1960. Mechanisms of removal of infected microorganisms from the American oyster *Crassostrea virginica* Gmelin. Biol. Bull. 119:273–282.

Tripp, M. R. 1961. The fate of foreign materials experimentally introduced into the snail *Australorbis glabratus*. J. Parasitol., 47:745–751.

Tripp, M. R. 1963. Cellular responses of mollusks. Ann. NY Acad. Sci. 113:467–474.

Tripp, M. R. 1966. Hemagglutinin in the blood of the oyster, *Crassostrea virginica*. J. Invert. Pathol. 8:478–484.

Tripp, M. R. 1969. General mechanisms and principles of invertebrate immunity. *In* G. J. Jackson, R. Herman and I. Singer (eds.) Immunity to parasitic animals. Appleton-Century-Crofts, New York. 292 p.

Tyler, A. 1946. Natural hemagglutinins in the body fluids and seminal fluids of various invertebrates. Biol. Bull. 90:213–219.

Warren, L., Glick, M. C. and Nass, M. K. 1966. The isolation of animal cell membranes. *In* B. D. Davis and L. Warren (eds.) The specificity of cell surfaces. Prentice-Hall, Englewood Cliffs, N.J. 290 p.

Wallach, D. F. H. 1966. Isolation of plasma membranes of animal cells. *In* B. D. Davis and L. Warren (eds.) The specificity of cell surfaces. Prentice-Hall, Englewood Cliffs, N.J. 290 p.

Weiss, L. 1967. The cell periphery, metastasis and other contact phenomena. John Wiley & Sons, New York. 388 p.

Experimental Studies on the Protection of Anemone Fishes from Sea Anemones

RICHARD N. MARISCAL

Department of Biological Science
Florida State University
Tallahassee, Florida

INTRODUCTION

Although an anemone fish was probably first described by Valentyn in 1726, and Collingwood first described the symbiosis between anemone fishes of the family Pomacentridae and tropical sea anemones in 1868, it remained for Davenport and Norris, in 1958, to conduct the first good experimental study of this association. In the last ten years or so a score of other papers have appeared which have helped to clarify (and in some cases confuse) our understanding of the procedure by which anemone fishes become protected from their sea anemone hosts.

This symbiosis consists of the association between colorful pomacentrid fishes of the genera *Amphiprion* (Fig. 1), *Premnas*, and *Dascyllus*, and giant sea anemones, mainly of the family Stoichactiidae, which are found throughout the tropical Indo-Pacific. Like all coelenterates, the tentacles of these giant sea anemones bear thousands of microscopic stinging capsules or nematocysts. Although small fishes are generally considered to be normal prey for sea anemones, anemone fishes are somehow protected from their host's nematocysts in spite of extensive tentacle contact.

The key to experimental investigation of the protection of anemone fishes from sea anemones lies in the interesting behavioral sequence known as acclimation (Fig. 2). Before acclimation, an anemone fish (e.g., *Amphiprion*) is stung by a sea anemone; following acclimation, the same fish is no longer stung by the same anemone. The obvious question then is just what is occurring during this period of acclimation which results in the protection of the fish? An understanding of this process should be the key to understanding the phenomenon of protection from coelenterate nematocysts for anemone fish in particular and for other organisms, perhaps even man, in general.

Fig. 1. Laboratory electronic flash photograph of *Amphiprion percula* hovering over the tentacles of the California sea anemone, *Anthopleura xanthogrammica*. Although this fish initially made attempts to acclimate to this anemone, it did not do so and was never seen to enter freely among the tentacles.

Fig. 2. Laboratory electronic flash photograph of an unacclimated *Amphiprion xanthurus* (orange morph) hovering to one side of *Anthopleura xanthogrammica*, while performing acclimation behavior. Occasionally, the fish would move in to touch the edge of its pelvic or anal fins to the anemone and then dart back following contact.

The various studies and/or hypotheses which have been conducted or put forth to explain this protection can be placed in either of two major categories. One hypothesis suggests that the anemone is responding to its symbiotic fish in such a way that it alters or inhibits its nematocyst discharge through the receipt of either behavioral, physical, or chemical stimuli from the fish (Gohar, 1948; Hackinger, 1959; Koenig, 1960; Graefe, 1963, 1964; Blösch, 1961, 1965). This hypothesis implies that nematocysts are not independent effectors, as they are classically considered to be, and instead postulates that the anemone is somehow able to control its nematocyst discharge, at least to some degree (see Mariscal, 1966a for discussion of nematocyst physiology in this regard).

The second hypothesis implies that the anemone is the passive member of the symbiosis and suggests that the fish is able to protect itself from its host's nematocysts by somehow altering its surface, presumably its epidermal mucous coat (Davenport and Norris, 1958; Eibl-Eibesfeldt, 1960; Mariscal, 1965, 1966b, 1967, 1969a, 1970a; Schlichter, 1967, 1968). Abel (1960b) also found that the surface mucus of the fish *Gobius* is important in protecting it from *Anemonia* in the Mediterranean, but did not find evidence for acclimation behavior in this case.

Since there are several ways an anemone fish might become protected from sea anemones through acclimation, Table 1 has been prepared to clarify the various types of acclimation phenomena. The two major categories of "unacclimated" and "acclimated" may be subdivided, depending on how this general condition was attained, into eight more or less distinct subcategories. Although theoretically all eight categories are open to investigation, only five of these are practicable for experimental purposes because of the difficulties in breeding and raising both fishes and sea anemones under well-controlled laboratory and field conditions over long periods of time.

Before going further, it might be well to define the terms associated with the general category of symbiosis as they are used in the present study. Because of widespread misusage in recent years, recent workers have attempted to standardize this often confusing terminology (e.g., Allee *et al.*, 1949; Noble and Noble, 1964; Henry, 1966). *Symbiosis* is used here as defined by DeBarey, who coined the term in 1879 (see Davenport, 1955; Henry, 1966; Mariscal, 1966a), simply to indicate the living together of two phylogenetically unrelated organisms. Under this encompassing term, the following major subdivisions are recognized: (1) *mutualism*, in which both partners of an association are benefited; (2) *commensalism*, in which one partner is benefited and the other neither harmed nor benefited; and (3) *parasitism*, in which one partner is benefited and the other is harmed (following Allee *et al.* 1949; Davenport, 1955; Noble and Noble, 1964; Henry, 1966). The present symbiosis between fishes of the genus *Amphiprion* (primarily) and sea anemones is probably best described as mutualism (see Mariscal, 1970d).

REVIEW OF THE LITERATURE

Gudger (1947) and Mariscal (1966a) have reviewed the literature on the anemone fish symbiosis in general, while Verwey (1930) and Mariscal (1970b,c), among others, have presented extensive field and behavioral observations of both partners. Since a number of interesting, and in some cases, conflicting, experimental studies on the protection of anemone fishes have been conducted in the last decade or so, it is believed that a critical review would be useful at this time in order to determine the direction which future experimental studies of this, and perhaps other symbioses, might take.

Gohar (1948) has made a number of observations on the association of *Amphiprion bicinctus*, *Amphiprion xanthurus* (= *A. clarkii*), and *Dascyllus* (= *Tetradrachmum*) *trimaculatus* with *Stoichactis* (= *Discosoma*) *giganteum* in the Red Sea. Unfortunately, nearly all of his observations regarding the protection of the fish are totally lacking in controls and thus alternative explanations are equally possible. For example, he reports that not all individuals of the symbiotic species of fishes are protected from *Stoichactis giganteum* and that as a rule, the association with *Stoichactis* anemones "is restricted to the very individuals that happen to live with them, whether from one, two or even the three species." Gohar goes on to say, "Other individuals that happen to come in touch with the tentacles will at once be caught, paralyzed and in due course also devoured and digested. These may belong to the same species as its own commensal fish, which may be seen at the same time lying safely and happily among the tentacles on the peristomial disc..." In attempting to explain these observations, Gohar further states (p. 39) that, "The anemones appear as if they recognize their partners by their mode of movement. Thus they would sting and seize their own fish if the latter are passively pushed against their tentacles." Gohar's initial observations that fish living with *Stoichactis* are protected while other individuals, even of the same species, may be stung and killed, could easily be explained by the fact that the fish living with *Stoichactis* were acclimated and thus protected while the others had been isolated and were not. Eibl-Eibesfeldt (1960) also points this out. Had the others been allowed to go through acclimation behavior, they too would probably have become protected. Since Gohar does not tell us the past history of the "other individuals" above, we have no way of evaluating this observation. Gohar, himself, states that, "Fish of the commensal species may develop partnership with such anemones as *Discosoma giganteum* by cautiously approaching it. The association is complete in one to a few days." This in fact is probably the first description of what Davenport and Norris (1958) later termed acclimation behavior.

In support of his idea that the anemones appear to recognize their symbiotic fish by their "mode of movement," Gohar relates that the association between the anemones and their symbiotic fish "lasts only as long as the fish are alive and healthy. Immediately the fish are dead or even moribund their partner devours them, if it happens to be hungry."

This statement is somewhat puzzling since subsequent experiments by the author and others (e.g., Davenport and Norris, 1958; Eibl-Eibesfeldt, 1960) have shown quite clearly that anemones do not sting, let alone ingest, dead acclimated fish or even pieces of *Amphiprion* so long as the undisturbed, epidermal mucous coat is in contact with the tentacles. Perhaps Gohar's dead or moribund fish had been injured or handled in such a way that the surface mucous coat had been damaged.

Although both Hackinger (1959) and Koenig (1960) have supported Gohar's idea that *Amphiprion* is protected by its behavior, neither author has presented any experimental evidence in this regard. Therefore, in view of the evidence which argues against Gohar's idea and the lack of any which supports it, we must conclude that there is no justification for the notion that an anemone is capable of "recognizing" its individual symbiotic fish by its behavior, while stinging others of the same species which approach it. Abel (1960a) similarly has argued against Gohar's idea on other grounds.

As mentioned earlier, the first substantial experimental study of the anemone fish symbiosis was that of Davenport and Norris (1958). Although they were unable to conduct reciprocal experiments between acclimated and unacclimated fish and anemones, since they had only a single anemone at their disposal, their work clearly demonstrated the importance of the surface of the anemone fish, and presumably the mucous coat, in its protection. Although Davenport and Norris (1958) do not specifically state in each case whether the fish were acclimated or unacclimated, it seems likely that they must all have been acclimated at the time of the experiments. They reported that a piece of plastic sponge coated with the mucus of a nonanemone fish (*Fundulus* or *Hypsypops*) caused very strong adhesion of the tentacles of *Stoichactis* (indicating nematocyst discharge), while clean pieces of plastic sponge (control) and pieces of sponge coated with *Amphiprion percula* mucus (presumably acclimated) caused essentially no adhesion of the tentacles. Further experiments in which a piece of *A. percula* skin and flesh, placed skin-side down on the oral disc of *Stoichactis*, was slowly worked to the edge of the disc and dropped, while a similar piece of the control fish, *Fundulus*, placed next to it was ingested, reveal again that some factor on the surface of the *Amphiprion* was protecting it from being stung and ingested.

I have conducted similar experiments aboard ship, comparing the responses of Seychelles *Radianthus ritteri* anemones to pieces of freshly collected, acclimated *Amphiprion akallopisos* skin and flesh with similar pieces from a control fish, the grouper *Epinelphus merra*. When both pieces (about 1 cm^2) were presented skin side down, there was massive adhesion of the *R. ritteri* tentacles to the grouper skin, but no adhesion to the acclimated *Amphiprion* skin, except around the cut edges. However, when the *Amphiprion* piece was turned over, there was massive clinging and contraction of the tentacles and the piece was ingested, as was the grouper flesh in the same experiment (see Table 5). These results are identical to those of Davenport and Norris (1958) and again indicate the importance of the undamaged skin surface in the protection of acclimated *Amphiprion* from the nematocysts of its sea anemone host.

Davenport and Norris (1958) also tested *Amphiprion percula* acclimated to *Stoichactis kenti* against the California sea anemone *Anthopleura xanthogrammica* and found that the tentacles of the latter strongly adhered to and stung the fish. I found that unacclimated *A. percula* also were stung by this anemone, and in fact appeared incapable of acclimating to it. After a few hesitant contacts and stingings by *Anthopleura xanthogrammica, A. percula* appeared to lose interest in acclimating to it (Fig. 1). Davenport and Norris (1958) also reported that an *Amphiprion frenatus*, although it had lived with their *Stoichactis kenti* for eight months prior to the experiments, was never seen to move into the tentacles. Similarly, *A. frenatus* and *A. ephippium* which were tested against California *Anthopleura xanthogrammica* and *Anthopleura elegantissima* by the author appeared to lose interest in these anemones after the initial contacts and stingings. Instead, they swam over and around the anemone, but were never seen to go through acclimation behavior nor did they ever directly enter the tentacles. Blösch (1965) has reported similar behavior for Indian Ocean *Amphiprion* when presented to Mediterranean anemones.

Such behavior appears to be in keeping with field observations made on these species by the author and others. For example, *A. percula* appears to be restricted to only certain sea anemone species and will be stung if forced into others, even in the same reef area; in addition it shows avoidance responses to some anemone species when placed in the same aquarium with them (Eibl-Eibesfeldt, 1960, 1965; Mariscal, 1966b, 1970c).

On the other hand, a very nonspecific species of *Amphiprion*, such as *A. xanthurus*, has been found by the author in seven different species of sea anemones from three different families (Mariscal, 1966b, 1970b,c). *Amphiprion xanthurus* showed no hesitation in acclimating to (and was able to become protected from) *Anthopleura xanthogrammica* in the aquarium (Fig. 2). *Amphiprion xanthurus* also repeatedly attempted to acclimate to *Anthopleura elegantissima* but only one fish out of three was ever successful in this, although the reason remains unclear (Fig. 3). In addition, *A. xanthurus* which were acclimated to and protected from *S. kenti* anemones (Fig. 4) were not protected from *Anthopleura xanthogrammica* anemones without re-acclimating directly to these anemones. However, an *A. xanthurus* acclimated

Fig. 3. Laboratory electronic flash photograph of an unacclimated *Amphiprion xanthurus* (orange morph) hovering above a California *Anthopleura elegantissima* anemone while attempting to acclimate to it. This particular fish appeared unable to become protected from this anemone and was still being stung following 119 hours of acclimation behavior, including hovering, settling into the tentacles, and contacting with the tentacles with the fins and body.

Fig. 4. Laboratory electronic flash photograph of a fully acclimated *Amphiprion xanthurus* deeply buried among the tentacles of *Stoichactis kenti*. Although massive contact and battering of the anemone's tentacles and oral disc occurred, such a fully acclimated fish was never observed to be stung by its anemone.

to and protected from *Anthopleura xanthogrammica* was also protected immediately from *Stoichactis kenti* (Mariscal, 1970a and see Table 4). This appeared to be correlated with the length of time required for acclimation, in that protection from the anemone which required the longest time of acclimation also seemed to provide protection for *A. xanthurus* from those anemones requiring correspondingly shorter times of acclimation (Mariscal, 1966b).

Hackinger (1959) and Koenig (1960) at one time believed that perhaps one aspect of the protection of *Amphiprion* from symbiotic sea anemones is that their nematocyst toxins were not strong enough to capture healthy fish in the field. Buhk (1939) was apparently the first to put forth this hypothesis and more recently Fishelson (1965) has said essentially the same thing. However, Gohar (1934, 1948), Herre (1936), and I have observed living fish captured and killed by stoichactiid anemones in aquaria and Eibl-Eibesfeldt (1960) has observed this under field conditions. Although there appear to be differences in the strength of nematocyst discharge and toxicity among the stoichactiid anemones, without some solid experimental evidence this will not suffice to explain the protection of anemone fishes from sea anemones. In addition, I have observed that some sea anemones tend to lose their nematocyst discharge capabilities under aquarium conditions, and thus attempts to explain anemone prey capture in the field, based strictly on aquarium observations, may be misleading.

As mentioned earlier, the main explanation for the protection of anemone fishes put forth by Hackinger (1959) and Koenig (1960), follows Gohar (1948) in postulating that the behavior of the fish is somehow important in preventing it from being stung by the host anemone. Unfortunately, they simply state that they have come to this conclusion based on their field and aquarium observations, but apparently have never conducted any carefully controlled experiments; at least they give no data in their papers to suggest this idea.

Although Eibl-Eibesfeldt (1960, 1965) was unable to use either isolated anemone fishes or sea anemones in his field studies conducted in the Indian Ocean, he has provided much interesting information concerning the behavior of anemone fishes, their protection, and the specificity of the association. He found that *Amphiprion xanthurus* was protected from all three species of sea anemones found in the vicinity of his study area—*Stoichactis* (= *Discosoma*), *Radianthus kuekenthali*, and *Radianthus ritteri*. However, two other species of anemone fish, *Amphiprion percula* and *A. akallopisos* (= *A. akallopisus*), were protected only from *Radianthus ritteri*, and not only avoided contact with both *Radianthus kuekenthali* and *Stoichactis* in the aquarium, but were stung by these anemones when forced into the tentacles. These results are comparable to mine in which unacclimated *A. percula* avoided contact with the California anemone, *Anthopleura xanthogrammica*, which subsequent tests revealed would sting this fish when forced into the tentacles. On the other hand, *A. xanthurus* would freely acclimate to *Anthopleura xanthogrammica* and became protected from it (Mariscal, 1965, 1970a).

Eibl-Eibesfeldt (1960) has postulated that perhaps *A. xanthurus* was protected from all three species of sea anemones because it possessed a greater concentration of protective substance on its surface while *A. percula* and *A. akallopisos* possessed less of this material. The other possibility is that the protective substance may be qualitatively different for the different species of anemones. Eibl-Eibesfeldt (1960) tends to feel that Davenport and Norris' idea of a protective substance on the surface of the fish best explains his own experimental results.

Graefe (1963, 1964), on the basis of his studies of *Amphiprion bicinctus* in the Red Sea, has hypothesized that nematocyst discharge against a symbiotic fish by a host anemone is a function of the size of the fish and of the consequent degree of tactile stimulation of the anemone's tentacles by this fish. For example, introduced *A. bicinctus*, which were larger than the resident fish, were stung by *Radianthus* (= *Antheopsis*) *koseirensis* (see Carlgren, 1949), while fish of the same size or smaller than the resident fish were not harmed. Since adult *A. bicinctus* were found only in *Stoichactis* anemones and since these anemones "stick" less to the hand of the observer than *Radianthus*, Graefe tends to believe that *Stoichactis* is less sensitive to mechanical stimuli than is *Radianthus*. He postulates that this may help to explain why the greater mechanical stimulation caused by the larger surface area of an adult *Amphiprion* is tolerated by *Stoichactis* while only smaller fish can live among the tentacles of *Radianthus*.

Although this idea is of interest and should allowed for in the controlling of experiments to analyze the protection of anemone fishes, certain difficulties in the controls and the method of handling the experimental fish have become apparent in attempting to evaluate Graefe's (1963, 1964) results. For example, Graefe does not always explain how he handled his experimental fish in transferring them between anemones, but he apparently used a net since he mentions this later in his 1964 paper. Although Graefe was certainly aware of the fact that a net might disturb the fish's surface mucous coating, as it indeed seemed to do by his own admission (experiment c. 4, p. 476), he appears to ignore the possibility that this might have influenced his experimental results. For example, in his first series of experiments with *Radianthus* (= *Antheopsis*) *koseirensis* (experiment a. 1, p. 472), he transferred a 32 mm long *A. bicinctus* from a 5 cm diameter *Radianthus* to a 6 cm diameter *Radianthus* containing an 11 mm long *A. bicinctus* (presumably by means of a net, although he does not say) and found that the larger fish was stung by the anemone containing the smaller fish. He states that such a stinging could not be explained by

Davenport and Norris' (1958) idea of a protective substance on the surface of the fish, without apparently considering the possibility that he could easily have damaged this coating during the transfer. Instead, he attributes the stinging of the new fish to the fact that it was a good deal larger than the original inhabitant and that the resulting increased mechanical stimulation of this anemone was triggering nematocyst discharge (p. 472). Nearly all of Graefe's subsequent translocation experiments are open to the same type of criticism. Graefe makes the statement later that injuries to the fish by the net do not appear likely because there was no loss of body fluids by the fish due to the abrasions by the net! He did attempt to scrape the mucus off the surface of the fish and found that it was only occasionally stung by a single tentacle; however, there is no way of knowing if he was successful in removing all the mucus present. In any case, experiments by Davenport and Norris (1958) and the author (Mariscal, 1966b, 1970a) clearly show that great care must be taken not to disturb the fish's surface mucous coating during the experiment, otherwise it is impossible to evaluate properly the results obtained. The fact that this was not carefully controlled leaves Graefe's (1963, 1964) experiments open to alternative explanations.

Even if it is assumed that there was no damage to the surface of the fish during the transfer from one anemone to another, other necessary controls appear lacking as well. For example, Graefe tested an adult *A. bicinctus* acclimated to one species of anemone (e.g., *Stoichactis*) against an entirely different anemone (e.g., *Radianthus*) and found that the adult fish normally found with the *Stoichactis* was stung by the *Radianthus* which normally harbors juvenile *A. bicinctus*. However, apparently Graefe did not run the reciprocal experiment of testing *A. bicinctus* acclimated to *Radianthus* against *Stoichactis*. Such reciprocal experiments are necessary in order to evaluate properly the results of experiments using more than one species of anemone, as well as in those using both acclimated and unacclimated fish and anemones of the same species, both with and without fish. Both Eibl-Eibesfeldt (1960, 1965) and the author have conducted experiments which revealed that *Amphiprion* which are acclimated to one anemone species are not protected from another species which may be very closely related.

Fraefe (1964) carries his arguments concerning the lack of a chemical protective substance further by his assumption that the constant bathing or "snuggling" in the tentacles of the anemone prevents any sustained mechanical pressure of the sides of the fish against the tentacles, thus somehow preventing nematocyst discharge. However, some *Amphiprion* may remain completely motionless on the oral disc of an anemone for periods lasting from seconds to minutes during the day and for many minutes and even hours at night when the fish remains in a single position on the oral disc completely surrounded by its host's tentacles. The "pressure" of the nematocysts against the fishes must be maximal at such times according to Graefe's hypothesis, yet the fishes were never observed to be stung. This nocturnal behavior has been observed in the shipboard aquaria for *A. nigripes* and *A. percula* in *Radianthus ritteri* anemones as well as in the laboratory for *A. xanthurus* in both *Anthopleura* and *Stoichactis kenti* (Mariscal, 1966b).

Finally, experiments were conducted by the author to control for Graefe's idea that the size of the fish is important in its protection. This was done by using *Amphiprion xanthurus* of approximately the same size in all experiments (Mariscal, 1966b, 1970a). However, to control for even slight size differences, the role of the fish was reversed from one experiment to the next. For example, the fish which was unacclimated in the first series of experiments became the acclimated fish in the subsequent experiments and vice versa. In no case was any difference ever detected in the results which could be attributed to even a few millimeters of difference in size between the fish. Recently, Schlichter (1967, 1968) has studied the same species of *Amphiprion* and anemones that Graefe used in the field in the Red Sea and has been unable to confirm Graefe's hypothesis concerning the protection of *A. bicinctus*.

Fishelson (1965) has concluded after seven years of study that *Stoichactis* is "not adapted to the catching of living fish, and that this is the basic condition for the association of fish with them." However, the capture of fishes by anemones harboring symbiotic fish has been observed by several workers under both aquarium and field conditions and there is no question that at least some stoichactiid anemones are capable of fish capture.

Fishelson finds no evidence for acclimation behavior in his experiments. Significantly, however, he appears to have been testing only fish which had been living with an anemone just prior to their being transferred to an uninhabited anemone of the same species (*Stoichactis*). Such fish were obviously already acclimated, and based on the experiments of Davenport and Norris (1958) and the author, there is no reason to expect already acclimated fish to go through acclimation behavior.

At the time that I initiated the experiments to be discussed later, Blösch's (1961) preliminary study appeared to present a fairly clear case for some kind of change occurring to the anemone during acclimation which resulted in the protection of the fish. However, as Davenport (1962, 1966a,b) cautioned in his short initial paper, Blösch did not give the details of his experimental procedure, his method of handling the fishes, nor the type of controls. This information was presented in his 1965 study and revealed that he took care to avoid damaging the mucous coat while handling his experimental fishes. Blösch (1965) was also well aware of the changes which may occur to anemones under aquarium conditions, especially regarding their capability for nematocyst discharge, and he tested this before each experiment. One difficulty in this connection is that Blösch differentiates between the adhesion of the tentacles or holding of the prey, supposedly by spirocysts, and the actual stinging or injection of toxin by the nematocysts. Although this idea is certainly plausible and worth investigating experimentally, Blösch presents no experimental evidence in favor of it. In addition, recent studies which attempted to demonstrate which coelenterate cnidae are involved in the behavioral patterns of other symbiotic sea anemones have revealed that in addition to whatever spirocysts might be utilized, nematocysts (microbasic p-mastigophores and basitrichous isorhizas respectively) are also involved in the respective adhesion response of the pedal disc of *Stomphia* and tentacles of *Calliactis* (Ellis *et al.*, 1969; Mariscal, 1969b). The *Anthopleura* and *Stoichactis kenti* tentacle nematocysts involved in the adhesion response to organic material appeared to be primarily the basitrichous isorhizas (Mariscal, 1966b).

Unfortunately, Blösch (1961, 1965) did not give the identifications of the anemones he used in his experiments. However, from his descriptions and photographs, it is possible to make at least tentative identifications based on field observations and collections.

1. Anemone A appears to be *Radianthus ritteri*
2. Anemone B also appears to be *R. ritteri* or a very close relative, and, in fact, Blösch says it scarcely differed in appearance from Anemone A. He separates it primarily on the basis of its reaction to anemone fishes.
3. Anemone C appears similar to *Radianthus kuekenthali*
4. Anemone D appears to be *Stoichactis kenti*

Although Blösch's behavioral observations seem comprehensive and accurate, his interpretation of some of these observations, as well as the supporting experiments regarding the protection of the fish, appear in some cases to be lacking in controls and thus open to alternative hypotheses. Even though it is difficult to analyze Blösch's 1965 study because his data and statements are self-contradictory in places, his basic premise is that acclimation is in reality an adaptation of the anemone to the fish (e.g., p. 41). However, he also states that, "*Zusammenfassend kann gesagt werden, dass der Nesselschutz der Anemonenfische in der besonderen chemischen Beschaffenheit ihrer äusseren Schleimhaut liegt*" (p. 34). He believes that the anemone is becoming "accustomed" to the surface mucous membrane ("*Schleimhaut*") of the fish during acclimation, rather than the fish acclimating to the anemone. This is because he believes that the protection of the fish generally does not vanish when the fish are isolated, but only during isolation of the anemones (p. 27). However, as Eibl-Eibesfeldt (1965: 128) points out, "It is worth noting that those of Blösch's anemones which first have to become accustomed to a fish, will accept anemone-accustomed fish which have come from an anemone of the same species, quicker than those which have had no contact with an anemone." For example, on page 13, Blösch (1965) states that acclimation required from 5 to 15 minutes for fish which had been isolated for some months, while acclimated fish penetrated the tentacles after only a few contacts. In addition, isolated fish take much longer before approaching an anemone for the first time (20–60 minutes), while acclimated fish head for the anemone after only 1–2 minutes. Later (p. 27), he similarly states, "*Immerhin muss aber festgehalten werden, dass ein an Anemonen gewöhnter Fisch von anderen Anemonen weniger stark und anhaltend genesselt wird als ein isolierter Artgenosse.*"

Although Blösch does not indicate what species of either fish or anemone he is describing, or whether the above times varied for the different anemone or fish species he studied, his observations are nearly identical to those of the author for the acclimation of *Amphiprion xanthurus* to the anemone *Stoichactis kenti*, these both being species that Blösch appears to have used. In the author's reciprocal experiments using both acclimated and unacclimated fish and anemones both with and without acclimated fish, the protection of the fish could not be attributed to an "accustoming" or acclimation by the anemone, but rather to an acclimation of the fish to the anemone (Mariscal, 1969a, 1970a, and see Table 3). Blösch appears to base a good deal of his hypothesis of what is occurring during acclimation on the results of the experiments reported in his table at the top of page 26 and his Table 5 (e.g., see his p. 27). For that reason it would be well to examine his results and experimental procedure in these instances.

For example, in the table at the top of page 26, if one compares the number of specimens used and the number of experiments run, it seems clear that the fishes were used more than once. However, it is not clear whether the fishes were re-isolated after each trial (which should have been done to prevent the effects of any partial acclimation on the results of subsequent experiments) or whether the same fishes were tested against all four anemones during the same general time period, following their initial isolation as a group. Furthermore, based on this same table, Blösch appears to have arrived at some major conclusions regarding the stinging behavior of the four species of anemones he used in relation to various species of anemone fish. These conclusions were conveyed to Eibl-Eibesfeldt (1965) by Blösch and are also found in the summary to Blösch's 1965 paper.

In the summary Blösch (1965) generalizes, apparently on the basis of this table, by saying that he could break up his experiments into four groups depending on the responses of the anemones to the fishes. For example, in his summary he says that Anemone A tolerates *A. percula* immediately but stings other species. However, a glance at the results in the table on page 26 indicates that 11 fish of three other species were also not stung by this anemone. Anemone B supposedly tolerates all *Amphiprion* species immediately with two exceptions. Anemone C supposedly tolerates *A. xanthurus* immediately but stings other species; however, the same table shows that 6 fish from 4 additional species were also tolerated by this anemone. In fact, 11 fish of 5 species total were tolerated as against only 7 fish of 3 species which were stung. Anemone D (probably *Stoichactis kenti*) supposedly stings all species of *Amphiprion* initially. In this case, Blösch tested only 2 out of the 8 species of anemone fish available and came to this conclusion because these two species (total of 9 specimens) were stung. Thus it seems clear that such generalizations are not only misleading when cited alone, but in fact may be unwarranted to begin with. In the table at the bottom of page 26, Blösch finds that Indian Ocean anemone fishes tested against Mediterranean anemones were stung on first contact with one exception. However, here he neglects to mention which species of anemone fish he used and it is not clear how these data fit into his hypothesis.

More important to Blösch's hypothesis, however, are the data reported in his Table 5 (see pp. 27 and 28 of his paper). Together with the results in the table at the top of page 26, the results from Table 5 suggest to Blösch that the fish's protection does not vanish during extended isolation, and thus acclimation cannot consist of the production or activation of a protective substance on the surface of the fish. The conflicting results and the confusing interpretation of the table at the top of page 26 have already been mentioned and they do not necessarily support this idea. The results in the table at the bottom of page 26 do not support this idea at all. In fact, Blösch appears to be interpreting the results in the table at the bottom of page 26 as demonstrating the presence of a protective mucous coat, in contradiction to the interpretation of his other data and to his theory in general (e.g., see last paragraph of p. 26).

In any case, it is clear that the lack of controls in the experiments reported in Table 5 misled Blösch. In this table, Blösch lists the results of 50 experiments which, according to him, show that acclimation of an anemone fish to a sea anemone does not protect it from being stung by isolated sea anemones; that is, the acclimation of the anemone to the fish was lost during the isolation of the anemones, resulting in the fish being stung. This is supposedly because in 43 out of 50 experiments (86 %), the fishes were stung by anemones other than the specific individual in which they had been living. However, even a casual perusal of his data shows that in many of these experiments he is testing fishes acclimated to one anemone species against an entirely different species of anemone. Such an experiment does not control for the well-known observation that different species of coelenterates may have markedly different strengths of nematocyst discharge. For example, in spite of the acknowledgment in his paper that the Mediterranean anemones sting more strongly than do the Indian Ocean forms, he tested fishes acclimated to the Indian Ocean species against the Mediterranean species without conducting the necessary reciprocal experiments. Interestingly enough, in only 7 of his experiments were acclimated fishes not stung when presented to other anemones and in 6 of these 7 cases, he was correctly testing fishes acclimated to Mediterranean anemones against the same species of Mediterranean anemone. Blösch does not attempt to explain these data which suggest the alternative hypothesis to his, namely that an acclimated fish may possess something on its surface which protects it. Furthermore, Davenport and Norris (1958), Eibl-Eibesfeldt (1960), and Mariscal (1970a) have all found that *Amphiprion* acclimated to one species of sea anemone are not *ipso facto* automatically protected from all other species of anemones, as Blösch apparently believes. For example, Eibl-Eibesfeldt has shown in his field and aquarium observations that acclimated fishes, even from the same reef area, are not protected from all species of sea anemones there. He found that *Amphiprion percula* and *A. akallopisos* living in and protected from *Radianthus ritteri* were stung and in some cases killed by *Radianthus kuekenthali* and *Stoichactis*. I have found that an *Amphiprion xanthurus* protected from *Stoichactis kenti* was stung by *Anthopleura xanthogrammica* but not vice versa. Similarly, an *A. xanthurus* acclimated to *Anthopleura xanthogrammica* was still stung by *Anthopleura elegantissima* but the reciprocal experiment revealed that a fish acclimated to *A. elegantissima* was not stung by *A. xanthogrammica* (see Table 4).

Thus, since one cannot meaningfully test an anemone fish acclimated to one species of sea anemone against an entirely different species of sea anemone (even closely related species) without conducting the necessary reciprocal experiments, we are forced to reevaluate Blösch's experiments in this light. Upon so doing, we find that in only 10 out of the 50 experiments (20 %) was a fish which was acclimated to one anemone still stung by another of the same species. Nine out of 10 of these experiments were carried out with anemones which appear to be of the genus *Radianthus* (anemones A and C), while one case occurred with the Mediterranean form, *Actinia equina*. Assuming no additional experimental errors, these data are of interest and deserve further study. However, in all 34 experiments conducted by the author in which respectively acclimated fish were tested against three different species of sea anemones it was found that the acclimated fish were not stung by the same species of sea anemone as that to which they were presently acclimated (Mariscal, 1966b, 1969a, 1970a, and see Table 3). Schlichter (1967, 1968) has reported a similar observation.

Davenport (1965: 53; 1966a: 424) has reported that Blösch (1961) stated that an anemone was capable of "recognizing" its own fish symbionts but stung other *Amphiprion* individuals of the same species. However, this interpretation is not supported by Blösch's (1961, 1965) data and conclusions.

In addition to the author's studies which will be mentioned later, another recent worker has been unable to verify Graefe's (1963, 1964) or Blösch's (1961, 1965) ideas concerning the protection of anemone fishes. Independently, Schlichter (1967, 1968) has come to the same conclusion as the author (Mariscal, 1965, 1966b), that during acclimation something is acquired on the surface of the fish which protects it. He finds no change in the anemone's pattern of nematocyst discharge occurring during this time. For example, he has been able to confirm the author's observations that anemone fish accli mated to one anemone species may be stung when tested against a different anemone species, while fish acclimated to one species are not stung by another anemone of the same species (either with or without acclimated fish). Similarly, Schlichter also found that an *Amphiprion* will lose its protection when isolated from sea anemones and will again be stung when reintroduced to them, regardless of whether the anemones contained acclimated fish at the time or not. Schlichter (1967, 1968) has also been able to confirm by reciprocal experiments Eibl-Eibesfeldt's (1960) and my own observations, that protection from one anemone species may provide protection from other species and that there appear to be different "levels" of acclimation for *Amphiprion*, this being dependent on the species of anemone and not the fish. Concerning the two possibilities which I have previously suggested (Mariscal, 1966b), that the fish may be physiologically altering its own mucous coat or picking up mucus from its anemone host during acclimation, Schlichter (1967, 1968) feels that the latter possibility best explains his observations. By picking up the anemone's mucus, he suggests that the fish's surface may become "anemone-like" and thus no longer capable of provoking nematocyst discharge. Unfortunately, however, sufficient experimental evidence is not yet available upon which to base a firm judgment.

A third possibility concerning the protection of anemone fishes from sea anemones has been mentioned by Gudger (1947) and Mariscal (1966a) in their reviews. This concerns the acquisition of some form of immunity to the nematocyst toxins of the coelenterate host, presumably by the ingestion of the anemone's tentacles and contained nematocysts during acclimation. Although Verwey (1930) appears to have been the first to make this suggestion, no one has yet conducted any experiments to examine this idea in connection with the protection of anemone fishes. Some work has been done, however, demonstrating that it is possible for vertebrates to develop an immunity to coelenterate toxins (Martin, 1967; Baxter *et al.*, 1968). Such studies show, as do others concerning the development of an immune response, that the acquisition of immunity to the toxin is probably much too slow to account for the relatively rapid protection of anemone fishes during acclimation (average of about ten minutes for *Amphiprion xanthurus* and *Stoichactis kenti.* See Table 2). Furthermore, there is no evidence for the very non-adaptive situation in which a symbiotic anemone might be continually discharging its nematocysts in response to an acclimated anemone fish which is protected by its physiological immunity. In any case, even a partial immunity might serve to protect an *Amphiprion* from slight or even massive accidental stinging during acclimation, and it would be of interest to examine this idea experimentally.

It was observed (Mariscal, 1966b) that *Amphiprion xanthurus* often hovered just to one side of the sea anemone to which it was acclimating for relatively long periods of time (minutes) and only occasionally moved in to touch the edge of its pelvic fin or other limited portion of its body against the tentacles (Fig. 2). Such behavior suggested that perhaps the fish was in some way bathing in and acquiring or altering its own mucous surface in response to anemone-released chemical compounds which may have been present in the vicinity of the anemone. However, at that time there was no evidence that sea anemones released large molecules which might in some way combine with the surface of an anemone fish or somehow cause the fish to respond physiologically to the anemone. Recently, Martin (1968) has obtained experimental evidence for the release of species-specific antigens into the surrounding seawater by sea anemones during contraction. He suggests that in addition its interesting implications for the study of allergenic phenomena, such a release of antigenic material may be pertinent to the protection of various organisms which associate with sea anemones, including anemone fish. It is to be hoped that this possibility will be studied experimentally in the near future.

Although the foregoing discussion has concerned itself almost entirely with the various species of *Amphiprion* or *Premnas,* there are several other species of pomacentrids which may commonly associate with sea anemones, especially as juveniles. One of these fish, *Dascyllus trimaculatus* (Fig. 5) has been found by the author to associate with a variety of anemone species throughout the Indo-Pacific, while its very close relative, *Dascyllus albisella* (Fig. 6) from Hawaii, also may associate with sea anemones (Stevenson, 1963; Mariscal, 1966b, 1970c). Both of these species must undergo acclimation similar to *Amphiprion,* but appear to be more facultative in their association, since they may also be found among branching coral heads as juveniles and appear to be largely free-living as adults. I have found juvenile *Dascyllus trimaculatus* living in large sea anemones with several species of *Amphiprion* in both the Indian Ocean and South Pacific, as well as among coral heads. No *Amphiprion* species or *Dascyllus trimaculatus* occur as far north as the Hawaiian Islands and here *Dascyllus albisella,* obviously a very closely related mutant form of *D. trimaculatus,* is found exclusively. I have similarly seen the juveniles of this species associating both with sea anemones and coral heads while again the adults appear to be free-living. The acclimation of both *Dascyllus* species is similar to that of *Amphiprion*, including nibbling of the tentacles, hesitant fin and body contact, and finally leaning or brushing up against the tentacles (Figs. 5, 6).

Fig. 5. Laboratory electronic flash photograph of two Indo-Pacific *Dascyllus trimaculatus* undergoing acclimation to *Stoichactis*. The fish would commonly make initial hesitant contacts with the tentacles and later lean up against individual tentacles or the folds of the oral disc as shown here.

Fig. 6. Laboratory electronic flash photograph of Hawaiian *Dascyllus albisella* acclimating to California *Anthopleura xanthogrammica*. Towards the end of acclimation, this species cautiously leaned up against individual tentacles, as did the very closely related *D. trimaculatus*.

MATERIALS AND METHODS

Field studies were conducted during Cruises I and V of Stanford University's (Hopkins Marine Station) *Te Vega* Expeditions. Cruise 1, under the direction of Dr. Rolf Bolin, sailed across the South Pacific and terminated in Singapore. Studies were conducted at Tutuila Island, East Samoa; Great Astrolabe Reef and Viti Levu Island, Fiji Islands; Vanikoro Island, Santa Cruz Islands; Guadalcanal and Bougainville Islands, Solomon Islands; New Britain Island and New Guinea, Territory of New Guinea; Green Island, Great Barrier Reef, Queensland, Australia; Oahu Island, Hawaiian Islands; and Tahiti Island, Society Islands.

Cruise V, under the direction of Dr. Donald P. Abbott, started in Mombasa, Kenya, East Africa and sailed across the Indian Ocean to terminate also in Singapore. Field work and shipboard experiments were conducted at Mombasa and Jadini, Kenya, East Africa; Mahé Island, Seychelles Islands; Malé Atoll, Maldive Islands; Ceylon; and Pipilek and Pipidon Islands off Phuket, West Thailand.

Extensive field studies were also carried out in the Galápagos Islands while participating in the University of California's Galápagos International Scientific Project, but no examples of the anemone fish symbiosis were found there.

In the field, underwater observations were made by both skin and scuba diving and were recorded on waterproof plastic sheets. Photographic records were obtained with Calypso and Nikonos underwater flash cameras as well as with an Exakta camera built into a watertight housing with a Rolleimarin flash attachment. All underwater photographs were taken on Kodachrome-X or Ektachrome-X 35mm color film. Laboratory photographs were taken on Kodachrome-II 35mm color film with a Nikon F camera and Micro-Nikkor lens with electronic flash.

Shipboard experiments were conducted in two gimbaled, fiberglass running seawater aquaria of about 20 gallons capacity each. Laboratory experiments were carried out in 15–20 gallon Plexiglas aquaria using natural seawater and both a subgravel and outside charcoal and glass wool filter. Daily records were kept of temperature, feedings, pH, density, and additions of water. The fishes lived on live crustaceans, algae, and a prepared flake food (Tetramarin and Tetramin). The anemones were fed on pieces of shrimp and fish, as well as live crustaceans.

To avoid disturbing the epidermal mucous coat of the fish, all fish were transferred by means of clean glass beakers containing enough water to immerse the fish in a normal upright swimming position. The anemones were maintained in numbered glass finger bowls to facilitate transfer without disturbance. No differences could be detected in the results depending on whether the fishes, the anemones, or both, were transferred during an experiment.

The main species of *Amphiprion* studied in the field included *A. akallopisos, A. nigripes, A. percula, A. ephippium*, and *A. xanthurus*. The primary species of anemones studied in the field were *Radianthus ritteri, Stoichactis kenti, Stoichactis giganteum*, and *Physobrachia ramsayi*. In the laboratory studies, the main fish studied was *Amphiprion xanthurus* which has been found in the field associating with *Stoichactis kenti*, this being the main tropical anemone used in the laboratory experiments. In addition, the tropical *Radianthus malu* and two California anemones, *Anthopleura xanthogrammica* and *Anthopleura elegantissima*, were used in laboratory experiments with *A. xanthurus*. Other *Amphiprion* observed in the laboratory included *A. frenatus, A. ephippium, A. percula, A. laticlavius, A. nigripes, A. akallopisos*, and *A. bicinctus*. In addition, observations were made of another anemone fish, *Premnas biaculeatus*, and of both *Dascyllus trimaculatus* and *Dascyllus albisella*.

Tests with live prey were conducted before and after each experiment to ensure that the anemones were fully capable of prey capture. In addition, tests on the fingers of the observer as well as on organic-coated coverslips were made to be certain that the anemones were capable of nematocyst discharge. Both light and repeated contact of the coverslips against the tentacles caused massive discharge and adhesion to the coverslip of thousands of basitrichous isorhizas. This was the only nematocyst present in large numbers on the coverslips.

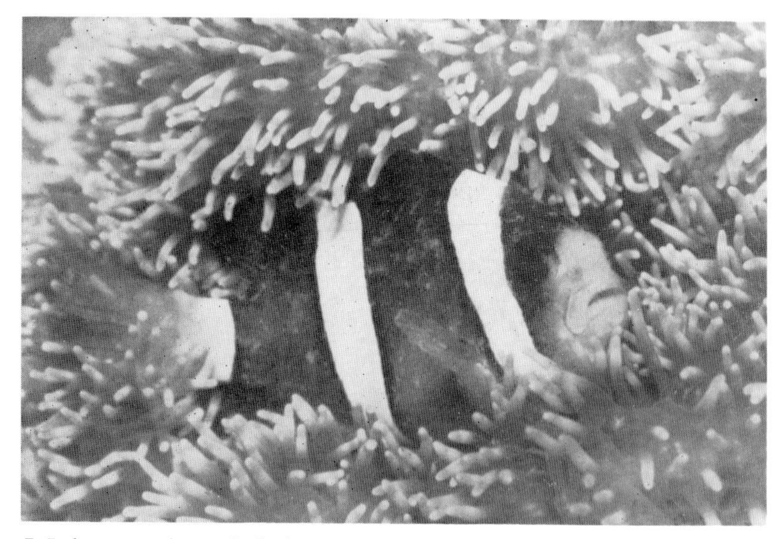

Fig. 7. Laboratory electronic flash photograph of an unacclimated *Amphiprion xanthurus* which had been directed into the tentacles of a *Stoichactis kenti* anemone, which at the time contained another, fully acclimated *A. xanthurus*. Note that the surface of this fish (now partially acclimated) is covered with small lesions and bits of *Stoichactis* tentacle where it was stung before being allowed to go through acclimation.

Four criteria were used to determine if the experimental fish was causing nematocyst discharge during acclimation: (1) adhesion of the tentacles to the surface of the fish, (2) the darting or jumping back of the fish immediately upon making tentacle contact, (3) the contraction of the anemone's tentacles after contacting the fish, and (4) the adherence of small tentacle fragments as well as the blistering of the fish's epidermis. The latter was especially evident with *Stoichactis* (Fig. 7).

RESULTS

Acclimation Behavior

In order to clarify the various types of acclimation phenomena, Table 1 has been prepared. Acclimation behavior (that is, "Reacclimated" in Table 1) is initiated by the fish (*Amphiprion xanthurus*) cautiously approaching an anemone and gently nosing or nibbling the tentacles (Fig. 8). Following this the tentacles can be seen to adhere strongly to the surface of an unacclimated fish. Upon contact, the tentacles generally contract and bend away, sometimes rather sharply. In the case of *Anthopleura xanthogrammica*, a wave of tentacle contraction away from the point of contact and encircling

Fig. 8. Laboratory electronic flash photograph of a partially acclimated *Amphiprion xanthurus* nibbling the tentacles of a *Stoichactis kenti* during acclimation to this anemone. Nibbling or mouthing the tentacles of the anemone was a common act at the start of and during acclimation. However, fully acclimated fish also engage in this behavior as well.

TABLE 1. *Definition of the types of acclimation phenomena*

General condition of fish or anemone	*Unacclimated*		*Acclimated*	
Means by which condition arrived at	Never acclimated	Deacclimated	Originally acclimated	Reacclimated
Fish	Anemone fish (*Amphiprion, Premnas, Dascyllus*) which have never before associated with sea anemones.	Anemone fish isolated from any contact with sea anemones (visual, physical, etc.) generally for five days or longer.	Anemone fish living with the original sea anemones to which they first became acclimated.	Anemone fish living with sea anemones capable of prey capture, other than those to which they first became acclimated, and which are no longer stung by them.
Anemone[a]	Sea anemones capable of prey capture which have had no prior contact with anemone fish	Sea anemones which have been isolated from any contact with anemone fish for one week or longer.	Sea anemones harboring their original anemone fish symbionts.	Sea anemones capable of prey capture, harboring other than their original anemone fish symbionts, but which no longer sting them.

[a] It should be mentioned that to date there is no good experimental evidence for "acclination" or a change in the anemone's nematocyst discharge pattern in response to a symbiotic fish: rather the change during acclimation which protects the fish appears to rest solely with the fish.

the oral disc occasionally occurred, along with rapid contractions of the oral disc.

Shortly after nosing the anemone, the fish generally began gently settling or moving into the tentacles until contact was made with the ventral fins or tail. The fish dashed up and out of the tentacles and then returned to repeat the same behavior. In other cases the fish would hover to one side of the anemone and slowly swim towards it until contacting the tentacles with the breast or pelvic fins. Although the overall acclimation procedure was similar, there was a good deal of individual variation in the case of *A. xanthurus*. Two specimens of *A. xanthurus* commonly backed tail-first into the tentacles, much as Stevenson (1963) has described acclimation for *Dascyllus albisella* and as I have observed for the closely related *Dascyllus trimaculatus*. As the *A. xanthurus* settled and backed into the tentacles on the oral disc, the tail would curve sharply to one side as if to delay making contact as long as possible and then with a flick of the tail, the fish would propel itself up and out of the tentacles.

Along with a deeper penetration of the anemone's tentacles, there was a decrease in the tentacle adhesion and contraction in response to the fish until finally the fish was able to bury itself among the tentacles without harm. When no further tentacle adhesion or contraction was elicited and the fish no longer responded to tentacle contact as if it had been "stung," acclimation was considered to be complete.

It was found that a separation of one hour from its anemone (either *Anthopleura* or *Stoichactis*) could be tolerated by *A. xanthurus* without being stung or having to undergo acclimation behavior upon reintroduction. Any separation of much longer duration than this resulted in the fish becoming partially deacclimated so that it was stung during approximately 50 % of the contacts with the anemone's tentacles. Between 3 and 10 hours following isolation of the fish, 2 out of 3 *A. xanthurus* became partially deacclimated while the third fish was completely deacclimated. However, the most striking finding in this regard was the unanimity of the results following a 20 hour isolation. In all cases, *A. xanthurus* which were isolated from sea anemones for 20 hours or more became completely deacclimated and were stung by both acclimated and unacclimated *S. kenti* and *A. xanthogrammica*. It should be pointed out that in all the experiments, both acclimated and unacclimated anemones were used with no observable effect on the results. In addition, the length of time that the acclimated anemones had fish living with them prior to the experiment, the length of time the fish had been acclimated, and the length of time the deacclimated anemones had been isolated seemed to have no bearing on the outcome of the experiment; only the length of time that the fish had been isolated was relevant to the experimental results.

Reacclimation Times for *Amphiprion xanthurus*

The acclimation (that is, reacclimation time) was immediate for an acclimated *Amphiprion xanthurus* which had just been removed from another anemone of the same species. However, the acclimation time for an unacclimated *A. xanthurus* varied, depending on the species of anemone to which it was acclimating (Table 2). Some anemones seemed "easier" to acclimate to than did others, based on the time required for acclimation. For example, the mean acclimation time of *A. xanthurus* to *Stoichactis kenti* was only 10.1 minutes whereas to *Anthopleura xanthogrammica* it was 59 minutes. Only one *A. xanthurus* out of three was able to acclimate to *Anthopleura elegantissima*. In this case, the acclimation time was approximately 45 hours. The other two fish remained unacclimated after 45 and 119 hours although they continued to perform acclimation behavior throughout this period (Fig. 3). Both acclimated and unacclimated anemones were tested and there was no correlation between the condition of the anemone and the acclimation time of the fish.

TABLE 2. *Times of acclimation for unacclimated* Amphiprion xanthurus *presented to sea anemones both with and without fish*

Anemone species	*No. of experiments*	*Range (minutes)*	*Mean time of acclimation*
Stoichactis kenti	8	5–20	10 minutes
Anthopleura xanthogrammica	8	25–140	59 minutes
Anthopleura elegantissima	3	——	~45 hours (1 fish)

Results of Experiments Presenting Unacclimated and Acclimated *Amphiprion xanthurus* to Anemones, both with and without Acclimated Fish

In view of the great similarity in morphology between *Anthopleura xanthogrammica* and *Anthopleura elgantissima* (C. Hand, per. comm.), it was interesting to note the great difference in acclimation times of *Amphiprion xanthurus* to these forms. Although Hand (1955) has pointed out their taxonomic differences, marked physiological differences may exist as well between these two species of *Anthopleura*.

As previously indicated, the period of acclimation is the focal point for any analysis of the protection of anemone fishes from sea anemones. At the start of acclimation, the anemone fish are stung; at the end of this period they are protected. Obviously something has happened to either the fish or anemone during acclimation which results in the fish becoming protected from the sea anemone's nematocysts.

Two hypotheses currently attempt to explain what occurs during acclimation. One hypothesis (e.g., Blösch, 1961, 1965) indicates that some change must be occurring in the anemone during acclimation. This is because Blösch found that an anemone which had acclimated anemone fish living with it prior to an experiment supposedly did not sting other anemone fish, regardless of whether the other fish had been isolated or living with sea anemones up until the time of the experiment. This observation implied that some change had taken place in the physiological state of the sea anemone which had *Amphiprion* living with it so that other *Amphiprion* of the same species which came into contact with this anemone were also protected.

The second hypothesis (e.g., Davenport and Norris, 1958) indicates that the surface mucous coating of the anemone fish is somehow altered during acclimation and that this change is responsible for the protection of the fish.

The rationale for the present experiments to test the above hypotheses was as follows: if an acclimated *Amphiprion* was not stung by both acclimated and unacclimated anemones, this would suggest that the fish had acquired some sort of protection during acclimation. If a deacclimated *Amphiprion* was stung by sea anemones both with and without acclimated fish, this would suggest that perhaps something was lacking from the fish which formerly had protected it. On the other hand, if both acclimated and unacclimated fish were stung by an isolated anemone, while an anemone containing acclimated fish stung neither, then this would be evidence for some sort of change occurring to the anemone during acclimation which resulted in its no longer stinging the fish.

In Table III are presented the results of experiments in which *Amphiprion xanthurus* (both acclimated and unacclimated) were presented to three different species of sea anemones (both with and without acclimated fish) and the responses recorded using the aforementioned controls.

It will be immediately noted that in all cases the unacclimated fish were stung by both acclimated and unacclimated anemones while the acclimated fish were stung by neither. This suggests that whatever change occurred during acclimation to protect the fish, lay with the fish and not the anemone in these cases. The outcome of these experiments was not influenced in any way by the condition of the anemone; both acclimated and unacclimated anemones responded in identical fashion to the fish.

If this protection acquired by the fish was due to some sort of change in or on its surface mucous coat, as suggested by the results of Davenport and Norris (1958), then one would expect that any disturbance in this mucous coating would result in an acclimated fish being stung. Going even further, if one were to carefully wipe this mucous coating off, one would expect the formerly acclimated fish to become deacclimated and thus be stung by the same anemone from which it had only seconds before been protected. As

TABLE 3. *Experiments presenting* Amphiprion xanthurus *(unacclimated or acclimated only to that species of anemone being tested) to three species of sea anemones, with and without acclimated* A. xanthurus

Amphiprion xanthurus	*Stoichactis kenti*		*Anthopleura xanthogrammica*		*Anthopleura elegantissima*			
			Condition of anemone					
Condition of fish	Without fish	With fish	Without fish	With fish	Without fish	With fish	*Total number of experiments*	*Summary of results*
Unacclimated	+[a] (9)[b]	+ (8)	+ (9)	+ (9)	+ (4)	+ (1)	40	All stung
Acclimated	—[c] (6)	— (7)	— (8)	— (10)	— (2)	— (1)	34	None stung
Total number of experiments	15	15	17	19	6	2	74	
After wiping off surface of acclimated fish	+ (2)	+ (2)	+ (2)	+ (2)	0[d]	+ (1)	9	All stung
After wiping off surface of unacclimated fish just as it was becoming acclimated	+ (1)	+ (1)	+ (1)	+ (2)	0	0	5	All stung
Total number of experiments	3	3	3	4	0	1	14	

[a] Fish stung
[b] Parentheses refer to the number of experiments
[c] Fish not stung
[d] No data

will be seen from the lower half of Table 3, this is what occurred. In all the experiments (14), a fully or partially acclimated fish which was wiped off, immediately became deacclimated and was stung upon every contact with the anemone from which it had just been removed. Interestingly, the behavior of an acclimated fish changed markedly after the wiping treatment. It became very reticent about approaching or entering the tentacles of its anemone and in fact did not do so until it had again gone through acclimation behavior to the same anemone.

To further control the experimental situation, some of the experiments summarized in Table 3 (termed "double" experiments) were run in which both acclimated and unacclimated fish and anemones were presented simultaneously in a single aquarium. This controlled for differences between the various experimental aquaria as well as between individual fish and anemones in single experiments. With either method, identical results were obtained. It proved quite striking to observe a single anemone sting a previously isolated, unacclimated fish (Fig. 7) while at the same instant its acclimated fish was buried deeply among its tentacles and completely protected from the nematocysts.

Other experiments were run in which the roles of the individual fish were reversed; that is, the deacclimated fish in one experiment became the acclimated fish in the subsequent one and vice versa. In this way, individual variation in size, shape, and behavior of the fish could be controlled. This is especially pertinent to Graefe's (1963, 1964) hypothesis that a size difference between anemone fish is involved in their protection. However, in all cases and regardless of the experimental situation, the presently unacclimated fish was always stung while the acclimated fish was not. Only anemones of the same species to which the fish had been acclimated were used in testing the acclimated fish. This control is extremely important as will be seen in the following series of experiments.

Results of Experiments with Acclimated *Amphiprion xanthurus* Presented to Different Species of Anemones, both with and without Acclimated Fish

Based on his field experiments in the Nicobar Islands, Eibl-Eibesfeldt (1960) believes that some sort of protective substance is present on the three species of anemone fish he studied. He found that *Amphiprion xanthurus* was protected from three species of anemones (*Radianthus ritteri, Radianthus kuekenthali,* and *Stoichactis = Discosoma*), while *Amphiprion percula* and *Amphiprion akallopisos* were protected from only one of these anemones (*R. ritteri*). He goes on to suggest that although either a qualitative or quantitative difference in this protective substance could account for these differences, if it were the latter, perhaps *A. xanthurus* simply had more of this substance on its surface. *A. percula* and *A. akallopisos*, with possibly a smaller concentration of this substance, were only protected from the more weakly stinging *Radianthus ritteri* and not the "clearly more sensitive" *R. kuekenthali* and *Stoichactis* (= *Discosoma*).

In order to obtain some experimental data which might be relevant to this idea, *Amphiprion xanthurus* which were acclimated to one species of anemone were reciprocally tested against another anemone species and the results compared with the mean time of acclimation of the same fish to each of these anemones. The idea was to see if a fish which was acclimated to an anemone which required a longer acclimation time was also protected from anemones which required shorter acclimation times.

From Table 4, it will be noted that a fish acclimated to and protected from *Anthopleura elegantissima* was also protected from *Anthopleura xanthogrammica* which did indeed require a much shorter time of acclimation (see Table 2). In turn, a fish acclimated to *A. xanthogrammica* was also protected from *Stoichactis kenti*, which also required a much shorter time of acclimation. In both of these cases the reciprocal experiments were consistent in that acclimation to an anemone requiring a shorter time of acclimation did not afford protection from an anemone which required a longer acclimation time.

TABLE 4. *Experiments presenting acclimated* Amphiprion xanthurus *reciprocally to three different anemone species (all without fish)*

Amphiprion xanthurus acclimated to:	*Anthopleura elegantissima*	*Anthopleura xanthogrammica*	*Stoichactis kenti*
Anthopleura elegantissima	—[a] (1)[b]	— (1)	0
Anthopleura xanthogrammica	+[c] (2)	— (6)	— (1)
Stoichactis kenti	0	+ (1)	— (5)

[a] Fish not stung
[b] Parentheses refer to the number of experiments
[c] Fish stung

Results of Shipboard Experiments on Isolated Pieces of *Amphiprion akallopisos* Skin

Experiments were conducted in the Seychelles Islands aboard the R/V *Te Vega* to test the response of *Radianthus ritteri* (containing acclimated *A. akallopisos*) anemones to the skin and flesh of *Amphiprion akallopisos* from another *R. ritteri*. Pieces of a control fish, the grouper *Epinelphus merra*, were tested simultaneously with the *Amphiprion* pieces (Table 5). Small pieces of skin and flesh (about 1 cm/) of both *A. akallopisos* and *Epinelphus* were first presented skin side down. This resulted in massive adhesion of the *R. ritteri* tentacles to the grouper skin but not to that of *Amphiprion* except around the cut edges. However, upon turning both pieces over and presenting them flesh side down, there was massive adhesion and contraction of the anemone's tentacles and both pieces were ingested. Repeated experiments provided identical results: only when the *Amphiprion* piece was presented skin side down was there no stinging and ingestion by the anemone. This suggested that the *Amphiprion* skin possessed some protective substance on its surface which did not elicit nematocyst discharge, while *Amphiprion* flesh and both grouper skin and flesh lacked this substance.

TABLE 5. *Shipboard experiments on the responses of Seychelles* Radianthus ritteri *anemones to 1 cm² pieces of skin and flesh from* Amphiprion akallopisos *(acclimated to* R. ritteri*) and the grouper,* Epinelphus merra

Fish	*Response to pieces presented skin side against tentacles*	*Response to pieces presented flesh side against tentacles*
Amphiprion akallopisos	–[a] (3)[b]	+ (3)
Epinelphus merra	+[c] (3)	+ (3)

[a] No response
[b] Parentheses refer to the number of experiments
[c] Strong adhesion of tentacles

Gohar (1948) has made the suggestion that the "mode of movement" of an anemone fish was recognized by its symbiotic anemone and, therefore, it was not stung. In order to test this hypothesis, several species of dead *Amphiprion* were tested against the anemones from which they had been captured to see if they elicited nematocyst discharge.

Four out of the five species of dead *Amphiprion* (*A. akallopisos, A. nigripes, A. percula, A. ephippium*) were not stung by their anemones when placed among the tentacles, nor was *Dascyllus trimaculatus.* In addition, *Amphiprion ephippium* was stung neither by its own anemone, *Physobrachia ramsayi*, nor by *Radianthus ritteri*, an anemone with which it was not found in the field. *Amphiprion xanthurus* was only slightly stung by its anemone, but its surface was abraded by the net when returned to the ship and thus its epidermal mucous coating could have been damaged, resulting in its being partially stung. In any case, no convincing evidence was found that tends to support Gohar's idea that the anemone recognizes its fish by its movement. Obviously dead fish are not noted for their motor activity, yet they were not stung by their anemones. Davenport and Norris (1958) and Eibl-Eibesfeldt (1960) have reported similar results.

DISCUSSION AND CONCLUSIONS

It seems clear from the present experiments that something is acquired by an anemone fish during acclimation, either on or in the epidermal mucous coat, which protects it from its host's nematocysts. This is attested by the fact that acclimated *Amphiprion xanthurus* were not stung by sea anemones either with or without acclimated fish, while unacclimated fish were stung by both.

I have therefore been unable to verify the hypotheses of Gohar (1948), Blösch (1961, 1965), Graefe (1963, 1964), or Fishelson (1965) concerning the protection of anemone fishes from sea anemones. Rather, my data support the hypothesis that something on the surface of an anemone fish (presumably associated with the mucous coat) is responsible for its protection from the anemone's nematocysts. The present study supports the conclusions of Davenport and Norris (1958), Eibl-Eibesfeldt (1960), and Abel (1960b) in this regard, although it should be mentioned that the latter author found no evidence of acclimation for the fish (*Gobius*) he studied.

The present study demonstrates that during acclimation a change occurs on the surface of an anemone fish which results in its protection. This change could occur in either of two ways. The fish might be either quantitatively or qualitatively altering its own mucous coat in response to a sea anemone by some physiological means. The second, equally likely, possibility is that the fish may be coating or complexing its own surface with the sea anemone's mucus during acclimation and in this way becomes protected by masking its own natural stimuli for nematocyst discharge.

SUMMARY

Various pomacentrid fishes, especially of the genus *Amphiprion*, live symbiotically with stoichactiid sea anemones in the tropical Indo-Pacific. The fishes are somehow protected from the nematocysts of their sea anemone hosts and are not stung in spite of massive tentacle contact.

The present paper reviews critically all of the studies to date which offer either suggestions or hypotheses based on experimental observations concerning the protection of anemone fishes from the nematocysts of the host.

The behavioral process known as acclimation, during which time a previously isolated anemone fish becomes protected from a sea anemone, serves as the focal point for such an investigation into the mechanism of anemone fish protection.

Based on the experiments conducted to date by various workers around the world, two main possibilities emerge concerning what occurs during acclimation to protect the fish: one suggests that the sea anemone may be altering its pattern of nematocyst discharge through the receipt of either behavioral, physical, or chemical stimuli from the fish; the second suggests that the fish is in some way altering its surface mucous coat in response to the anemone, which results in the fish no longer provoking nematocyst discharge.

A careful study of the literature which suggests that the anemone may be altering its nematocyst discharge in response to the fish during acclimation, has revealed either a lack of experimental evidence or some serious difficulties in the controls of the experiments which were interpreted as supporting this idea.

On the other hand, controlled experimental studies by others as well as by the author strongly suggest that, at least for the species of fishes and anemones so far investigated, some change occurs on the surface of the fish, presumably associated with the mucous coat, which results in its protection.

This change could be due either to a physiological secretory process on the part of the fish, or it might somehow involve the acquisition of the anemone's mucus on the fish's surface so that the fish no longer triggers nematocyst discharge.

ACKNOWLEDGMENTS

The field observations which serve as a background for this paper were conducted during two Stanford University (Hopkins Marine Station) *Te Vega* Expeditions, one across the South Pacific under Dr. Rolf L. Bolin and the second across the Indian Ocean under Dr. Donald P. Abbott. These expeditions were supported by NSF Grant G 17465. Portions of this work were supported under NSF Grant GB 1237 to Dr. Cadet Hand of the University of California at Berkeley, and in part by Public Health Service predoctoral fellowship 5-Fl-Gm-22, 391-02 from the National Institute of General Medical Sciences.

REFERENCES

Abel, E. F. 1960a. Zur Kenntnis des Verhaltens und der Ökologie von Fischen an Korallenriffen bei Ghardaqa (Rotes Meer). Z. Morphol. Ökol. Tiere 49:430–503.

Abel, E. F. 1960b. Liason facultative d'un poisson (*Gobius bucchichii* Steindachner) et d'une anémone (*Anemonia sulcata* Penn.) on Méditerranée. Vie et Milieu 11:517–531.

Allee, W. C., Park, O., Emerson, A. E., Park, T. and Schmidt, K. P. 1949. Principles of animal ecology. W. B. Saunders Co., Philadelphia. 837 p.

Baxter, E. H., Marr, A. G. and Lane, W. R. 1968. Immunity to the venom of the sea wasp *Chironex fleckeri*. Toxicon 6:45–50.

Blösch, M. 1961. Was ist die Grundlage der Korallenfisch-symbiose: Schutzstoff oder Schutzverhalten? Naturwissenschaften 48:387.

Blösch, M. 1965. Untersuchungen über das Zusammenleben von Korallenfischen (*Amphiprion*) mit Seeanemonen. Inaugural Dissertation, Eberhard-Karls-Universität zu Tübingen. 44 p.

Bowman, T. E. and R. N. Mariscal. 1968. *Renocila heterozota*, a new cymothoid isopod, with notes on its host, the anemone fish, *Amphiprion akallopisos*, in the Seychelles. Crustaceana 14(1):97–104.

Buhk, F. 1939. Lebensgemeinschaft zwischen Riesenseerose und Korallen-fischen. Wochensch. Aquar. Terrarienk. 46:672–674.

Carlgren, O. 1949. A survey of the Ptychodactiaria, Corallimorpharia and Actiniaria. Kungl. Svenska Vetens. Handling, Fjarde Serien 1(1):1–121.

Collingwood, C. 1868. Note on the existence of gigantic sea-anemones in the China Sea, containing within them quasiparasitic fish. Ann. and Mag. Natur. Hist., Ser. 4(1):31–32.

Davenport, D. 1955. Specificity and behavior in symbioses. Quart. Rev. Biol. 30:29–46.

Davenport, D. 1962. Physiological notes on actinians and their associated commensals. Bull. Inst. Oceanogr. (Monaco) 1237:1–15.

Davenport, D. 1965. E. H. Thorpe and D. Davenport (eds.) *In* Learning and associated phenomena in invertebrates. Anim. Behav. Suppl. No. 1:53.

Davenport, D. 1966a. The experimental analysis of behavior in symbioses. *In* S. M. Henry (ed.) Symbiosis, Vol. 1. Academic Press, New York. 478 p.

Davenport, D. 1966b. Cnidarian symbioses and the experimental analysis of behavior. In W. J. Rees (ed.) The Cnidaria and their evolution (Symp. Zool. Soc. London, no. 16). Academic Press, London. 449 p.

Davenport, D. and Norris, K. S. 1958. Observations on the symbiosis of the sea anemone *Stoichactis* and the pomacentrid fish *Amphiprion percula*. Biol. Bull. 115:397–410.

Eibl-Eibesfeldt, I. 1960. Beobachtungen und Versuche an Anemonenfischen (*Amphiprion*) der Malediven und der Nicobaren Z. Tierpsychol. 17:1–10.

Eibl-Eibesfeldt, I. 1965. Land of a thousand atolls. MacGibbon and Kee, London. 195 p.

Ellis, V. L., Ross, D. M. and Sutton, L. 1969. The pedal disc of the swimming sea anemone *Stomphia coccinea* during detachment, swimming, and resettlement. Can. J. Zool. 47(3):333–342.

Fishelson, L. 1965. Observations on the Red Sea anemones and their symbiotic fish *Amphiprion bicinctus*. Bull. Sea Fish. Res. Sta. (Haifa) 39:1–14.

Gohar, H. A. F. 1934. Partnership between fish and anemone. Nature, 134:291.

Gohar, H. A. F. 1948. Commensalism between fish and anemone. (With a description of the eggs of *Amphiprion bicinctus* Rüppell). Fouad I. Univ. Publ. Marine Biol. Sta. Ghardaqa (Red Sea) 6:35–44.

Graefe, G. 1963. Die Anemonen-Fisch-Symbiose und ihre Grundlage—nach Freilanduntersuchungen bei Eilat/Rotes Meer. Naturwissenschaften 50:410.

Graefe, G. 1964. Zur Anemonen-Fisch-Symbiose, nach Freilanduntersuchungen bei Eilat/Rotes Meer. Z. Tierpsychol. 21:468–485.

Gudger, E. W. 1947. Pomacentrid fishes symbiotic with giant sea anemones in Indo-Pacific waters. J. Roy. Asiatic Soc. Bengal 12:53–76.

Hackinger, A. 1959. Freilandbeobachtungen an Aktinien und Korallenfischen. Mitteil. Biol. Sta., Wilhelminenberg, Wien 2:72–74.

Hand, C. 1955. The sea anemones of central California. Part II. The endomyarian and mesomyarian anemones. Wasmann J. Biol. 13:37–99.

Henry, S. M. (ed.). 1966. Symbiosis, Vol. 1. Academic Press, New York. 478 p.

Herre, A. W. 1936. Some habits of *Amphiprion* in relation to sea anemones. Copeia 1936: 167–168.

Koenig, O. 1960. Verhaltensuntersuchungen an Anemonenfischen. Die Pyramide 8:52–56.

Mariscal, R. N. 1965. Observations on acclimation behavior in the symbiosis of anemone fish and sea anemones. Amer. Zool. 5:694.

Mariscal, R. N. 1966a. The symbiosis between tropical sea anemones and fishes: a review. *In* R. I. Bowman (ed.) The Galápagos. Univ. Calif. Press, Berkeley. 318 p.

Mariscal, R. N. 1966b. A field and experimental study of the symbiotic association of fishes and sea anemones. Doctoral Dissertation, Univ. Calif., Berkeley. University Microfilms, Ann Arbor. 262 p.

Mariscal, R. N. 1967. A field and experimental study of the symbiotic association of fishes and sea anemones. Diss. Abstr. 28(1):388–B.

Mariscal, R. N. 1969a. The protection of the anemone fish, *Amphiprion xanthurus*, from the sea anemone, *Stoichactis kenti*. Experientia 25:1114.

Mariscal, R. N. 1969b. The tentacle nematocysts involved in the shell adhesion response of the symbiotic sea anemone, *Calliactis tricolor*. Amer. Zool. 9:1140.

Mariscal, R. N. 1970a. An experimental analysis of the protection of *Amphiprion xanthurus* Cuvier & Valenciennes and some other anemone fishes from sea anemones. J. Exp. Marine Biol. Ecol. (In press)

Mariscal, R. N. 1970b. The symbiotic behavior between fishes and sea anemones. *In* How-. ard E. Winn and Bori Olla (*eds*.) Behavior of marine animals—recent advances. Plenum Publishing Corp., New York. (In press)

Mariscal, R. N. 1970c. A field and laboratory study of the symbiotic behavior of fishes and sea anemones from the tropical Indo-Pacific. Univ. Calif. Pub. Zool. 91:1-33 (4 plates).

Mariscal, R. N 1970d The nature of the symbiosis between lndo-Pacific anemone fishes and sea anemones. Mar. Biol. 6(1):58–65.

Martin, E. J. 1967. Antitoxin against a coelenterate poison. Int. Arch. Allergy Appl. Immunity 32:342–348.

Martin, E. J. 1968. Specific antigens released into sea water by contracting anemones (Coelenterata). Comp. Biochem. Physiol. 25:169–176.

Noble, E. R. and Noble, G. A. 1964. Parasitology: The Biology of Animal Parasites. Second edition. Lea & Febiger, Philadelphia. 724 p.

Schlichter, D. 1967. Zur Klärung der "Anemonen-Fisch-Symbiose". Naturwiss enschaften 54:569.

Schlichter, D. 1968. Das Zusammenleben von Riffaaemonen und Anemonenffschen. Z. Tierpsychol. 25:933-954.

Stevenson, R. A. 1963. Behavior of the pomacentrid reef fish *Dascyllus albiesella* Gill in relation to the anemone *Marcanthia* [sic] *cookei*. Copeia 1963:612–614.

Valentyn, F. 1726. Oud en nieuw Oost-Indien, etc., Vol. 3. Omftandig Verhaal van de Geschiedenissen en Zaaken het Kerkelyke ofte den Godsdienst Betreffende, zoo in Amboina, etc. Joannes Van Braam, Dodrecht.

Verwey, J. 1930. Coral reef studies. 1. The symbiosis between damselfishes and sea anemones in Batavia Bay. Treubia 12:305–366.

Index

INDEX